Hybrid Formulation of Wave Propagation and Scattering

NATO ASI Series

Advanced Science Institutes Series

A Series presenting the results of activities sponsored by the NATO Science Committee, which aims at the dissemination of advanced scientific and technological knowledge, with a view to strengthening links between scientific communities.

The Series is published by an international board of publishers in conjunction with the NATO Scientific Affairs Division

A	**Life Sciences**	Plenum Publishing Corporation
B	**Physics**	London and New York
C	**Mathematical and Physical Sciences**	D. Reidel Publishing Company Dordrecht and Boston
D	**Behavioural and Social Sciences**	Martinus Nijhoff Publishers Dordrecht/Boston/Lancaster
E	**Applied Sciences**	
F	**Computer and Systems Sciences**	Springer-Verlag Berlin/Heidelberg/New York
G	**Ecological Sciences**	

Series E: Applied Sciences – No. 86

Hybrid Formulation of Wave Propagation and Scattering

edited by

L.B. Felsen

Department of Electrical Engineering and Computer Science
Polytechnic Institute of New York
Farmingdale, NY 11735
USA

1984 **Martinus Nijhoff Publishers**
Dordrecht / Boston / Lancaster
Published in cooperation with NATO Scientific Affairs Division

Proceedings of the NATO Advanced Research Workshop on Hybrid Formulation of Wave Propagation and Scattering, IAFE, Castel Gandolfo (Rome), Italy, August 30-September 3, 1983

Library of Congress Cataloging in Publication Data

NATO Advanced Research Workshop on Hybrid Formulation of Wave Propagation and Scattering (1983 : Castel Gandolfo, Italy)
Hybrid formulation of wave propagation and scattering.

(NATO ASI series. Series E, Applied sciences ; no. 86)
"Proceedings of the NATO Advanced Research Workshop on Hybrid Formulation of Wave Propagation and Scattering, IAFE, Castel Gandolfo, (Rome), Italy, August 30-September 3, 1983"--T.p. verso.
"Published in cooperation with NATO Scientific Affairs Division."
Includes bibliographical references.
1. Beam optics--Congresses. 2. Electromagnetic waves--Transmission--Congresses. 3. Electromagnetic waves--Scattering--Congresses. 4. Waves--Congresses.
I. Felsen, Leonard B. II. North Atlantic Treaty Organization. Scientific Affairs Division. III. Title.
IV. Series.
QC389.N38 1983 535'.32 84-20651
ISBN 90-247-3094-5

ISBN 90-247-3094-5 (this volume)
ISBN 90-247-2689-1 (series)

Distributors for the United States and Canada: Kluwer Academic Publishers, 190 Old Derby Street, Hingham, MA 02043, USA

Distributors for the UK and Ireland: Kluwer Academic Publishers, MTP Press Ltd, Falcon House, Queen Square, Lancaster LA1 1RN, UK

Distributors for all other countries: Kluwer Academic Publishers Group, Distribution Center, P.O. Box 322, 3300 AH Dordrecht, The Netherlands

Printed in The Netherlands

Preface

The Workshop on Hybrid Formulations of Wave Propagation and Scattering underwent a sequence of iterations before emerging in the format recorded here. These iterations were caused by various administrative and logistical problems which need not be detailed. However, its direction being set initially, the iterations led to modifications of the original concept so that the final form was arrived at through an indirect approach. This circumstance may explain some possible deficiencies which might have been removed, had the final concept been implemented directly.

The motivation arose from a perception that the newly restored interest, coupled with new developments, in hybrid methods employing progressing wave fields and oscillatory wave fields for time-harmonic and transient guided propagation in manmade or general geophysical environments, and for scattering by targets and irregularities, merits exposure to the wider scientific community. Accordingly, a meeting with highly tutorial content was envisaged. For administrative reasons, related to sponsorship and organizational structure, this objective could not be realized but, eventually, there emerged the possibility of convening an Advanced Research Workshop (ARW) under the auspices of the NATO Advanced Study Institute Series. The original concept was then modified to accommodate a Workshop, wherein state-of-the-art science is discussed by a relatively small group of specialists, instead of tutorial presentations of more basic material. Moreover, the concept was broadened to provide interaction between researchers in different disciplines for the purpose of exposing specialists in one area to the methods and techniques employed for similar problems by workers in other areas. The designation "hybrid" was now interpreted more broadly to encompass an approach to difficult wave problems that involves various combinations of techniques leading to a more effective solution. Much of the discussion at the Workshop centered in fact on what is meant by "effective". Here, the constraints imposed by differing requirements and objectives in various research organizations (academic, governmental and industrial) turned out to play an essential role. A major accomplishment of the Workshop was to impress upon groups motivated by one objective the different viewpoints of groups motivated by other objectives.

A prerequisite for establishing interaction among specialists from different disciplines is the ability of each to understand the language of the other. Although no formal mechanism was employed to achieve this goal, it turned out that the "language barrier" was overcome rather quickly so that intensive interplay characterized the discussions after the first day of presentations. Here, the informality of a Workshop, where speakers are interrupted for on-the-spot clarification, sets an effective framework. At the conclusion of the Workshop, there was agreement that everyone had learned to listen to everyone else. While this may not lead to a change of the modus operandi of those who already feel to be operating in the most "effective" manner, it will at least have helped them to understand the operational procedure of others having different requirements and constraints.

The disciplines included in the Workshop were colored by the background of the Director who, from personal knowledge, attempted to select areas (and individuals representing them) so as to optimize the likelihood of interdisciplinary dialogue. A special effort was made to include in each discipline an emphasis on analysis as well as numerical implementation. The analysis-computer interface was, in fact, a principal item for lively debate. Deliberately excluded were statistical phenomena in the propagation environment as well as the consideration of inverse problems. The scope of these unquestionably important problem areas was regarded to be so broad as to dilute the concentration on the other topical themes. Invitations were extended to workers in electromagnetics, optics, tropospheric propagation, underwater acoustics, elasticity and seismology. The final program had gaps in some of these areas because of last minute perturbations caused by the inability of several invitees to attend. Nevertheless, the strong representation in electromagnetics, guided optics, elasticity and underwater acoustics included enough diversity to demonstrate that a Workshop with strong interdisciplinary flavor can be an enriching experience. Should the exercise be repeated in the future, one may draw on the present experience to broaden and improve.

The papers included in this volume are based on preprints submitted by participants at the time of the Workshop and were solicited by the Director who also made suggestions concerning their length and contents. By an informal internal review scheme, most papers were subjected to a reading by at least one of the other contributors. Thereafter, only a minimal attempt was made to edit the final submitted manuscripts and to standardize the notation. This route was followed to expedite the processing of this volume, but it resulted in a lack of uniformity and in certain linguistic transgressions. On the other hand, this style preserved some of the spontaneity and international flavor that characterized a Workshop attended by participants from many countries.

The contributions are grouped into four sections: 1. Rays and modes; 2. Rays and beams; 3. Transient propagation and scattering; and 4. Numerical modeling. The hybrid viewpoint is exploited most specifically in section 1, which contains several contributions emphasizing the interplay between time-harmonic ray fields and modal fields in guiding configurations, the role of adiabatic invariants and of spectral considerations in defining local modes in waveguides with weak longitudinal and lateral variation, and, via two introductory papers, the hybrid mix of progressing (ray) and oscillatory (mode) fields. Section 2 begins with ray methods for time-harmonic diffraction problems involving edges on planar and curved boundaries, and with the uniformization of conventional ray techniques required there. Also illustrated is the use of ray methods for propagation in complicated inhomogeneous media and for propagation in tapered guiding structures. Conventional ray theory is then generalized to include complex rays and their utilization for the description of beam type fields. This is followed by several examples of beam tracing in different propagation and diffraction configurations. The construction of transient field solutions in terms of progressing (wavefront) and oscillatory (resonance) constituents is taken up in section 3, and is applied there to diffraction problems and to propagation in weakly and strongly dispersive environments. Here, the requirement of causality leads to formulations that effectively employ hybrid wavefront-resonance combinations. While numerical implementation is of concern throughout these presentations, the contributions in section 4 focus attention on various algorithms that have been devised for solution of time-harmonic and transient propagation under quite general conditions. Here, one may gain an appreciation of the accuracy, complexity and efficiency of computer codes derived from models based on rays, modes and beams, and on layered approximations for inhomogeneous environments.

The contributions in each section demonstrate both the diversity of wave problems in different disciplines and the commonality of techniques to cope with them. While a specialist in one discipline may not be familiar with the terminology in another discipline, he should recognize the relevance of what is done elsewhere by the similarity of the mathematical formulation and by the physical content. Cross fertilization was achieved rather well by direct contact at the Workshop, and it is hoped that this feature has not been lost through the more formal structure of papers in this volume. Especially those readers unfamiliar with propagation in elastic media and its application to seismology should not be discouraged by the complicated structure of the basic equations since the techniques employed in the seismological examples here deserve to be carefully explored in electromagnetics, optics and underwater acoustics. Conversely, electromagnetic diffraction, both time-harmonic and transient, has been developed to a degree not yet incorporated into other types of fields.

As noted earlier, much time at the Workshop was spent in discussion periods. Although the discussions were taped, it was impractical either to incorporate the voluminous verbatim transcripts, or to have the Director edit them, subject to approval of the participants. The possible gains obtained by these options would have been offset by substantial delays in publication. It is hoped that the included Summary, based on the reports from the Working Groups, will provide the principal perspectives and assessments that emerged from the exchanges of views during the Workshop.

The Workshop would not have been possible without the assistance of the Organizing Committee, which carried the burden of coping with logistical and organizational problems in Italy and coordinating these problems with a non-resident, the Director. I express my sincere gratitude especially to Professors Laura Ronchi Abbozzo and Annamaria Scheggi at the Institute for Electromagnetic Wave Research (IROE) in Florence for their dedication. My appreciation is also extended to Professor V. Cappellini, Director of IROE, for his generous support in making available administrative services for typing, preparation of preprints and other conference materials. A generous financial contribution from SELENIA, Rome, is also gratefully acknowledged. Finally, it has been a pleasure to interact with Dr. M. DiLullo, the Director of the NATO ARW Program, who has extended friendly advice on a host of administrative details.

Leopold B. Felsen
Workshop Director

TABLE OF CONTENTS

PART I
RAYS AND MODES

PROGRESSING FIELDS, OSCILLATORY FIELDS, AND HYBRID COMBINATIONS

L.B. Felsen

Department of Electrical Engineering and Computer Science/
Microwave Research Institute
Polytechnic Institute of New York
Route 110
Farmingdale, NY, 11735 USA

ABSTRACT

Progressing and oscillatory waves provide alternative time-harmonic or time-dependent descriptions of propagation in layered media, and of scattering phenomena. A progressing formulation characterizes the wave motion in terms of direct and multiple wavefronts or rays, and an oscillatory formulation in terms of resonances or modes. Each description is convenient and physically incisive when it requires few elements but inconvenient and physically less transparent when it requires many elements. In the latter event, it is desirable to express many inconvenient elements collectively as fewer convenient ones. For a variety of propagation and scattering environments, this can be done by expressing rays (wavefronts) collectively in terms of modes (resonances), or modes (resonances) collectively in terms of rays (wavefronts). When performed selectively, there emerges a hybrid representation combining progressing and oscillatory fields in uniquely defined proportions. Rigorous for coordinate separable guiding or scattering environments, the progressing-oscillatory equivalence can be extended approximately also to environments with weak departures from separability. The theory, based on Poisson summation and on alternative treatments of wave spectra, is summarized here, with references to diverse applications in the recent published literature.

I. INTRODUCTION

Two principal methods of analyzing wave propagation involve a description in terms of either progressing (traveling wave) or oscillatory (standing wave) events. In the former, the wave motion is organized along wavefronts that travel away from the source region along trajectories called rays. In an isotropic environment, the rays are perpendicular to the wavefronts. Under conditions admitting multiple reflections, for example, in layered media, an observer is reached by wavefronts coming directly from the source and also after repeated interaction with the boundaries. Analogous considerations apply to scattering where the wavefronts undergo multiple traversals of the scattering object. A wavefront conveys local information about the environment, gathered along the ray path that connects source and receiver. Changes in source and receiver locations generally change the ray path and therefore the local information corresponding to it. The wavefront description is useful at short observation times since, due to causality, their number is then relatively small and the fields carried along them can be readily resolved. At long observation times, the number of wavefront arrivals is large, their individual isolation difficult, and their summation inconvenient. It would then be desirable to treat their contribution collectively in terms of other types of wave processes.

In multiple interaction environments, wave motion can alternatively be synthesized by oscillatory events that are phase coherent so that, in the absence of dissipative or radiation damping, they maintain themselves even without continuous excitation. These modal or resonant fields encompass the environment as a whole, if it is bounded, or transverse portions thereof, if one (longitudinal) dimension is unbounded, as in a layer. Modal fields therefore convey global information independent, except for excitation amplitude, of source and receiver arrangements. They generally are effective at long observation times when the environment has fully responded to the incident disturbance. At short observation times, when part of the environment is still at rest, they are inconvenient because the destructive interference of many of them must then produce the quiescent conditions in the as yet unexcited portion. Their collective contribution is then better expressed by other wave phenomena, for example, the wavefronts mentioned above. Thus, progressing and oscillatory descriptions are complementary in the sense that one is convenient when the other is awkward. These observations may be supplemented by noting that a "convenient" description usually also provides a better physical understanding of the wave phenomena. The same considerations apply to the time-harmonic regime where equiphase surfaces replace wavefronts and normal modes replace resonances. Collective treatment of one or the other is again desirable when the relevant contributions from either become excessively large.

The problems posed above may be phrased as requiring the transformation of poorly into more rapidly convergent representations. For wavefronts and resonances, or rays and modes, the transformation from one to the other is accomplished by Poisson summation which expresses a group of elements as a sum of its Fourier transforms. Rays and modes are essentially one another's Fourier transforms.[1] Thus, the Poisson sum transforms a series of ray fields into a series of modal fields, and vice versa. This equivalence has been long recognized and widely used in the theory of guided propagation. However, since rays (wavefronts) alone or modes (resonances) alone are not convenient for <u>all</u> observation ranges of interest, one has in practice employed either the one or the other, with a switching to the more convenient one in an overlapping domain where both are still acceptable. This circumstance has motivated the search for a hybrid formulation that combines rays (wavefronts) and modes (resonances) within a self-consistent framework wherein the need for switching is eliminated, and the number of elements retained in the one uniquely determines the number retained in the other.[2,3] Viewed more broadly, a given finite spectral wavenumber interval can be filled either with ray fields or with mode fields plus a remainder to take care of any "spectral voids" near the edges. These alternative descriptions express two distinctive orderings of the constructive interference between the spectral constituents.

The interplay between rays (wavefronts) and modes (resonances), while applicable most directly to coordinate separable configurations, can also be exploited for non-separable structures. For weak departures from separability, one can define adiabatic invariants that ensure that field constituents adapt smoothly to the changing environment without coupling, in a lowest order approximation, to other "local" constituents.

The hybrid method was first developed for guided electromagnetic and acoustic fields[4-7], and thereafter for SH motion in elastic media.[8-10] A recent application is to transient scattering by a smooth convex object.[11] The principles of the method are here reviewed, with references to the published literature for details and examples.

II. COLLECTIVE TREATMENT AND HYBRID FORMULATION OF POORLY CONVERGENT REPRESENTATIONS

II.1. Collective Treatment by Poisson Summation

II.1.1 The Poisson sum formula. The ray-mode equivalence may be regarded within the general framework of converting a poorly convergent representation into another that expresses the collective effect of either more succinctly. Among the various mathematical techniques addressing this problem, the Poisson summation formula, which expresses an infinite series of elements f_q as a series over the Fourier transforms F_p of these elements, is especially relevant, viz.[12]

$$\sum_{q=-\infty}^{\infty} f_q = \sum_{p=-\infty}^{\infty} F_p \tag{1}$$

where

$$F_p = \frac{1}{2\pi} \int_{-\infty}^{\infty} f(\tau)\, e^{ip\tau}\, d\tau \tag{1a}$$

In (1), $f(\tau)$ is obtained from f_q on replacing the discrete index q by the continuous variable $(\tau/2\pi)$, it being required thereby that f_q is continuable into a function $f(\tau)$ which reduces to f_q when $(\tau/2\pi)$ takes on integer values. We shall assume that f_q satisfies these requirements. That the series on the right-hand side of (1) has convergence properties different from that on the left-hand side may be anticipated from the observation that a process wide (poorly convergent) in configuration space has a narrow Fourier spectrum. The relevance of the formulation in (1) to the ray-mode equivalence arises from the fact, shown subsequently, that ray fields and mode fields are one another's Fourier transforms.

For a hybrid formulation, one seeks to transform not all, but only a portion, of the elements f_q into the elements F_p. This requires use of the finite Poisson summation formula[12]

$$\sum_{q=q_1}^{q_2} f_q = \frac{1}{2}\left(f_{q_1} + f_{q_2}\right) + \sum_{p=-\infty}^{\infty} F_p(q_1,q_2) \tag{2}$$

where

$$F_p(q_1,q_2) = \frac{1}{2\pi} \int_{\tau_1}^{\tau_2} f(\tau)\, e^{ip\tau}\, d\tau = F_p + \hat{F}_p(q_1) + \hat{F}_p(q_2) \tag{2a}$$

with F_p defined in (1a) and

$$\hat{F}_p(q_{\frac{1}{2}}) = \pm\frac{1}{2\pi} \int_{\mp\infty}^{\tau_{\frac{1}{2}}} f(\tau)\, e^{ip\tau}\, d\tau \ , \ \tau_{1,2} = 2\pi q_{1,2} \tag{2b}$$

The alternative form

$$\sum_{q=q_1}^{q_2} \varepsilon_q f_q = \sum_{p=-\infty}^{\infty} F_p(q_1,q_2) \tag{3}$$

where $\varepsilon_q = 1/2$ for $q = q_{1,2}$ and $\varepsilon_q = 1$, $q \neq q_{1,2}$, highlights the effect of truncating the f_q series by half-amplitude weighting of the first and last retained elements.

II.1.2 Asymptotic approximations. While the formulas in (2) or (3) are exact, subsequent interpretation of the results for the present purpose is facilitated by recourse to asymptotic approximations based on the assumption that the elements f_q behave in a manner that permits identification of a slowly varying amplitude a_q and a rapidly varying phase $k\psi_q$, where k is a (large) reference wave number:

$$f_q = a_q \exp(ik\psi_q) \tag{4}$$

This identification pertains to the local plane waves that form the basis of the wave phenomena to be studied here. More generally, the local plane waves may be combined into a "spectral integral," with f_q occurring in the integrand. In that event, the integration remains in the corresponding transition to $f(\tau)$ in (2a). This aspect is pursued later on. When (4) is continued to $f(\tau)$ and substituted into (2a), the integral may be evaluated asymptotically by the method of saddle points or stationary phase. Stationary points τ_p of the composite phase $k\Psi_p$ are solutions of

$$\frac{d\Psi_p}{d\tau} = 0 \text{ at } \tau_p, \quad \Psi_p = \psi(\tau) + p\tau/k \ . \tag{5}$$

Principal contributions to the integral arise from those τ_p which lie on the interval $\tau_1 < \tau_p < \tau_2$ and also from the endpoints $\tau_{1,2}$. When the saddle points are well separated and none lies near an endpoint nor near a singularity (if any) of $f(\tau)$, each may be treated in isolation. The relevant integral for saddle point evaluation is F_p in (1a) taken along the steepest descent path $\bar{C}'$ (SDP) in the complex τ-plane (see Fig. 1) and yields the result[13]

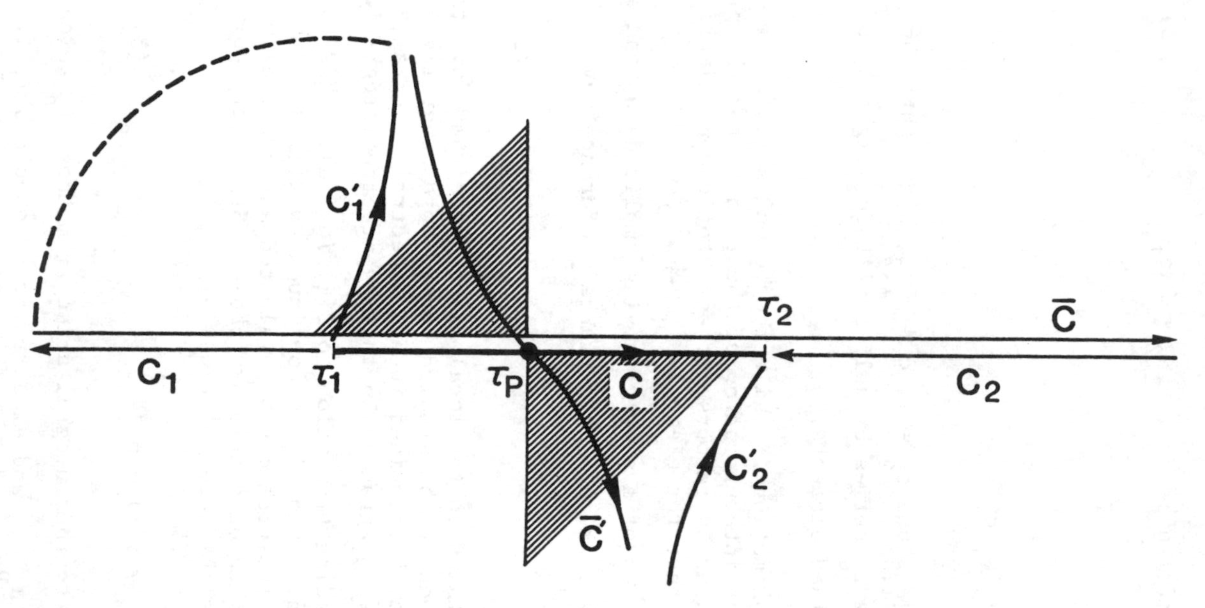

Fig. 1. Contours for spectral integral in complex τ-plane. The initial contour C is expressed as $C = \bar{C} + C_1 + C_2$, with subsequent deformation into $\bar{C}'$ (SDP), C_1' and C_2' into valleys corresponding to the saddle point τ_p. For several saddle points in the inverval $\tau_1 < \tau_p < \tau_2$, there is a corresponding multiplicity of valley regions.

$$F_p(\tau_p) \sim \frac{1}{2\pi}\sqrt{\frac{2\pi i}{k\Psi''_p(\tau_p)}}\; a_p(\tau_p)\, e^{ik\Psi_p(\tau_p)}, \quad \tau_1 < \tau_p < \tau_2 \tag{6}$$

with a prime denoting the derivative with respect to the argument. Equation (6) may be written in a form analogous to (4),

$$F_p(\tau_p) \sim A_p(\tau_p)\, e^{ik\psi(\tau_p)} \tag{7}$$

where

$$A_p(\tau) = \frac{1}{2\pi}\, a_p(\tau_p)\sqrt{\frac{2\pi i}{k\Psi''_p(\tau_p)}} \tag{7a}$$

Endpoint contributions to the integral in (2a) are always present. When isolated from a saddle point and from singularities (if any) of $f(\tau)$, the endpoints yield, asymptotically, by integration along C'_2 and C'_1 in Fig. 1,[13]

$$F_p(q_{2,1}) = \frac{1}{2\pi}\int_{C'_{2,1}} f(\tau)e^{ip\tau}\, d\tau \sim \pm\, \frac{f(\tau_{2,1})\exp(ip\tau_{2,1})}{2\pi i k\Psi'_p(\tau_{2,1})}$$

$$= \mp\, \frac{i/2\pi}{k\psi'(\tau_p)+p}\, f_{q_{2,1}} \tag{8}$$

i.e., the limiting elements of the <u>original</u> series, but with a modified amplitude. The p-series in (2), with (7), can be summed into closed form via the formula

$$\sum_{p=-\infty}^{\infty} \frac{1}{\nu-p} = \pi\cot(\nu\pi) \tag{9}$$

whence

$$\sum_{p=-\infty}^{\infty} F_p(q_{2,1}) \sim \pm\frac{1}{2}\,\Delta_{q_{2,1}}\, f_{q_{2,1}}, \quad \Delta_{q_{2,1}} = i\cot[\pi k\psi'(\tau_{2,1})] \tag{10}$$

When these results are collected, one finds

$$\sum_{q=q_1}^{q_2} f_q \sim \sum_{p=p_1}^{p_2} F_p U(\tau_2-\tau_p)\, U(\tau_p-\tau_1) + \bar{f}_{q_1} + \bar{f}_{q_2} \tag{11}$$

where

$$\bar{f}_{q_{2,1}} = \frac{1}{2}(1 \pm \Delta_{q_{2,1}}) f_{q_{2,1}} \tag{11a}$$

and $U(\tau) = 0, \tau < 0$, but $U(\tau) = 1, \tau > 0$. The formulation in (11) expresses the original group of elements f_q in terms of a series of its Fourier transformed elements F_p plus elements $\bar{f}_{q_{2,1}}$, which are the collectively weighted last and first elements, respectively, in the original group. These "collective" elements, expressive of the truncation of the original set, account for the spectral voids $(\tau_{p_1} - \tau_1)$ and $(\tau_2 - \tau_{p_2})$ left unfilled by the F_p elements, where τ_{p2} and τ_{p1} denote the saddle points closest to τ_2 and τ_1, respectively, and contained within the interval $\tau_1 < \tau_p < \tau_2$.

The elements f_q and F_p generally depend also on other parameters which determine the distribution and location of the saddle points τ_p. Should it happen that the f_q set and F_p set fill the spectral interval $\tau_1 < \tau < \tau_2$ completely (i.e., $\tau_{p1} = \tau_1$, $\tau_{p2} = \tau_2$), implying coincident saddle points and endpoints, then the endpoint contributions $F_p(q_{2,1})$ equal (1/2) of the saddle point contributions $F_p(\tau_{2,1})$ [13] so that (10) may be rewritten in the symmetric form

$$\sum_{q=q_1}^{q_2} \varepsilon_q f_q \sim \sum_{p=p_1}^{p_2} \varepsilon_p F_p \tag{12}$$

which exhibits the bilateral equivalence of the truncated sets f_q and F_p. Alternatively, when an endpoint lies approximately midway between the last included and the first excluded saddle point, then $\cot[\pi k\psi'(\tau_1)]$ and (or) $\cot[\pi k\psi'(\tau_2)]$ in (10) is small. By setting $\Delta_{q_1} \approx 0$ and (or) $\Delta_{q_2} \approx 0$ in (10), one thereby obtains a different simplification of (11). Finally, when the saddle point τ_{p_2} approaches the endpoint τ_2, the simple asymptotic formulas in (6) and (7) must be replaced by the uniform formula involving a Fresnel integral [13]

$$F_p(q_2) \sim \frac{e^{ik\Psi(\tau_{p_2})}}{2\pi} a(\tau_{p_2}) h_p Q(s_e) + \frac{e^{-s_e^2}}{2s_e}[a(\tau_2)h_e - a(\tau_{p_2})h_p] \tag{13}$$

$$Q(y) = \int_y^\infty e^{-x^2}\,dx, \quad s_e = [ik(\Psi(\tau_{p_2}) - \Psi(\tau_2))]^{1/2},$$

$$h_e = \frac{-s_e}{ik\Psi'(\tau_2)}, \quad h_p = \sqrt{\frac{-2}{ik\Psi''(\tau_{p_2})}} \tag{13a}$$

where $\arg h_p = \arg(d\tau)\tau_{p_2}$, $d\tau$ being an element along the steepest descent path SDP through τ_{p_2}, and $\arg s_e$ is defined so that $h_e \to h_p$ as $s_e \to 0$. With (13), passage of the saddle point τ_{p_2}, and therefore the corresponding element F_{p_2}, out of the spectral interval, or transition of a previously external saddle point τ_p and therefore the corresponding element F_p into the spectral interval, may be tracked smoothly and continuously.

To demonstrate the bilateral equivalence between the f_q and F_p, we repeat the steps in (2)-(10), now starting with the elements F_p:

$$\sum_{p=p_1}^{p_2} F_{p_1} = \frac{1}{2} F_{p_2} + \frac{1}{2} F_{p_1} + \sum_{q=-\infty}^{\infty} \hat{f}_q(p_1,p_2) \tag{14}$$

where with $p \to \mu/2\pi$,

$$\hat{f}_q(p_1,p_2) = \frac{1}{2\pi}\int_{\mu_1}^{\mu_2} F(\mu)\, e^{i\mu q} d\mu \tag{14a}$$

$$= \hat{f}_q + \hat{f}_q(p_1) + \hat{f}_q(p_2) \tag{14b}$$

with contours in the complex μ-plane analogous to those in the complex τ-plane of Fig. 1,

$$\hat{f}_p \to \int_C, \quad \hat{f}_q(p_1) \to \int_{C_1'}, \quad \hat{f}_q(p_2) \to \int_{C_2'} \tag{14c}$$

Substituting for $F(\mu)$ from (2a), with p replaced by the continuous variable τ, inverting the orders of the μ and τ integrations, and recognizing that the μ-integration yields the delta function (2π) $\delta(\tau + \mu)$, one verifies from (14a) that

$$\hat{f}_q = f_{-q} \tag{14d}$$

Then referring to (11) and (6),

$$\sum_{p=p_1}^{p_2} F_p = \sum_{q=\hat{q}_1}^{\hat{q}_2} \hat{f}_q + \bar{F}_{p_1} + \bar{F}_{p_2} \tag{15}$$

with the collective elements

$$\bar{F}_{p_{1,2}} = \frac{1}{2}(1 \pm \bar{\Delta}_{p_{1,2}})F_{p_{1,2}}, \quad \bar{\Delta}_{p_{1,2}} = i \cot[\pi k \Psi'(\mu_{1,2})] \tag{15a}$$

and the μ_q determined from the saddle point condition (see (5))

$$\frac{d\hat{\Psi}_q}{d\mu} = 0 \text{ at } \mu_q, \quad \hat{\Psi}_q(\mu) = \Psi(\mu_q) + \mu q/k \tag{15b}$$

In (15), $\mu_{\hat{q}_1}$ and $\mu_{\hat{q}_2}$ denote the saddle points μ_q closest to the end-points μ_1 and μ_2, respectively, and contained within the interval $\mu_1 < \mu_q < \mu_2$. Although the elements f_q in (15) and in (2) are identical (except for the reflection $q \to -q$), the number of elements in the set $\hat{q}_1 \le q \le \hat{q}_2$ differs in general from that in the original set $q_1 \le q \le q_2$. The difference is accounted for by the different collective elements in (11) and (15). If the integral in (14a) had been defined with $\exp(-i\mu q)$ in the integrand, there would have been a complete equivalence $\hat{f}_q = f_q$, with the implication that F_p in (2a) and $\hat{f}_q$ in (14a) (with the integration paths extended to infinity) are Fourier transform pairs.

It follows that a set of elements f_q can be expressed by a set of its Fourier transforms F_p (and vice versa) in two different ways, depending on whether one wishes to characterize the truncation effect in terms of "collective f_q" or "collective F_p" elements. These alternative formulations are contained in (11), repeated here for convenience,

$$\sum_{q=q_1}^{q_2} f_q = \sum_{p=p_1}^{p_2} F_p + \bar{f}_{q1} + \bar{f}_{q2}, \quad F_p \sim A_p e^{ik\Psi_p}, \quad f_q \sim a_q e^{ik\psi_q}, \tag{16}$$

$$\bar{f}_{q_{1,2}} = \frac{1}{2}(1 \pm \Delta_{q_{1,2}})f_{q_{1,2}}, \quad \Delta_{q_{1,2}} = i \cot[\pi k \psi'(\tau_{1,2})] \tag{16a}$$

and in (15), rewritten as

$$\sum_{q=\hat{q}_1}^{\hat{q}_2} \hat{f}_q = \sum_{p=p_1}^{p_2} F_p - \bar{F}_{p_1} - \bar{F}_{p_2} \tag{17}$$

$$\bar{F}_{p_{1,2}} = \frac{1}{2}(1 \pm \bar{\Delta}_{p_{1,2}}) F_{p_{1,2}} , \quad \bar{\Delta}_{p_{1,2}} = i \cot[\pi k \Psi'(\tau_{1,2})] \tag{17a}$$

Since the formulation in (17) (see (15)) is based on the F_p spectral interval $\tau_1 \leq \tau \leq \tau_2$ $(p \to \tau/2\pi)$, the number of elements f_q with $q_1 \leq q \leq q_2$ cannot be arbitrarily specified as in (16) but is subject to the constraint on τ_q noted after (15b). Dual considerations apply to (16) when one seeks to replace a set of F_p elements by f_q elements with collective f_q remainders.

II.1.3 Spectral integrals vs asymptotic approximations: example of of a longitudinally homogeneous guiding environment.

a. Spectral integral representation for generalized rays. When the asymptotic form of the element f_q in (4) can be generated from an exact integral representation, application of the Poisson summation formulas in (1)-(3) leads to exact instead of asymptotic equivalence statements. This will be illustrated by assuming that f_q is a multiple reflected scalar ray field in a homogeneous plane parallel layer of width $\underline{a}$, with the index q denoting the number of reflections encountered between a y-directed harmonic line source at S = (x',z') = (0,z') and an observer at P = (x,z). The totality of rays is grouped into the familiar four categories distinguishing departure at S toward, and arrival at P from, the top and bottom boundaries, respectively. The ray field f_q is taken to refer to one of these categories. Such a ray field is generated by a spectral integral (a "generalized" ray), typically of the form [14]

$$f_q^g = \frac{i}{4\pi} \int_{-\infty}^{\infty} \exp[ik\psi_q^g(\zeta)]\frac{d\zeta}{\xi} , \quad \xi = (1 - \zeta^2)^{1/2} \tag{18}$$

where ζ is the spectral variable (normalized wavenumber) along the z-direction parallel to the layer boundaries, ξ (with Im $\xi \geq 0$ on the spectral Riemann sheet) is the wavenumber along the x-direction perpendicular to the boundaries, and $k = \omega/v$, with ω denoting the angular frequency in the suppressed time dependence $\exp(-i\omega t)$ and v denoting the wave propagation speed in the medium. The phase ψ_q^g has the form

$$k\psi_q^g(\zeta) = k\Omega(\zeta) + q\phi(\zeta) \tag{19}$$

where

$$\Omega(\zeta) = \zeta z + \xi(x_> - x_<) \tag{19a}$$

accounts for translational effects from S to P while

$$\phi(\zeta) = 2ka\xi(\zeta) + \phi_o(\zeta) + \phi_a(\zeta) \tag{19b}$$

accounts for reverberation effects between the boundaries. Here, $\phi_o(\zeta)$ and $\phi_a(\zeta)$ denote, respectively, the phases of the spectral plane wave reflection coefficients at the lower ($x = 0$) and upper ($x = a$) boundaries, and the form of $\Omega(\zeta)$ in (19a) refers to one of the ray categories with equal number of reflections at top and bottom. The integrand in (18) has a saddle point ζ_q determined by

$$\frac{d\psi_q^g}{d\zeta} = 0 \text{ at } \zeta_q,$$

the solution of which defines the ray path between S and P according to asymptotic ray theory. Application of the saddle point formula along the SDP through ζ_q then yields

$$f_q^g \sim f_q = a_q \exp(ik\psi_q),\ a_q \equiv a_q(\zeta_q) = \frac{1}{4}\frac{1}{\xi(\zeta_q)}\left[\frac{2i}{\pi k\psi_q''}\right]^{1/2}, \tag{21}$$

with a prime denoting the derivative with respect to the argument. Subject to the identification $\psi_q(\zeta_q) = \psi_q$, (21) is evidently of the same form as (4). The amplitude variation a_q is here expressed in terms of the wavefront curvature ($\propto\psi_q''$) but can equivalently be expressed in terms of ray tube spreading.[14,15] The result in (21) could have been constructed directly by asymptotic ray theory without recourse to the exact spectral integral in (18). When (18) is substituted into (2), one obtains for the transformed function $F_p(q_1,q_2)$, after a permissible interchange of the orders of the ζ and τ integrations

$$F_p^g(q_1,q_2) = \frac{i}{8\pi^2}\int_{-\infty}^{\infty}\frac{d\zeta}{\xi}\exp[ik\Omega(\zeta)]\int_{\tau_1}^{\tau_2}\exp\{i\tau[\frac{\phi(\zeta)}{2\pi} + p]\}d\tau \tag{22a}$$

$$= F_p^g(q_1) - F_p^g(q_2) \tag{22b}$$

where, after performing the integration over τ and recognizing that $\exp(ip\tau_{1,2}) = 1$,

$$F_p^g(q_{1,2}) = \frac{-i}{8\pi} \int_{-\infty}^{\infty} \frac{d\zeta}{\xi} \frac{\exp[ik\psi^g_{\tau_{1,2}}(\zeta)]}{i[\frac{\phi(\zeta)}{2\pi} + p]} \tag{23}$$

These integrals have common pole singularities ζ_p specified by

$$\frac{\phi(\zeta)}{2\pi} + p = 0 \text{ at } \zeta_p, \tag{24}$$

which is the modal resonance condition defining the eigenvalues for the guided modes. The integrals have individual saddle points at $\zeta_{1,2}$, where

$$\psi'^g_{\tau_{1,2}}(\zeta_{1,2}) = 0 \tag{25}$$

and therefore, from (20), identify rays that undergo q_1 and q_2 reflections, respectively. By deforming the integration paths for $F_p^g(q_1)$ and $F_p^g(q_2)$ into the steepest descent paths through these saddle points (see Fig. 1), one obtains rapidly decaying integrands. This facilitates asymptotic or direct numerical evaluation. During the path deformation, one has to extract residues at any poles ζ_p traversed in this process. Since common intercepted poles in $F_p^g(q_1)$ and $F_p^g(q_2)$ have identical residues (from (24), $\tau_{1,2}\ \phi(\zeta_p)$ is an integer multiple of 2π), the only contributing poles are those lying between ζ_1 and ζ_2. The residue contributions are the partial guided mode fields contributed by the ray category in (18), with (19)

$$F_p^g = -\ 2\pi i \frac{i}{4\pi} \frac{\exp[ik\Omega(\zeta_p)]}{i\ \phi'(\zeta_p)\xi_p}, \quad \zeta_1 < \zeta_p < \zeta_2. \tag{26}$$

The p-summation in (2), which remains intact for the integrals (23) along the deformed contours, can be done in closed form on the integrands, using (9). Collecting these results, one obtains for the group of ray integrals in (18):

$$\sum_{q=q_1}^{q_2} f_q^g = \sum_{p=p_1}^{p_2} F_p^g + (\Delta f)^g_{q_1} - (\Delta f)^g_{q_2} + \frac{1}{2} f^g_{q_1} + \frac{1}{2} f^g_{q_2} \tag{27}$$

where

$$(\Delta f)^g_{q_{1,2}} = (\frac{-1}{8\pi}) \int_{C_{1,2}} d\zeta \exp\left[ik\ \Psi^g_{q_{1,2}}(\zeta)\right] \cot[\frac{\phi(\zeta)}{2}] \tag{27a}$$

The equivalence in (27), with (27a), is exact. If $\cot[\frac{\phi(\zeta)}{2}]$ is slow varying near ζ_1 and ζ_2 (i.e., none of the poles lie near the saddle points of the integrands in (23)), this factor may be approximated by its value at $\zeta_{1,2}$ and removed from the integrand. When this is done, one has

$$(\Delta f)^g_{q_{1,2}} \sim \frac{1}{2} \Delta_{q_{1,2}} f^g_{q_{1,2}}, \quad \Delta_{q_{1,2}} = i \cot\left[\frac{\phi(\zeta_{1,2})}{2}\right] \tag{28}$$

whence (27) then has the same form as (16), except for replacement of the ray field f_q by the generalized ray field f^g_q. In view of the asymptotic approximation in (28)), consistency requires that, also, $f^g_q \sim f_q$. When a pole of the integrand in (27a) lies near a saddle point, a uniform asymptotic approximation is appropriate. This leads to a Fresnel integral as in (13) (see [2]).

b. Direct asymptotics. It may be verified that the asymptotic approximations leading to (28) et. seq. via the exact spectral integral route agree with those obtained from (16) when (2) is applied to the asymptotic ray field in (21), which could have been constructed directly by asymptotic ray theory without recourse to the exact ray integral in (18). Details may be found in references 2, 6, 16 and 17.

II.2 Hybrid representation. Assuming that a desired wave field G is initially expressed in terms of the set f_q,

$$G = \sum_{q=q_1}^{\infty} f_q, \tag{29}$$

let us partition this set into N subgroups

$$G = \left(\sum_{q_1}^{q_2} + \sum_{q_2+1}^{q_3} + \cdots + \sum_{q_N+1}^{\infty} \right) f_q \tag{30}$$

Subgroups with favorable convergence properties are to be retained intact while subgroups with unfavorable convergence properties are to be converted into F_p groupings. The conversion may take place in terms of any of the alternative options listed in (16)-(17). This leads to a hybrid representation for G that involves alternating groupings of f_q and F_p elements, with collective elements $\bar{f}_q$ or $\bar{F}_p$ added to fill any spectral voids between adjacent groups. A hybrid formulation appears typically as

$$G = \sum_{q_1}^{q_2} f_q + \sum_{p_1}^{p_2} F_p + \sum_{q_3+1}^{q_4} f_q + \cdots + \sum_{p_j}^{\infty} F_p + R(q_2,P_1)$$
$$+R(p_2,q_3+1) + R(q_4,p_3+1) + \cdots + R(q_N,p_j) \tag{31}$$

where $R(q,p)$ denotes collective elements pertaining to the spectral interval between τ_q and τ_p (see Fig. 2). The hybrid representation for a source-excited field G may be obtained not only by the Poisson summation route but also by manipulation of a spectral integral (see references 5,8,9).

III RAYS AND LOCAL MODES IN LONGITUDINALLY VARYING (NON-SEPARABLE) WAVEGUIDES

III.1 Spectral Scaling

When the guiding environment varies in the longitudinal direction in a manner that renders the wave equation non-separable, it is no longer possible to define translationally invariant plane wave or local plane wave congruences which can be used, by superposition, to synthesize the cross-sectional self-consistency required of modal fields. However, for sufficiently slow deviations from separability, one may attempt to satisfy modal consistency for each mode locally along the propagation direction, with a z-dependent modal resonance condition. The resulting modal fields, so-called local or adiabatic modes, adapt continuously to the changing guiding environment in a manner that retains intact the individual features of each mode and avoids coupling to other modes. In spectral terms, each mode in a longitudinally non-varying waveguide is characterized by a spectral invariant, the function $-\frac{\phi(\zeta_p)}{2\pi}$ in (24), which is required, for the p-th mode, to have the constant value p. In a longitudinally varying environment, the spectral invariant is locally adjusted so that its functional value $-\frac{\phi[\zeta_p(z)]}{2\pi}$ (=p) remains constant. This implies a corresponding "scaling" of the spectral variable ζ along the longitudinal coordinate z. For details, see reference 18.

III.2 Ray-to-Mode Conversion

Asymptotic ray fields in a longitudinally changing waveguide are not hampered by the need for spectral scaling as in Sec. III.1 because they describe ab initio a local propagation process. Ray fields can be constructed without constraints of separability, provided only that conditions change sufficiently slowly to validate local plane wave propagation along each trajectory defined by the ray equations. The ability to define local modes in a non-separable configuration can be tied to the ability to treat multiple reflected ray fields collectively by Poisson summation as in Sec. II.1.3.b. Details may be found in reference 19.

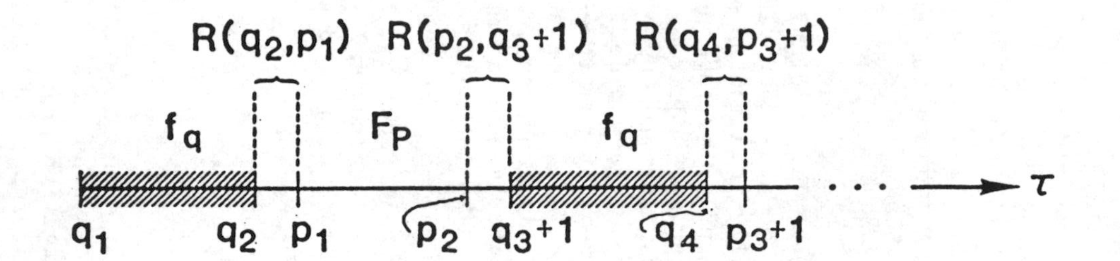

Fig. 2. Partitioning of spectral range for hybrid formulation. q_i and p_i denote spectral values delimiting intervals occupied by f_q and F_p elements, respectively. $R(q,p)$ and $R(p,q)$ are remainders accounting for spectral voids.

IV. HYBRID TREATMENT OF MULTIPLE REFLECTED RAY FIELDS TRANSMITTED THROUGH A LAYER

In Section III, it has been noted that collective treatment of multiple reflected ray fields can be employed to define local modes propagating inside a tapered layer with penetrable boundaries. The collective approach may be used as well to describe transmission through a layer. While Poisson summation (Section II.1) furnishes the necessary mathematical framework, the major contribution in the guiding problem arises from the stationary points inside the Fourier spectral interval whereas in the transmission problem, the major contribution arises from the "collective elements" due to the end-points of the spectral interval, provided that source and observer lie sufficiently far away from the layer boundaries. In that event, possible trapped modes inside the layer are excited and observed via evanscent tunneling and therefore have negligibly small amplitudes. These considerations have been illustrated for a (coordinate separable) circular layer and also for a (non-separable) tapered layer (reference 20).

V. SUMMARY

Progressing and oscillatory building blocks represent alternative options for structuring representations of source-excited wave fields. These alternative formulations are generally useful in complementary parameter ranges and furnish local and global information, respectively, about the propagation or scattering process. Being one another's spectral Fourier transforms, they can be converted one into the other by application of the Poisson sum formula. The hybrid technique seeks to exploit the favorable convergence properties of each within a single self-consistent framework through elimination, by collective summation, of the poorly convergent portions in these representations. A selected list of references indicates the variety of problems in various areas of wave propagation and scattering, to which the concepts have already been applied. Extension to elastic media with p-SV coupling at the boundaries is presently being developed. [21]

The exact ray-mode equivalence that may be established for coordinate separable configurations can be generalized to environments with weakly non-separable characteristics, as in a tapered open guiding channel. By use of adiabatic invariants that maintain mode shapes and ray trajectories locally intact, one may generate a scaled spectrum that accommodates to the range dependence. The resulting approximate spectral integral then contains all of the information about the propagation process, just like the exact spectral integral for the range independent case. The collective summation technique, applied to multiple internal reverberations, has also proved useful for transmission through non-plane-parallel layers. Thus, the hybrid point of view incorporates a degree of flexibility that may serve to enhance the understanding and simplify the calculation of complicated wave phenomena in various disciplines.

VI. ACKNOWLEDGEMENT

The research summarized here was sponsored in part by the Office of Naval Research under Contract No. N00014-79-C-0013, and in part by the Joint Services Electronics Program under Contract No. F-49620-82-C-004.

VII. REFERENCES

1. Gao, T.F. and E.C. Shang, "The Transformation Between the Mode Representation and the Generalized Ray Representation of a Sound Field," J. Sound Vib. 80 (1982), pp. 105-115.

2. Felsen, L.B., "Hybrid Ray-Mode Fields in Inhomogeneous Waveguides and Ducts", J. Acoust. Soc. Am., 69 (1981), pp. 352-361.

3. Kamel, A. and L.B. Felsen, "On the Ray Equivalent of a Group of Modes," J. Acoust. Soc. Am. 71 (1982), pp. 1447-1452.

4. Ishihara, T., Felsen, L.B. and A. Green, "High Frequency Fields Excited by a Line Source Located on a Perfectly Conducting Concave Cylindrical Surface," IEEE Trans. on Antennas and Propagation, AP-26, (1978), pp. 757-767.

5. Felsen, L.B. and A. Kamel, "Hybrid Ray-Mode Formulation of Parallel Plane Waveguide Green's Functions," IEEE Trans. on Antennas and Propagation, AP-29 (1981), pp. 637-649.

6. Niver, E., Cho, S.H., and L.B. Felsen, "Rays and Modes in an Acoustic Channel with Exponential Velocity Profile," Radio Science 16 (1981), pp. 963-970.

7. Migliora, C.G., Felsen, L.B. and S.H. Cho, "High-Frequency Propagation in an Elevated Tropospheric Duct," IEEE Transactions on Antennas and Propagation, AP-30 (1982) pp. 1107-1120.

8. Kamel, A. and L.B. Felsen, "Hybrid Ray-Mode Formulation of SH Motion in a Two-Layer Half Space," Bull. Seismol. Soc. Am., 71 (1981), pp. 1763-1781.

9. Kamel, A. and L.B. Felsen, "Hybrid Green's Function for SH Motion in a Low Velocity Layer," Wave Motion 5 (1983), pp. 83-97.

10. Niver, E., Kamel, A. and L.B. Felsen, "Modes to Replace Transitional Ray Fields in a Vertically Inhomogeneous Earth Modal," in preparation.

11. Heyman, E and L.B. Felsen, "Creeping Waves and Resonances in Transient Scattering by Smooth Convex Objects," IEEE Transactions on Antennas and Propagation, AP-31 (1983), pp. 426-437.

12. Papoulis, A., "Signal Analysis," McGraw-Hill, New York, 1977. Sec. 3.3.

13. Felsen, L.B. and N. Marcuvitz, "Radiation and Scattering of Waves," Prentice Hall, Englewood Cliffs, New Jersey, 1973. Secs. 4.1 and 4.6.

14. Ibid, Secs. 1.6 and 1.7.

15. Deschamps, G.A., "Ray Techniques in Electromagnetics," Proc. IEEE 60 (1972), pp. 1022-1035.

16. Felsen, L.B., "Progressing and Oscillatory Waves for Hybrid Synthesis of Source-Excited Propagation in Layered Media," to be published in Geophys. J. Roy. Astron. Soc.

17. Felsen, L.B., "Progressing and Oscillatory Fields for Propagation and Scattering," invited paper, to be published in IEEE Transactions on Antennas and Propagation.

18. Kamel, A. and L.B. Felsen, "Spectral Theory of Sound Propagation in an Ocean Channel with Weakly Sloping Bottom," J. Acoust. Soc. Am., 73 (1983), pp. 1120-1130.

19. Arnold, J.M. and L.B. Felsen, "Rays and Local Modes in a Wedge-Shaped Ocean," J. Acoust. Soc. Am., 73 (1983), pp. 1105-1119.

20. Einziger, P.D. and L.B. Felsen, "Ray Analysis of Two-Dimensional Radomes," IEEE Transactions on Antennas and Propagation, AP-31 (1983), pp. 870-884.

21. Lu, I.T., Felsen, L.B. and A. Kamel, "Eigenreverberations, Eigenmodes and Hybrid Combinations: A New Approach to Propagation in Layered Multiwave Media," to be published in Wave Motion.

TRANSFORM RELATIONS BETWEEN RAYS AND MODES

E. C. Shang*

University of Wisconsin- Madison
Department of Geology and Geophysics
Madison, WI 53706 USA

I. INTRODUCTION

The sound field produced by a point source in a stratified wave-guide can be expressed in three different representations based on different approaches to solve the boundary value problem, namely[1]:

1. Integral representation (spectral representation):

$$G(r,z,z_o) = \int_{\Gamma_\mu} F(z,z_o,\zeta)\, H_o^{(1)} (r\sqrt{k_o^{\,2} - \zeta^2})d\zeta \qquad (1)$$

2. Normal mode representation:

$$G(r,z,z_o) = \sum_m^\infty \phi_m \qquad (2)$$

3. Multiple scattering (generalized ray) representation:

$$G(r,z,z_o) = \sum_\ell^\infty B_\ell \qquad (3)$$

Each representation has advantages and disadvantages in different cases. Conventionally, wave propagation in a stratified waveguide is analyzed either in terms of normal modes or in terms of multiple reflected rays. When the number of modes or rays is excessively large, either method is inconvenient. Furthermore, the asymptotic (geometric) ray method runs into difficulties in transition regions near caustics and cups of the various ray systems. Since the FFT techniques were introduced to underwater acoustic field calculation (FFP)[2], calculation of the propagation field by using the integral representation directly becomes a useful numerical approach. But a pure numerical method lacks the physical interpretation that leads to a better understanding of the relevant propagation mechanisms.

To develop an "information concentrated" representation is an attractive goal for the wave propagation (scattering) investigators, because an "information concentrated" representation allows a more compact physical interpretation of the mechanism and saves computing time.

All of these three representations are formulated as a summation of its "member",--"m" member in a mode representation, "ℓ" members in a ray representation. Each "member" of the representation possesses a certain degree of "information concentration" for describing the field. The question is how to establish a suitable criterion justifying the degree of "information concentration"?

The concept of the "interval" of member, proposed by Weston in 1963[3], is a useful one. When the "interval" is measured by the horizontal component of wave number, we have[4]:

$$\Delta\mu_m = -2\pi/S_m \qquad \text{(for mode representation)} \tag{4}$$

$$\Delta\mu_\ell = S_\ell^2/r\dot{S}_\ell \qquad \text{(for ray representation)} \tag{5}$$

$$\Delta\mu = 1/\Delta L \qquad \text{(for spectral representation)} \tag{6}$$

where S_m is the "skip distance" of the modal ray, S_ℓ is the "skip distance" of the ray, and ΔL is a certain distance specified by a characteristic scale of the sound velocity profile[5].

According to the magnitude of different $\Delta\mu$, one can judge which representation is more "economical." For instance, if

$$\Delta\mu_m << \Delta\mu_\ell \tag{7}$$

this means that the information of the field carried by one of the "ℓ" members will be carried by a large number of "m" members (the number is $M \approx \Delta\mu_\ell/\Delta\mu_m$).

Furthermore, it is interesting to see whether the field superposed by a large number of "m" members in the interval (magnitude of $\Delta\mu_\ell$) is the same as given by one of the "ℓ" members in the same interval.

Early in 1966, Tolstoy and Clay obtained a fuzzy ray picture given by two mode interference (see Fig. 1)[6].

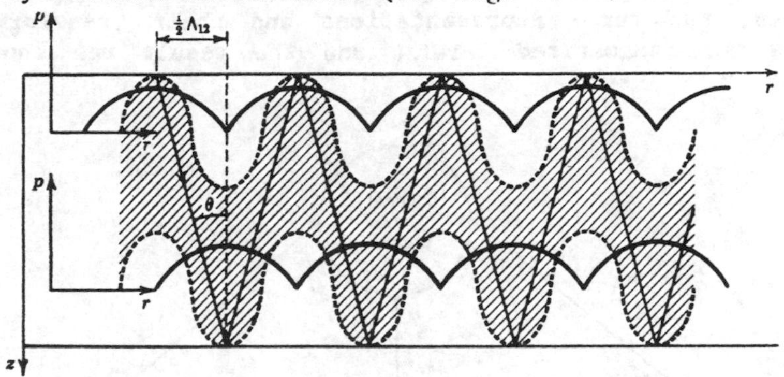

Fig. 1

It was shown by Tindle and Guthrie[7,8] that a cluster of modes will produce an interference maximum along a trajectory (emanating from the source) equivalent to the path of the modal ray for the central mode in the group (see Fig. 2).

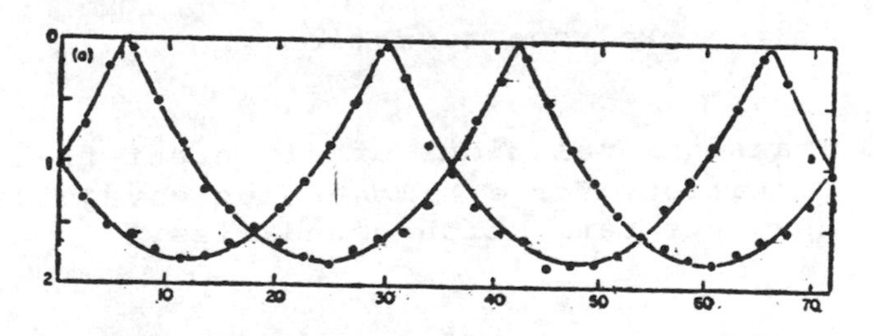

Fig. 2

These interesting features lent insight into the behavior of mode (ray) fields, which individually possess global (local) properties but collectively become strongly localized (globalized). It also motivates to research the transformation relation between modes and rays in a more rigorous framework.

Recently, the research on transformation between modes and rays has been progressing in two directions: (1) total transformation based on mathematical formula of Poisson summation; and (2) partial transformation based on mathematical formula of partial

Poisson summation. We will discuss them in part II and part III, respectively.

II. TOTAL TRANSFORMATION

For a homogeneous waveguide with perfect reflection **boundaries, the three representations and their transformation relations were summarized** in ref. (1) **and the result was shown in Fig. 3**

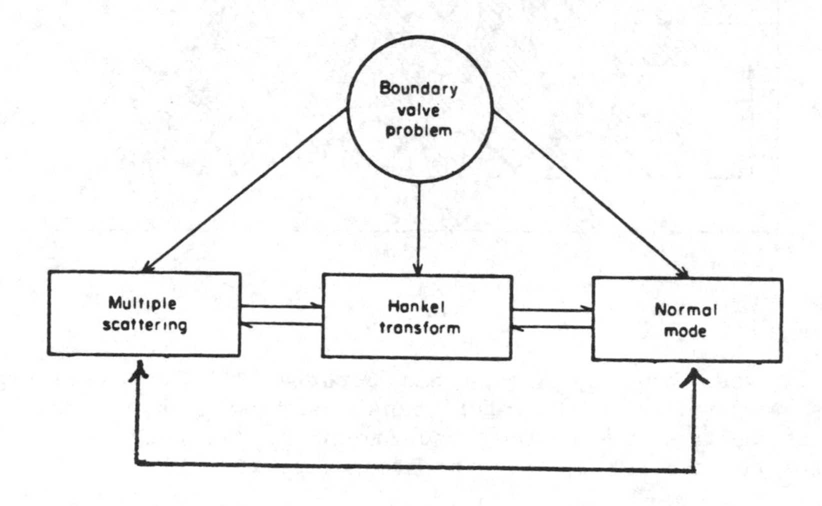

Poisson summation

Fig.3 Transform relations of different field representations for a homogeneous waveguide with absolute reflection boundaries.

For a general inhomogeneous waveguide with an ocean bottom, the case is somewhat more complicated. In 1971, the Poisson summation formula connecting the asymptotic representation (WKB) of the modes and rays was established by Batorsky and Felsen[9]. A rigorous proof that the mode representation and the "generalized ray" representation are connected by Poisson summation formula in **arbitrary stratified waveguide was given by Gao and Shang in 1980[4].**

In (4), the "mode generation function" $\phi(v)$, and "generalized ray generation function" B(u) were introduced for a stratified

waveguide. The mode ϕ_m in the mode representation is the sampling of its generation function $\phi(v)$, and the sampling period is 2π:

$$\phi_m = \phi(m\, v_o) \tag{8}$$

$$v_o = 2\pi \tag{9}$$

The generalized ray B_ℓ in the ray representation is the sampling of its generation function $B(u)$, and the sampling period is 1:

$$B_\ell = B(\ell u_o) \tag{10}$$

$$u_o = 1 \tag{11}$$

It was proved rigorously that $\phi(v)$ and $B(u)$ are connected by Fourier transformation, provided lateral waves do not appear:

$$\phi(v) = \int_{-\infty}^{\infty} B(u)\, e^{-iuv}\, du \tag{12}$$

$$B(u) = \frac{1}{2\pi} \int_{-\infty}^{\infty} \phi(v)\, e^{iuv}\, dv \tag{13}$$

The rigorous transformations (12) and (13) show that each particular member can be constructed by the superposition of all members (total) of the other representation.

It is very interesting that a member of one representation can actually be approximated by some of the members of the other representation, instead of all the members (total). It was found that by using the asymptotic expressions of $\phi(v)$ and $B(u)$, the result of the Fourier transformation was mainly contributed by a local region around the stationary phase point of the integrand[4,10]. This local conversion problem has been discussed quantitatively in a recent paper by Kamel and Felsen[11], which we will review in part III.

In what was discussed above, the effect of the branch-cut on the transformation between modes and rays was omitted. This problem has been discussed in the recent papers (references (12 and 13).

The first task is to choose a branch-cut in the field representation. It was known that there are two kinds of branch-cut, namely "E-J-P" cut and "Pekeris" cut[2]. For the purpose of transformation between modes and rays, the leaky modes must be included in the mode series. So, a "Pekeris" branch-cut (or something equivalent) was taken. Then the mode representation is:

$$G(r,z,z_o) = \sum_{m}^{\infty} \phi_m + \oint_b \text{(Pekeris)} \tag{14}$$

The second problem is what happens to the ray representation when a branch-cut is relevant? There seems to be nothing if the integral contour in the generalized ray representation keeps the same path $\Gamma_u(\zeta)$ as in the original integral representation in Eq. (1):

$$G(r,z,z_o) = \sum_{\ell=0}^{\infty} B_\ell^{\Gamma_\mu} \tag{15}$$

But $B_\ell^{\Gamma_\mu}$ and ϕ_m can not be connected by Poisson summation formula, unless we deform the contour from Γ_u to real ζ axis; then we get:

$$G(r,z,z_o) = \sum_{\ell=0}^{\infty} B_\ell + \sum_{\ell=0}^{\infty} \oint_{b\ell} \tag{16}$$

It was shown that B_ℓ and ϕ_m are connected by Poisson summation formula. We may call the transformable part, B_ℓ in Eq. (16) as the "canonical" part. In other words, the "canonical" generalized rays and the modes with "Pekeris" branch-cut are transformable. The sum of the "lateral" term in generalized ray representation is exactly equal to the "lateral wave" (Pekeris) in the mode representation:

$$\oint_b \text{(Pekeris)} = \sum_{\ell=0}^{\infty} \oint_{b\ell} \tag{17}$$

In Fig. 4, what we have achieved on the total transformation relations is demonstrated.

III. PARTIAL TRANSFORMATION

The ray (mode) equivalent of a group of modes (rays), treated qualitatively heretofore, can be quantified within a rigorous framework (11, 14, 15). By using the partial Poisson summation formula:

$$\sum_{m=M+1}^{\infty} F_m = -\frac{1}{2} F_M + \frac{1}{2\pi} \sum_{\ell=-\infty}^{\infty} \int_{2\pi M}^{\infty} F(t)\, e^{i\ell t} dt \tag{18}$$

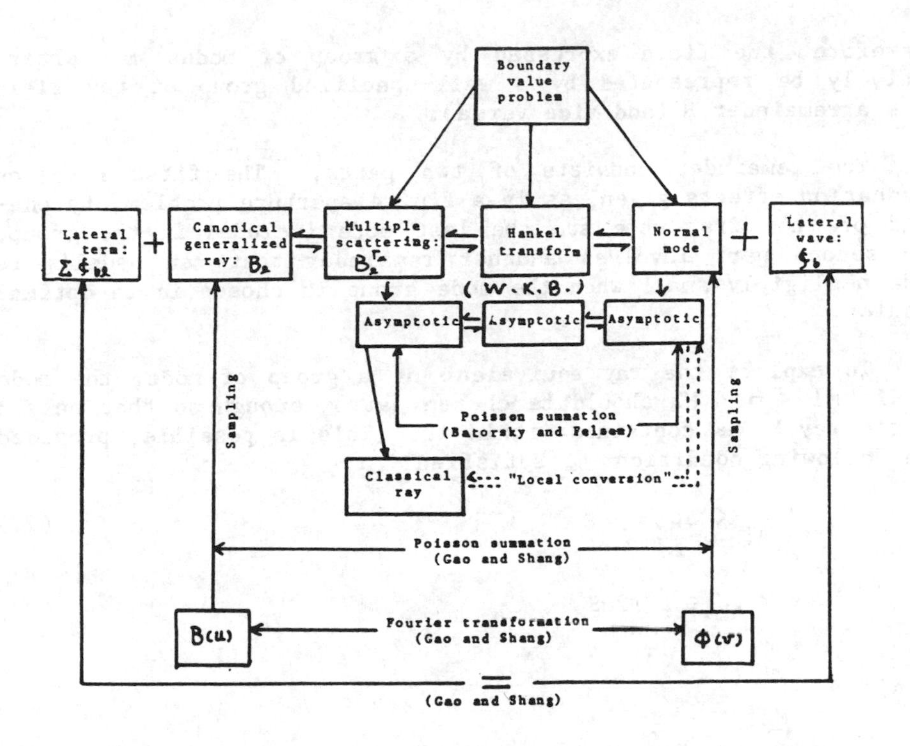

Fig. 4: Transformation relation between modes and rays in a stratified medium waveguide.

For a group of modes:

$$\sum_{m=M1}^{M2} \phi_m = \frac{1}{2}\,\phi_{M1} + \frac{1}{2}\,\phi_{M2} + \frac{1}{2\pi} \sum_{\ell=-\infty}^{\infty} \int_{2\pi M1}^{2\pi M2} \phi(v) e^{i\,\ell v} dv \qquad (19)$$

Substituting the asymptotic (WKB) approximation of $\phi(v)$ into Eq. (19) and by using saddle point technique to evaluate the integral, one gets:

$$\sum_{m=M1}^{M2} \phi_m = \sum_{\ell=L1}^{L2} \widetilde{B}_{\ell} + R \qquad (20)$$

$$R = \frac{1}{2}\,\widetilde{\phi}_{M1}\,(1 + \overline{\Delta}_{M1}) + \frac{1}{2}\,\widetilde{\phi}_{M2}\,(1 - \overline{\Delta}_{M2}) \qquad (21)$$

Therefore, the field expressed by a group of modes may alternatively be represented by a well-specified group of ray fields plus a remainder R (and vice versa).

The remainder consists of two parts. The first involves truncation effects given, as in a finite aperture problem, by one-half of the first mode and the last mode included in the group. The second part involves another remainder that can usually be made negligibly small when the mode group is chosen in an optimal fashion.

To explore the ray equivalent of a group of mode, the mode bundle $M1 \leq m \leq M2$ should be chosen narrow enough so that only a single ray $\tilde{B}_L$ is contained within it. This is possible, provided the following condition is satisfied[4]:

$$\Delta\mu_m << \Delta\mu_\ell \tag{22}$$

or

$$r < r_k \equiv s^3/2\pi\dot{S} \tag{23}$$

Then,

$$\sum_{m=M1}^{M2} \phi_m = \tilde{B}_L + R_L \tag{24}$$

On the other hand, if

$$r >> r_k \tag{25}$$

then, the mode equivalent of a group of rays is in effect[1]:

$$\sum_{\ell=L1}^{L2} \tilde{B}_\ell = \tilde{\phi}_M + R_M \tag{26}$$

IV. CONCLUSION

In this paper the state of the art on transformation relations between modes and rays was reviewed. Different representations of the field can be justified by a criterion of the degree of "information concentration." Then the transformation between different representations was discussed.

For total transformation, it was found that in a general inhomogeneous stratified waveguide, the modes series corresponds to the "Pekeris" branch-cut and the "canonical" generalized rays are transformable and they are connected by the Poisson summation formula in a rigorous manner. Likewise, the sum of the "lateral terms" of the "canonical" generalized rays was exactly equal to the "lateral wave" of the "Pekeris" branch-cut. Furthermore, the "local conversion" characteristic was found by using the stationary phase approach to estimate the Fourier transform integration.

For partial transformation, a quantitative analysis was obtained by using the partial Poisson summation formula for a group of modes (rays), which procedure incorporates systematically not only the transition to the continuum, but also the effect of truncation of the mode (ray) bundle. As a result, a mode (ray) cluster may be shown to be completely equivalent, in phase and amplitude, to a ray (mode) field plus a remainder, under the condition $r \ll r_k$ ($r \gg r_k$).

These developments have led to a rigorous hybrid formulation as an attractive alternative representation of the field. This incorporates the advantages of each representation within a single concise framework, the grouping of rays and modes can be made so as to simplify the overall numerical calculation, remove "singularities" from a given representation, and to predict the constructive interference properties of one representation in terms of a few members of the other. In other words, a hybrid representation was formed by the more "information concentrated" and more "convenient" members selected from each representation.

*On leave (1982) from the Institute of Acoustics, Academia Sinica, Peking, China. Current address: NOAA/AOML, Ocean Acoustics Division, Miami, Florida 33149 USA

REFERENCES

1. Keller, J. B. and J. S. Papadakis. Wave Propagation and Underwater Acoustics, Ch. 2 (Spring-Verlag, New York, 1977).
2. DeSanto, J. A. Ocean Acoustics, Ch. 3 (Spring-Verlag, New York, 1979).
3. Weston, D. E. Propagation of Sound in Shallow Water. Radio and Electronic Engineer 26 (1963) 329-337.
4. Gao, T. F. and E. C. Shang. The Transformation Between the Mode Representation and the Generalized Ray Representation of a Sound Field. Journal of Sound and Vibration 80(1) (1982) 105-115.
5. DiNapoli, F. R. A Fast Field Program for Multilayered Media. Tech. Rept. 4130 (NUSC° New London, CT, 1972) or see Ref. 2.
6. Tolstoy, I. and C. S. Clay. Ocean Acoustics, Ch. 4 (Spring-Verlag, New York, 1979).
7. Tindle, L. T. and K. M Guthrie. Ray as Interfering Mode in Underwater Acoustics. Journal of Sound Vibration 34 (1974) 291-295.
8. Guthrie, K. M. and C. T. Tindle. Ray Effects in the Normal Mode Approach to Underwater Acoustics. Journal of Sound Vibration 47 (1976) 403-413.
9. Batorsky, D. V. and L. B Felsen. Ray-Optic Calculation of Mode Excited by Sources and Scatterers in a Weakly Inhomogeneous Duct. Radio Science (1971) 911-923.
10. Wang, T. C. and E. C. Shang. Underwater Acoustics, Ch. 3 (Science Press, Peking, China, 1981).
11. Kamel A. and L. B. Felsen. On the Ray Equivalent of a Group of Modes. JASA 71(6) (1982), 1445-1452.
12. Gao, T. F. and E. C. Shang. Effect of the Branch-Cut on the Transformation Between Modes and Rays. JASA 73(5) (1983) 1551-1555.
13. Kamel, A. and L.B. Felsen. Hybrid Ray-Mode Formulation of SH Motion in a Two-Layer Half-Space. Bull. Seis. Soc. Am. 1763, 1981
14. Felsen, L.B. Hybrid Ray-Mode Fields in Inhomogeneous Wave Guides and Ducts. JASA 69(2) (1981) 352-361
15. Felsen, L.B. and T. Ishihara. Hybrid Ray-Mode Formulation of Ducted Propagation. JASA 65(3) (1979) 595-607.

SPECTRAL SYNTHESIS AND ADIABATIC INVARIANTS FOR NON-UNIFORM WAVEGUIDES

J. M. Arnold

Department of Electrical and Electronic Engineering,
University of Nottingham, University Park, Nottingham, NG7 2RD,
England.

INTRODUCTION

Wave propagation problems in which a waveguide undergoes slow transformations of its cross sectional geometry abound in the natural and the manufactured environment. In fact, the notion of a translation invariant waveguide is an idealisation which describes exceptional, rather than generic, cases. In many cases the non-uniformity of the waveguide determines much of the physics of wave propagation in those environments; examples can be found in optical fibre theory, underwater acoustics, synthetic seismology, tropospheric radio propagation and a host of related applications. For idealised waveguides, Fourier transformation along the longitudinal direction effectively reduces the number of dimensions of the problem, and a relatively complete account of wave propagation in such translation invariant waveguides can be formulated. This procedure is one example of the method of separation of variables, whereby some symmetry of the geometry permits the coordinate associated with that symmetry to be integrated out of the problem.

By definition, non-uniform waveguides break the symmetry leading to the application of separation of variables, and the loss of symmetry entails the absence of any 'exact' method to solve a general class of problem; inevitably, approximate methods must be sought.

The most favoured method for studying non-uniform waveguides has been the framework of coupled mode theory, systematically used to study time dependent quantum mechanics in the 1930s and probably originated, like almost all other good ideas, by Lord Rayleigh. It has since been applied by countless authors to a wide range of problems with a similar mathematical basis (weak perturbation of linear operators in Hilbert space). In its lowest order form (the adiabatic approximation) it has a direct physical appeal and a simplicity of calculation that renders it very attractive.

However, when this theory is applied to open waveguides, where radiation is as much a part of the physics as guidance, then serious defects are encountered. These are described fully in sections 2 and 3, and we will not dwell on them here, except to say they are concerned with singularities which appear in the radiation parts of the field description; for example, in the complete disappearance of an adiabatic mode from the equations, when it should smoothly convert into radiation at a critical cut off transition beyond which it cannot propagate as a bound mode. In fact, the proper treatment of radiation fields in this theory is fraught with difficulties which have hardly been acknowledged in the literature, let alone effectively addressed.

With these comments in mind we shall consider in this paper one such phenomenon wherein the adiabatic mode theory fails to give sensible results, namely, the disappearance of an adiabatic mode during upslope propagation in a wedge with one penetrable boundary. We show that this problem can be solved by constructing a plane wave spectral integral which reduces to an adiabatic mode where that is defined but which continues to be a good representation even where the adiabatic mode theory fails, beyond the cut off transition from bound to radiative behaviour of the local mode. Thus, while the adiabatic theory remains applicable away from these critical transitions, transformation to a more uniform representation is required to traverse the singular transition.

COUPLED MODES IN LONGITUDINALLY VARYING WAVEGUIDES

The conventional treatment of longitudinally varying (non-separable) waveguides is provided by coupled-mode theory [1]. This proceeds by first defining transverse cross-sections in the waveguide for all points along its axis (Fig.1).

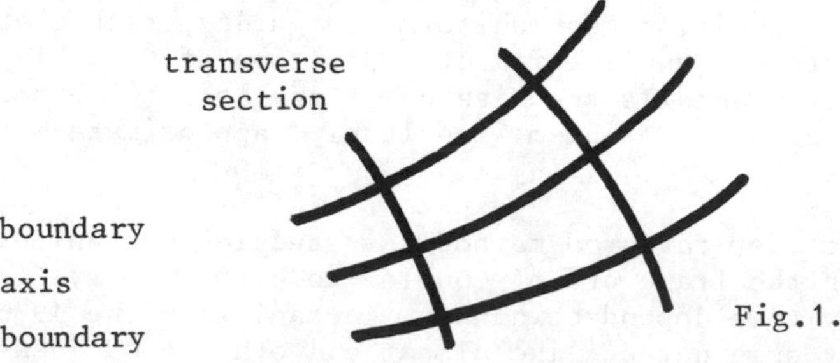

Fig.1.

The precise forms taken by the 'axis' and the 'cross sections can be rather arbitrary but it is desirable that cross sections should intersect normally both the axis and those penetrable boundaries where media of different refractive indices contact [2]. In particular, cross sections need not be flat. To illustrate the general development of the theory we shall refer to the wedge-shaped waveguide of Fig.2, which has been rather extensively studied recently [3].

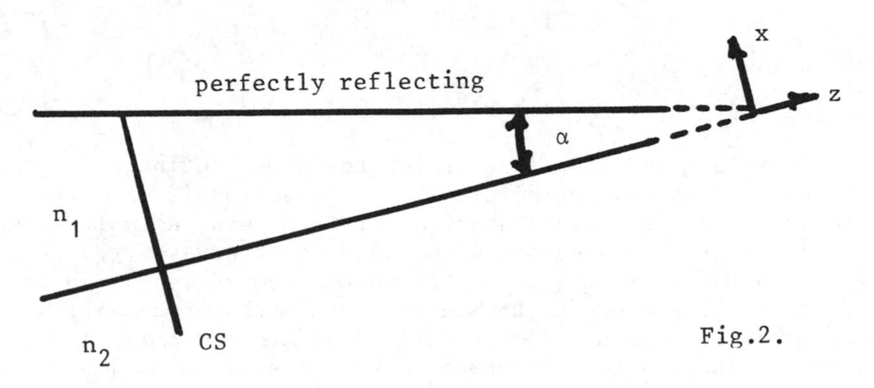

Fig.2.

In the wedge configuration, the refractive indices n_1 and n_2 are independent of position, the 'axis' is x = 0, and 'transverse cross sections' are z = constant. If the angle α is small the waveguide can be regarded as slowly varying in the axial (z) direction. Scalar waves $u(\underline{x})$ in this, or any other, waveguide must satisfy the Helmholtz equation

$$\nabla^2 u + n^2 k^2 u = 0 \tag{2.1}$$

(where n is the refractive index variation with position, k is the wavenumber in a medium n = 1) along with appropriate boundary conditions. The Laplacian ∇^2 can be split into 'longitudinal' and 'transverse' parts

$$\nabla^2 = \partial_z^2 + \nabla_T^2 , \tag{2.2}$$

where $\partial_z = \partial/\partial_z$ and the subscript T denotes restriction to the transverse coordinates, transforming (2.1) to

$$\partial_z^2 u + (\nabla_T^2 + n^2 k^2)u = 0 \tag{2.3}$$

In the most general case, where the axial and transverse coordinates are not Cartesian, metric coefficients will appear with the partial derivatives in (2.2); in the special case of Fig.2 this does not concern us.

In (2.3) the partial differential operator $\nabla_T^2 + n^2 k^2$ acts on u only through its transverse coordinates; its dependence on the axial coordinate z is merely as a parameter. Thus we can write

$$L \equiv L(z) = \nabla_T^2 + n^2k^2 \tag{2.4}$$

and

$$\partial_z^2 u + Lu = 0 \tag{2.5}$$

for the linear operator L.

The operator L is diagonalised by eigenfunctions ϕ satisfying

$$L\phi = \beta^2\phi \tag{2.6}$$

or, using (2.4),

$$[\nabla_T^2 + (n^2k^2-\beta^2)]\phi = 0 \tag{2.7}$$

wherein it is assumed that ϕ satisfies the same continuity boundary conditions as u at discontinuities in n. Equation (2.7) is recognisable as the differential equation for transverse normal modes in an axially invariant waveguide with the same transverse geometry as the given waveguide at the parameter value denoted by z. These eigenfunctions are known to form a complete orthonormal set, with a spectrum of eigenvalues β^2 partially discrete and partially continuous. The discrete spectrum corresponds to bound modes indexed by an integer, q, the continuous to radiation modes indexed by a variable, τ. The orthogonality relations then read

$$\begin{aligned} &\int_{CS}\phi_p^*\phi_q dS = \delta_{pq};\ \int_{CS}\phi^*(\tau)\phi(\tau')dS = \delta(\tau-\tau') \\ &\int_{CS}\phi_p^*\ \phi(\tau)dS = \int_{CS} \phi^*(\tau)\ \phi_p dS = 0 \end{aligned} \tag{2.8}$$

where dS is the area measure on CS, * denotes complex conjugation and δ_{pq} and $\delta(\tau-\tau')$ are the Kronecker and Dirac delta functions respectively.

The completeness of the eigenfunctions $\{\phi_p, \phi(\tau)\}$ implies the existence of the expansions

$$u = \sum_p A_p\phi_p + \int_C A(\tau)\phi(\tau)d\tau \tag{2.10a}$$

$$\partial_z u = \sum_p B_p\phi_p + \int_C B(\tau)\phi(\tau)d\tau \tag{2.10b}$$

where C denotes the range of values covered by τ.

Substitution of (2.10) in (2.5) leads, after use of (2.8), to the system of equations

$$\partial_z A_p + \sum_q \kappa_{pq}A_q + \int_C \kappa_p(\tau)A(\tau)d\tau = B_p \tag{2.11a}$$

$$\partial_z B_p + \sum_q \kappa_{pq} B_q + \int_C \kappa_p(\tau) B(\tau) d\tau = -\beta_p^2 A_p \qquad (2.11b)$$

$$\partial_z A(\tau) + \sum_q \overline{\kappa}_q(\tau) A_q + \int_C \kappa(\tau,\tau') A(\tau') d\tau' = B(\tau) \qquad (2.11c)$$

$$\partial_z B(\tau) + \sum_q \overline{\kappa}_q(\tau) B_q + \int_C \kappa(\tau,\tau') B(\tau') d\tau' = -\beta^2(\tau) A(\tau) \qquad (2.11d)$$

which must be satisfied by the coefficients in (2.10). The numbers $\{\beta_p^2, \beta^2(\tau)\}$ are the eigenvalues of the operator L, and both these numbers and the coefficients depend parametrically on z. The remaining quantities in (2.11) are defined by

$$\kappa_{pq} = \int_{CS} \phi_p^* \frac{\partial \phi_q}{\partial z} dS \qquad (2.12a)$$

$$\kappa_p(\tau) = \int_{CS} \phi_p^* \frac{\partial \phi}{\partial z}(\tau) dS \qquad (2.12b)$$

$$\overline{\kappa}_q(\tau) = \int_{CS} \phi^*(\tau) \frac{\partial \phi_q}{\partial z} dS \qquad (2.12c)$$

$$\kappa(\tau,\tau') = \int_{CS} \phi^*(\tau) \frac{\partial \phi}{\partial z}(\tau') dS \qquad (2.12d)$$

Equations (2.11) are the basic coupled mode equations. (For fuller derivations under a variety of conditions see [1,2,4,5].

Further analytical progress can be made after simplifying the notation of (2.11); we ignore the distinction between discrete and continuous parts of the spectrum. This is a temporary measure for convenience only and the correct notation can be restored at any time. Thus (2.11) will become

$$\partial_z A_p + \sum_q \kappa_{pq} A_q = B_p \qquad (2.13a)$$

$$\partial z B_p + \sum_q \kappa_{pq} B_q = -\beta_p^2 A_p \qquad (2.13b)$$

where the summation over q also implies integration over τ, and the index p similarly implies extension to continuous τ. These equations can now be given a simple operator formulation

$$\partial_z A + \kappa A = B \qquad (2.14a)$$

$$\partial_z B + \kappa B = -\beta^2 A \qquad (2.14b)$$

with $(A)_p = A_p$, $(B)_p = B_p$, $(\kappa)_{pq} = {}_{pq}$, $(\beta)_{pq} = \beta_p \delta_{pq}$. By eliminating B from (2.14) the second order coupled mode system results :

$$\partial_z^2 A + 2\kappa\, \partial_z A + (\kappa^2 + \partial_z \kappa + \beta^2) A = 0 \tag{2.15}$$

3. THE ADIABATIC SOLUTION

The lowest order approximation that can be made in equations (2.14) or (2.15) is to neglect the intermode coupling specified by the coupling operator κ, but retain the z-dependence of the phase factor β. Thus (2.15) reduces to

$$\partial_z^2 A + \beta^2 A = 0 \tag{3.1}$$

which has the WKB solution

$$A \sim \beta^{-\frac{1}{2}} \left(e^{i\int^z \beta dz'} A^+ + e^{-i\int^z \beta dz'} A^-\right) \tag{3.2}$$

where $A^\pm$ are arbitrary z-independent vectors fixed by the initial conditions at $z = 0$ and the radiation conditions for $z \to \pm\infty$, and it is recalled that β is a diagonal operator.

To this order of approximation, (2.10a) becomes

$$\begin{aligned} u \sim & \sum_q \beta_q^{-\frac{1}{2}} \left(A_q^+ e^{i\int^z \beta_q dz'} + A_q^- e^{-i\int^z \beta_q dz'}\right)\phi_q \\ & + \int_C \beta^{-\frac{1}{2}}(\tau)\left(A^+(\tau) e^{i\int^z \beta(\tau) dz'} + A^-(\tau) e^{-i\int^z \beta(\tau) dz'}\right)\phi(\tau)d\tau \end{aligned} \tag{3.3}$$

where we have restored the distinction between discrete and continuous spectra. The basis functions

$$\psi_q^\pm \sim \beta_q^{-\frac{1}{2}}\, e^{\pm i\int^z \beta_q dz'}\, \phi_q \tag{3.4}$$

appearing in (3.3) are called adiabatic modes, and (3.3) is called the adiabatic representation. The concept of an adiabatic mode is well defined for the discrete part of the spectrum, but this is not so for the continuous part. This is because arbitrary changes of the variable τ can be made for each z, and so the manner in which the expression

$$\psi^\pm(\tau) \sim \beta^{-\frac{1}{2}}(\tau) e^{\pm i\int^z \beta(\tau) dz'}\, \phi(\tau) \tag{3.5}$$

sections through all possible 'adiabatic radiation modes' depends on the choice of τ at each z. This is a primary example of circumstances in which the continuous part of the spectrum cannot be treated in the same way as the discrete part, despite the formal similarities in their notation which we have previously used to simplify the equations of coupled mode theory. In general, calculations which involve this continuous spectrum must be regarded as purely formal, and the meaning of expressions containing the continuum spectral coefficients must be carefully assessed at each stage. The key to the resolution of this problem probably lies in the concept of the adiabatic invariant, to be discussed shortly, but currently a satisfactory non-formal treatment of continuum modes in coupled mode theory must be regarded as an open problem.

Further limitations of the adiabatic mode theory are revealed when its interpretation on the discrete part of (3.3) is considered more deeply. Each adiabatic mode (3.4) consists of a local normal mode ϕ_q on each cross section CS, with a propagation factor

$$\beta_q^{-\frac{1}{2}} e^{i\int^z \beta_q dz'}$$

to connect up the modes on different cross sections into a single wavefunction defined over a range of z-values; the qth local normal mode on one cross-section translates smoothly into the qth local mode on another cross-section. The required continuity of this flow makes ambiguity impossible, and hence the discrete adiabatic modes are well defined. However it is an elementary fact that, in translation invariant open waveguides, the total number of discrete modes, N, is finite and depends on the transverse geometry of the waveguide and the operating frequency (proportional to k). Since these modes form part of the local normal mode basis, ϕ_q, it is quite possible for the dimension N of the adiabatic mode set to be different on different cross sections, reflecting different transverse geometries. In the wedge example, this happens as an adiabatic mode based on a bound local mode ϕ_q propagates upslope; eventually a critical transition is reached where ϕ_q is at cut off, beyond which it is part of the radiation field spanned by continuum modes (i.e. a leaky mode). The corresponding adiabatic mode simply disappears from the discrete part of (3.3), the dimension of which is reduced by 1. In addition, the amplitude of this mode tends to zero everywhere on the critical cross-section, and fails to be a differentiable function of z.

There is no known resolution of this mode cut off problem within coupled mode theory. Clearly, the assumption that all coupling may be neglected ($\kappa \equiv 0$) is wrong; at the critical range a bound mode converts <u>entirely</u> to radiation modes, a very strong coupling! However the ambiguities in the treatment of the radiation spectrum have so far prevented a rigorous treatment of this phenomenon using equations (2.14). Nevertheless, three recent papers (Pierce, [6], Arnold and Felsen [3], Kamel and Felsen [7] have successfully constructed solutions to the wave equation (2.1) for a tapered waveguide from which the correct behaviour of an adiabatic mode can be deduced, even beyond the transition where it disappears in conventional coupled mode theory. These analyses have used a variety of techniques, none of which is based on an attempt to solve (2.14). In the following section we concentrate on just one of these techniques (the plane wave spectral synthesis of Arnold and Felsen [3]) applied to the wedge geometry of Fig.2.

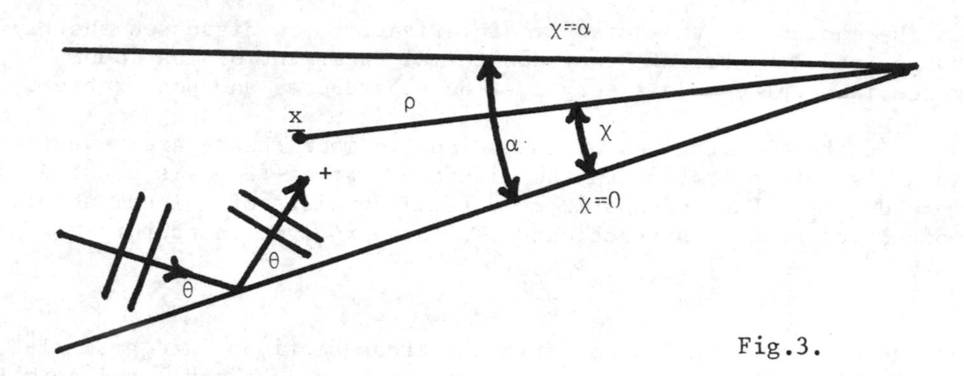

Fig.3.

3. PLANE WAVE SPECTRAL SYNTHESIS

The basic idea here is to try to construct a plane wave spectrum which satisfies the boundary conditions on the upper and lower surfaces (in Fig.3) and which maintains itself self-consistently without support from sources. A plane wave spectrum in the interior of the wedge may be written in general as

$$u(\underline{x}) = u^+(\underline{x}) + u^-(\underline{x}) \tag{4.1}$$

$$u^{\pm}(\underline{x}) = \int e^{-in_1 k\rho\cos(\theta\pm\chi)}\, \tilde{u}^{\pm}(\theta)\, d\theta \tag{4.2}$$

where $\tilde{u}^{\pm}(\theta)$ are undetermined spectral coefficients, and the integration contours are similarly undetermined; (ρ,χ) are polar coordinates with respect to the wedge apex and the lower boundary. The + (resp.-) sign denotes waves travelling away from (resp. towards) the lower boundary $\chi = 0$, and θ is the angle between a plane wave's direction of propagation and this boundary. The local boundary conditions on $\chi = 0$ can be satisfied by a superposition of the two spectra (4.2) if each downward plane wave in u^- reflects into the corresponding plane wave in u^+ with the appropriate Fresnel reflection coefficient; this requires

$$\tilde{u}^+(\theta) = e^{i\phi(\theta)}\, \tilde{u}^-(\theta) \tag{4.3}$$

where $e^{i\phi(\theta)}$ is the Fresnel coefficient. The superposition (4.1) then satisfies the local continuity conditions on $\chi = 0$.

On reflection of the upward wave u^+ in the top boundary $\chi = \alpha$, a new wave

$$v^-(x) = \int \tilde{u}^+(\theta)\, e^{i\pi}\, e^{-ikn_1\rho\cos(\theta+2\alpha-\chi)}\, d\theta \tag{4.4}$$

is produced, and we require for self-consistency that

$$u^-(\underline{x}) = v^-(\underline{x}) \tag{4.5}$$

From (4.3), (4.4) and (4.5) we deduce that

$$\tilde{u}^-(\theta+2\alpha) = \tilde{u}^-(\theta)\, e^{i[\phi(\theta)+\pi]} \tag{4.6}$$

If $\tilde{u}(\theta)$ is represented by

$$\tilde{u}(\theta) = e^{iS(\theta)} \tag{4.7}$$

then (4.6) implies

$$S(\theta+2\alpha) - S(\theta) = \phi(\theta) + \pi - 2q\pi \tag{4.8}$$

where q is an arbitrary integer.

Equation (4.8) is formally solved by the Euler-Maclaurin formula. If (4.8) is summed m times, then

$$S(\theta_m) - S(\theta) = -\phi(\theta_m) + \sum_{\ell=0}^{m} \phi(\theta_\ell) + m\pi - 2qm\pi \tag{4.9}$$

where

$$\theta_\ell = \theta + 2\ell\alpha \;, \quad \theta_o = \theta$$

The Euler-Maclaurin formula (see, for example, Olver [8] transforms the sum on the right of (4.9) into an integral :

$$S(\theta_m) - S(\theta) = -\frac{1}{2}[\phi(\theta_m) - \phi(\theta)] + \frac{1}{2\alpha}\int_{\theta}^{\theta_m} \phi(\theta')d\theta' + m\pi - 2qm\pi + E \tag{4.10}$$

where the remainder E can be shown to vanish as $\alpha \to 0$ for all θ. Equation (4.10) is derived assuming that $\theta_m-\theta$ is an integral multiple of 2α. It can be continued to arbitrary values of θ_m and θ by replacing m by $(\theta_m-\theta)/2\alpha$ and neglecting E. Then

$$S(\theta) \sim -\frac{1}{2}\phi(\theta) + \frac{1}{2\alpha}\int_{\theta_c}^{\theta} \phi(\theta')d\theta' + \frac{\pi\theta}{2\alpha} - \frac{q\pi\theta}{\alpha} \tag{4.11}$$

is a solution of (4.10) valid for arbitrary values of θ. The parameter θ_c is an arbitrary constant in (4.11); it is most convenient to choose it to be the critical angle for total internal reflection at the bottom interface $\chi = 0$.

We abbreviate (4.11) to

$$S(\theta) \sim \Phi(\theta) - \frac{q\pi\theta}{\alpha} \tag{4.12}$$

with

$$\Phi(\theta) = -\frac{1}{2}\phi(\theta) + \frac{1}{2\alpha}\int_{\theta_c}^{\theta} \phi(\theta')d\theta' + \frac{\pi\theta}{2\alpha} \tag{4.13}$$

With (4.12), (4.7) and (4.3), the spectral coefficients $\tilde{u}^{\pm}(\theta)$ can be written explicitly as

$$\tilde{u}^{\pm}(\theta) = e^{i(\frac{1}{2}\pm\frac{1}{2})\phi(\theta)}\, e^{i\Phi(\theta)}\, e^{-iq\pi\theta/\alpha} \tag{4.14}$$

We denote $(2\alpha)^{-\frac{1}{2}}$ times the integral in (4.2) by $W_q^{\pm}$, with the explicit form from (4.14) :

$$W_q^{\pm}(\underline{x}) = (2\alpha)^{-\frac{1}{2}} \int_C e^{iQ^{\pm}(\underline{x};\theta)}\, e^{-iq\pi\theta/\alpha}\, d\theta \tag{4.15}$$

and

$$Q^{\pm}(\underline{x};\theta) = -\, n_1\, k\rho \cos(\theta\pm\chi) + \Phi(\theta) + (\tfrac{1}{2}\pm\tfrac{1}{2})\,\phi(\theta) \tag{4.16}$$

The integrals in (4.16) can be summed according to (4.1) to give

$$W_q(\underline{x}) = W_q^{+}(\underline{x}) + W_q^{-}(\underline{x}) \tag{4.17}$$

The function W_q so defined is called an <u>intrinsic</u> mode. It has properties similar to the adiabatic mode defined in section 3, but is more general, as we show in the next section.

5. THE ADIABATIC INVARIANT

We now investigate the possibility of evaluating the integral in (4.15) by the saddle point method. Saddle points of the integrand are located by expanding the phase up to leading order in α, then finding zeroes of the derivative (with respect to θ) of this leading order term. Thus we have

$$Q^{\pm}(\underline{x};\theta) \sim \alpha^{-1}\, Q_o(\underline{x};\theta) + O(1) \tag{5.1}$$

with

$$Q_o(\underline{x};\theta) = -kh\cos\theta + \frac{1}{2}\int_{\theta_c}^{\theta} \phi(\theta')d\theta' + \frac{\pi\theta}{2} \tag{5.2}$$

and

$$h = \lim_{\alpha\to 0}(\rho\alpha)\ ,\ \rho \sim O(\alpha^{-1}) \tag{5.3}$$

Note in equation (5.3) that the ρ coordinate is scaled by α^{-1}; thus, h, is the 'local width' of the wedge waveguide at the observation point $\underline{x}$, and $\alpha \to 0$ in such a way as to make $h \sim O(1)$. Then h depends only on the axial coordinate z in Fig.2, and the entire leading order term Q_o is independent of the x-coordinate of the observation point.

Locating the saddle points of the phase of the integrand in (4.15), we require

$$\frac{\partial Q_o(\underline{x};\theta)}{\partial\theta} = q\pi \tag{5.4}$$

Expressing the required derivative of Q_o from (5.2), we have the saddle point condition

$$khsin\theta + \frac{1}{2}\phi(\theta) + \frac{\pi}{2} = q\pi \tag{5.5}$$

Equation (5.5) is precisely the same condition as that obtained in a translation invariant waveguide of constant width h as the condition that two plane waves at an angle θ to the lower boundary fit together self consistently to form a mode. The only difference is that here h is the local width at the observation point. Equation (5.5) is therefore the same adiabatic approximation as described in section 3, applied to the eigenvalue equation for the local normal modes rather than the local normal modes themselves.

The right hand side of (5.5) is constant for all observation points, and for this reason the function on the left hand side is called an adiabatic invariant. It is actually slightly more convenient to define the adiabatic invariant as π^{-1} times the left hand side of (5.5), as this quantity then takes integer values only.

Accordingly we have

$$\nu(z;\theta) = \pi^{-1}\{khsin\theta + \frac{1}{2}\phi(\theta) + \frac{\pi}{2}\} \tag{5.6}$$

as the invariant, where the first variable in the argument of ν indicates its dependence only on the axial coordinate; for each z there is a value of θ which makes ν equal to an integer, and this value of θ is the required solution of the saddle point equation (5.5), which is denoted by $\theta = \theta_q$.

Symmetry considerations applied to (5.6) reveal that $\pi-\theta_q$ is also a saddle point; θ_q and $\pi-\theta_q$ are the only saddle points in the range $0 \leqslant Re(\theta) \leqslant \pi$. For the time being we are assuming θ_q to be real, so that the local normal mode which corresponds to the angle θ_q is bound; later, we allow θ_q to become complex.

Since the saddle points of the integrand in (4.15) are now well defined, we may define the integration contour C in (4.15) with respect to these saddle points. Any contour which passes from $\theta = -i\infty$ to $\theta = \pi/2 + i\infty$, then to $\theta = \pi - i\infty$ can be deformed to lie along the steepest descent paths through θ_q and $\pi-\theta_q$, provided that θ_q is real and less than θ_c. In general the Fresnel phase $\phi(\theta)$ has a branch point at θ_c so, in moving θ_q from one side of the branch point to the other, difficulties with branch cuts might be anticipated, but these do not arise under the stated conditions on θ_q (Fig.4). This is the assumed path for the contour C.

The saddle point integration may now be carried out, with the result that, for $\alpha \to 0$,

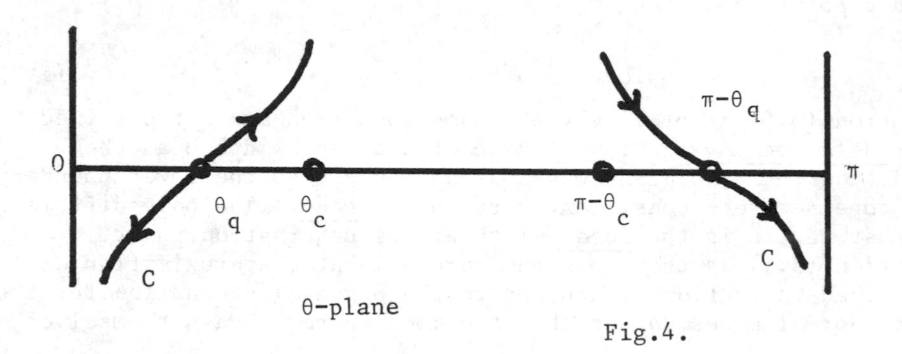

Fig.4.

$$W_q \sim -(2\pi)^{\frac{1}{2}} e^{-i\pi/4} e^{iq\pi} A_q e^{iS_q} \sin[k(h-x)\sin\theta_q] \tag{5.7}$$

with

$$A_q = \left\{\frac{1}{2}\left[kh\cos\theta_q + \frac{1}{2}\frac{d\phi}{d\theta}\bigg|_{\theta=\theta_q}\right]\right\}^{-\frac{1}{2}} \tag{5.8a}$$

$$S_q = \alpha^{-1}\left[Q_o(\underline{x};\theta_q) - q\pi\theta_q\right] \tag{5.8b}$$

It is quite easy to show that equation (5.7) for W_q agrees precisely with the adiabatic mode as constructed in section 3, and this identification motivates our first conclusion: integration of the intrinsic mode integral by the saddle point method yields the adiabatic mode approximation.

The above analysis was carried out assuming θ_q real and less then θ_c. As z increases upslope, the angle θ_q increases due to the tapering of the waveguide. As θ_q approaches the critical angle θ_c the local normal mode indexed by θ_q approaches cut off, and the adiabatic mode theory breaks down, as we have already observed. However, the integral (4.15) from which the intrinsic mode W_q is constructed remains well defined, although the saddle point method can no longer be used to approximate it, because of the confluence of θ_q and θ_c. Nevertheless, more uniform integration techniques can be used instead, and the result is an expression for $W_q(\underline{x})$ in terms of Airy functions. Pierce [6] also obtains a similar result by a completely different method.

Finally the observation point can be allowed to move through and beyond the critical region where $\theta_q \sim \theta_c$. Then θ_q, as defined by the adiabatic invariant (5.6), becomes complex with $\mathrm{Re}(\theta_q) > \theta_c$ corresponding to a 'leaky' local mode. In tracking the motion of θ_q around θ_c as the observation point moves upslope, a branch cut integral must be added to the (analytically continued) saddle point approximation. The saddle point contribution retains its interpretation as a local mode, which is now leaky, and the branch cut contribution is interpreted as a lateral wave excited at the critical transition when $\theta_q = \theta_c$. All the calculations referred to in this section are carried out and thoroughly discussed by Arnold and Felsen [3].

6. SUMMARY AND CONCLUSIONS

Conventional adiabatic mode theory for longitudinally varying waveguides has serious limitations if radiation must be correctly described. In particular, the disappearance of an adiabatic mode as it traverses cut off introduces singularities in the description of wavefields. A better description of these wavefields can be found using a spectral representation; in the example of a wedge studied here the spectral elements are plane waves, but more complicated problems might require other spectral elements. In essence, the spectral representation reduces to the adiabatic mode representation wherever that is well defined, but it also extends uniformly into regions where the adiabatic mode cannot be defined, to describe the radiative behaviour of the local mode beyond cut off.

Though we have not attempted it here, representations of source excited fields in terms of superpositions of intrinsic modes can be found, and the plane wave spectral representation of the intrinsic mode can be obtained for more general waveguide environments than the simple wedge. In this way the singularities of adiabatic mode theory can be traversed uniformly and systematically for a broad range of problems.

7. REFERENCES

1. Pierce, A. D., 'Extension of the method of normal modes to sound propagation in an almost stratified medium', J. Acous. Soc. Am., vol.37, (1965), pp.19-27.

2. Rutherford, S. and Hawker, K., 'Consistent coupled mode theory for a class of non-separable problems', J. Acous. Soc. Am., vol.70, (1981), pp.554-564.

3. Arnold, J. M. and Felsen, L. B., 'Rays and local modes in a wedge-shaped ocean', Jour. Acous. Soc. Am., vol.73, (1983), pp.1105-1119.

4. Arnold, J. M. and Allen, R., 'Microbending loss in optical fibres', IEE Proc. (H), vol.130, (1983), pp.331-339.

5. Sporleder, F. and Unger, H. G., Waveguide tapers, transitions and couplers, IEE Electromagnetic waves, Series 6, (Peter Peregrinus Ltd., 1979).

6. Pierce, A. D., 'Guided mode disappearance during upslope propagation in variable depth shallow water overlying a fluid bottom', J. Acous. Soc. Am., vol.72, (1982), pp.523-531.

7. Kamel, A. and Felsen, L. B., 'Spectral theory of sound propagation in an ocean channel with weakly sloping bottom', J. Acous. Soc. Am., vol.73, (1983), pp.1120-1130.

8. Olver, F. W. J., Asymptotics and special functions, (Academic Press, 1974).

ACKNOWLEDGEMENT

This work was partially supported by the US Office of Naval Research contract No. N00014-79-C-0013.

RAYS, MODES AND FLUX

D E Weston

Admiralty Underwater Weapons Establishment
Portland, Dorset DT5 2JS, England

ABSTRACT

The ray approach, the mode approach and where appropriate a flux approach are briefly reviewed for many guided wave concepts or applications. These comprise eigenvalues, transforms in the angle domain, mode filtering, cycle distance and time, phase and group velocity, attenuation, arrival time, transmission loss, interference patterns, range-dependent environments and horizontal refraction. For example the interference patterns show fuzzy rays, from both the source and the complementary source position, which combine to produce a series of focal points. The duality or trinity of descriptions is shown to be very widely useful and by no means restricted to the angle domain. It is helpful to distinguish between source distances that are finite and those that are infinite. There are many interesting symmetries in definitions, eg phase and group velocity correspond to different kinds of beam displacement at boundaries.

1. INTRODUCTION

This paper concerns guided propagation and is written mainly from the point of view of acoustics, with its usual scalar potential, and in particular concerns underwater acoustics. But the ideas do carry over to other mechanical waves (including gravity effects) and to electromagnetic propagation.

For a given angle domain it is possible to switch between ray and mode descriptions of the field, and this has received much attention lately. The main point of this paper is to stress that this duality occurs in a wide variety of applications besides the angle domain transforms. The paper starts by describing the two approaches, and then adds flux to make the duality into a trinity. Applications which are in some sence hybrid are then briefly reviewed for 13 concepts.

2. THE APPROACHES

2.1 Ray Approaches

There is a variety of usages for each approach, and the longer one looks the hazier become the boundaries between approaches. Ray concepts may be introduced by recognising three stages, though again the distinctions are not always sharp.

Stage 1 The transitive concept has a specific source and a specific receiver, with a number of line paths connecting them, sometimes called "eigenrays". Each path corresponds to a local extremum in travel time, and is associated with a number of reflections or turning-point refractions. In one version of this system we have a source, plus a number of image sources effectively reflected through the boundaries.

Stage 2 The broadcast concept has a specific source but we look at the whole field distribution rather than a single receiver. Neighbouring rays define a ray tube, and due to source directivity or multipath interference we may speak of a "beam".

Stage 3 The infinite or infinite-distance concept has its source far off, or perhaps it has no specific source at all, nor receiver. The term "plane wave" may be encountered, with so-called homogenous and inhomogeneous types. This is all in contrast to the finite concepts above, to which one must return for any real or thought experiment.

In all these stages the rays obey the laws of refraction, according to Snell, Bouguer or ray invariant ideas as appropriate. There is a host of possible wave-theory corrections.

2.2 Mode Approach

For normal modes, both propagating and non-propagating, it is convenient to take the same three stages, but natural to arrange them in the reverse order.

Stage 3 The normal modes or characteristic functions of a system may be calculated without any specific reference to sources or receivers, they are basically infinite-distance concepts.

Stage 2 By calculating the coupling of a source into each mode, and then summing, we can reach the field distribution, or broadcast concept.

Stage 1 By calculating both the source and receiver coupling for each mode, and then summing, we obtain the transmission loss, ie the transitive concept.

2.3 Flux Approach

Of course an energy flux may be directly associated with rays or modes. We wish to concentrate here on a rather special use of the term flux, concerning the averaged energy flux density per unit angle. The averaging in range extends effectively over one cycle distance, and to avoid wave effects is taken in the high-frequency limit. This leads to an approximate method which avoids the discreteness effects that come with rays and with modes.

Stage 3 With the infinite concept one can relate angle and depth distributions of energy.

Stage 2 The broadcast concept with a known source depth gives particular angle and depth distributions. The approximation predicts a smooth $(\text{range})^{-1}$ dependence of intensity. In one version all the energy within a critical angle is trapped, and the method automatically gives the depth distribution of the total energy flux density within the insonified channel.

Stage 1 In the transitive concept or application the flux approach gives very simple integral forms for the transmission loss.

Since the flux approach is the least-known of the three, and only slowly receiving more attention, it is worth giving a few representative references from underwater acoustics, where the method has been developed furthest (1-7). In addition it has received just a little use in electromagnetic propagation (8).

3. APPLICATIONS OF THE APPROACHES

3.1 General

A format will be attempted in which for each application rays, modes and flux are always mentioned, even if the mention is negative. The first nine items all correspond to a single angle or a small range of angles, ie a single ray or a single mode or a small group of either. The next two items are for a wide, perhaps complete angular range. We finish with two unclassified or miscellaneous entries.

3.2 Eigenvalues

Let us start with the stage 3 or infinite-distance concept of a mode. It is well known that this may be regarded as the sum of a down-going and an upgoing plane wave or ray (9, 10). See figure 1 for the first transverse mode. I regard this as the archetypal case of the duality. A rays-with-phase approach may be used to find the mode eigenvalues. (This is usually but not necessarily along a vertical.) This is exact for isovelocity water and uses the WKB approximation where there is a continuous stratification in velocity:

$$2n\pi = 2\int\gamma dz + \phi_1 + \phi_2. \quad (1)$$

Here n is mode number, γ is vertical component of wave number, z is depth and ϕ_1 and ϕ_2 are phase delays at the turning or reflection points.

But the phase change is equivalent (11) to reflection having taken place at a free surface lying beyond the real boundary by a vertical displacement distance

$$\delta_v = (\phi + \pi)/2\gamma. \quad (2)$$

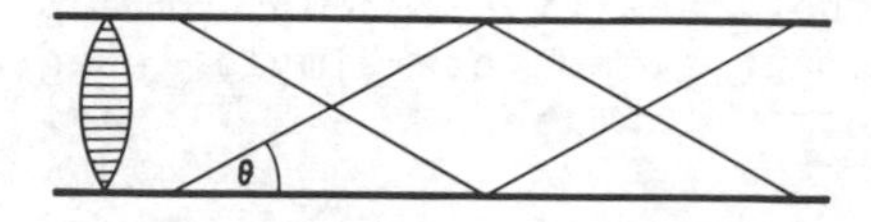

Fig 1. Ray-Mode Duality for the First Mode

The eigenvalue equation becomes

$$2(n+1)\pi = 2\int \gamma dz + 2\gamma_1 \delta_{v1} + 2\gamma_2 \delta_{v2}, \qquad (3)$$

or if the integral is considered to extend over δ_{v1} and δ_{v2} it is simply

$$(n+1)\pi = \int \gamma \, dz. \qquad (4)$$

We will return to displacement parameters later, and merely note now that there are related quantities measured along the wavefront or in the horizontal direction. The latter is

$$\delta_h = 2\delta_v \cot\theta = (\phi + \pi)/h. \qquad (5)$$

where θ is grazing angle and h is horizontal component of wave number. Of course another set of displacements applies to an equivalent hard surface, but this choice is less convenient.

The *flux* method is not applicable for eigenvalues, but exists as a parallel approach in stage 3 generally.

3.3 Transforms in the Angle Domain

We will jump over stage 2 broadcast (but see section 3.12) to the stage 1 transitive concept. Transforms between complete sets of *rays* and complete sets of *modes* have long been of interest (12). More recently it has been shown how this may be done within restricted angular regions (13), allowing use of rays in one region, and, at the same time, modes in another. This possibility is identified as a central feature of hybrid formulations, but because of the attention it is receiving it will not be pursued here.

Generally *flux* methods have been a parallel activity, but it is quite feasible to switch between rays, modes and flux in the angle domain (14).

3.4 Mode Filtering

It is possible to select or match to a given *mode* by tailoring the amplitude response of a vertical line of closely-spaced transducers. One alternative is to steer the line to the upward or downward angle of the equivalent plane wave or *ray*, though typically here only half the energy goes into the intended mode. The *flux* approach is inappropriate.

3.5 Cycle Distance

Cycle distance is an infinite or stage 3 concept, and the problem turns out to be that of finding a definition which always works. Weston (15) provides five definitions, including ones based on rays and modes. There is a small calculation which delights the writer and which brings out beautifully the relation between the two approaches in this context, though it involves some mathematical liberties. Start with the eigenvalue equation (1), differentiate with respect to -h (frequency constant), and make some minor adjustments using the relation between h and γ.

$$- 2\pi \frac{\partial n}{\partial h} = 2h \int \frac{dz}{\gamma} - \frac{\partial \phi_1}{\partial h} - \frac{\partial \phi_2}{\partial h} . \tag{6}$$

This may be rewritten or interpreted as

$$D_n = D_g + \Delta_{h1} + \Delta_{h2} . \tag{7}$$

Here D_n is the mode interference theory for cycle distance (16), D_g is the geometrical range theory formula for cycle distance, and the horizontal beam displacement terms Δ_h are corrections to the ray theory. They are the same as the Goos-Hänchen shifts in optics and related to the δ_h displacements; we will return to this later. Figure 2 illustrates the result for isovelocity water with displacement only at the bottom.

A good early reference for the basic calculation without the Δ terms is Tolstoy and Clay (16), but versions of it are continually cropping up in the literature out of context, often to the apparent astonishment of the new discoverer. One form (1) even goes in effect backwards from equation (6) to equation (1). The full calculation with Δ terms was first reported by Murphy (17).

Flux ideas appear here by courtesy of cycle time, as next discussed.

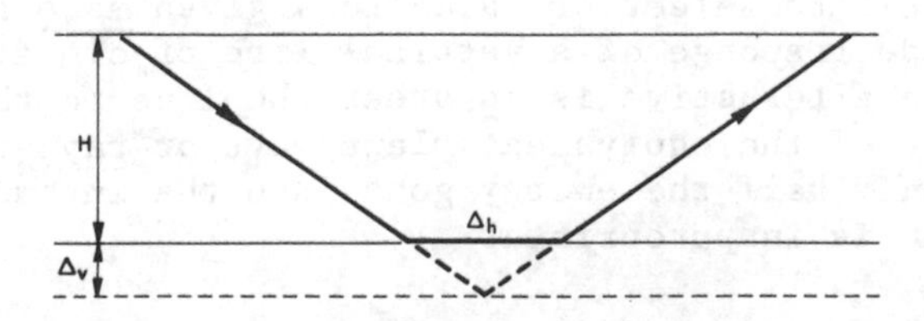

Fig 2. Cycle Distance with Displacement

3.6 Cycle Time

Cycle time as a parameter is encountered surprisingly rarely. It is of course another stage 3 infinite idea, and obtained by dividing the cycle distance by the group velocity. Starting with cycle time ideas Weston (18) reaches a definition of cycle distance as the ratio of the total horizontal energy flux to one vertical component of the flux density. Thus for cycle distance and time together we have the full complement of ray, mode and flux approaches.

3.7 Phase Velocity

Phase velocity c_p (a stage 3 infinite concept) is given by c sec θ where c is free space velocity. This is of course a constant independent of depth and in effect can be obtained from considerations of modes, rays or flux. For isovelocity water it may be written in the more specific form, which follows from equation (3),

$$c_p = c\left[1 - \left\{\frac{(n+1)\lambda}{2(H + \delta_{v1} + \delta_{v2})}\right\}^2\right]^{-\frac{1}{2}} . \qquad (8)$$

This form is presented in order to stress that there is a dependence on the δ displacements. We may imagine it applied to shallow water with δ_{v1} zero for the sea surface and δ_{v2} finite for a semi-infinite sediment bottom.

3.8 Group Velocity

Group velocity c_g (stage 3 infinite again) is well known as having a dual nature. We can start off with one definition based on modes and their frequency dispersion, and one equating c_g with the mean rate of horizontal energy flux. But sections 3.5 and 3.6 and figure 2 suggest we may also try a ray treatment. For the isovelocity case discussed the mean rate of advance along the ray path is

$$c_g = \frac{2H \cot\theta + \Delta_h}{2H \operatorname{cosec}\theta/c + \Delta_h/c \sec\theta} . \qquad (9)$$

Note that the speed along the (semi-infinite) bottom or displacement segment is equal to the mode phase velocity. For an account of this see especially Tindle and Bold (19). It is pleasing that formula (9) is exact for this Pekeris model case, and good approximations can be obtained for layered water by amending the first terms of numerator and denominator.

Note now some interesting and meaningful parallels in our various displacement and velocity terms δ_h, Δ_h, c_p and c_g. They respectively depend in a linear manner on ϕ/h, $\partial\phi/\partial h$, ω/h and $\partial\omega/\partial h$. Among other matters this leads on to the point (15) that δ and Δ can be defined for any stratum, not just the bottom or boundary. In the limit Δ_h can equal the cycle distance and the parallels become even closer.

3.9 Attenuation

One cause of mode attenuation in the sea (stage 3 infinite concept) is the lossiness of the material of the ocean bottom. Recent work has shown that the same answer may be reached by a number of different routes (20, 21). One set of routes is equivalent to a perturbation of the mode equations for lossless propagation. This corresponds closely to a ray approach in which there is a certain bottom loss per cycle distance. The cycle distance is calculated without including any beam displacement, and to compensate for this it is found necessary to reduce the normal bottom loss according to the inhomogeneity in the incident wave.

Alternatively we can use another ray approach where the cycle distance does include the displacement Δ_h, and the usual bottom loss for a homogeneous wave is appropriate. It is interesting here that two quite different ray approaches are possible, each one self-consistent. The approach using flux, in its special sense, is inapplicable.

Rayleigh (22) has pointed out a link between group velocity and attenuation, encountered in explaining why the same group velocity value arises from a dispersion calculation (mode) and from an energy transport equation (ray, mode or flux). He does this by considering increments in wavelength which are either real or imaginary, one gives dispersion and the other gives attenuation.

3.10 Arrival Time

The time pattern of arrivals (a transitive or stage 1 idea) may be modelled by rays or modes, and it is natural to use the former at short range and the latter at long range (see also section 3.11). Rays seem particularly appropriate at short range if one wishes to model the effect of source or receiver depth changes. Perhaps we may have the best of both worlds by building on the section 3.8 association of group velocity with a specific ray path. In another method we also have the best of both worlds through the simultaneous display of ray and mode effects (23, 24). The flux approach is inapplicable here.

3.11 Transmission Loss

It is well known that transmission loss (basically a stage 1 transitive idea) may be calculated either by summing all the ray contributions or all the mode contributions. In general neither is quite exact, assuming we restrict ourselves to propagating modes. A useful discriminant between the approaches, based on a preference for the smallest num'er of contributions, is $R\lambda/2H^2$ where R is range. This is shown as a dotted line in the universal range-frequency diagram of figure 3 (25). But provided there are at least a couple of rays in the ray description and at least a couple of modes in the mode description we can for many purposes scrap both ideas. Instead of the summation, of either type, we can use an integration of the flux. In fact all three approaches may be regarded as integrations over angle. Neglecting losses the relative intensity is

$$F = \frac{4}{R}\int \frac{d\theta_A}{\theta_B \, D_B} \tag{10}$$

where A and B refer to either source or receiver position (see section 2.3). With suitable definitions this will apply to multiple wave types, such as co-existing longitudinal and shear waves, and also to range-dependent media.

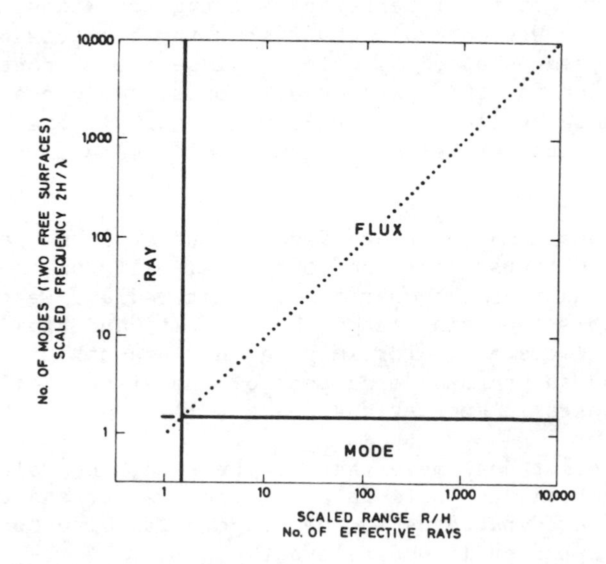

Fig 3 Diagram Showing Choice of Propagation Concepts for Constant-Depth Isovelocity Water with Lossless Boundaries

3.12 Interference Patterns, Beaming and Focussing

Consider now the field distribution or interference pattern for the stage 2 broadcast condition. With, say, two allowed modes the pattern defines a "fuzzy" ray (11, 16, 26-28). With a large number of modes the patterns develop in a way which the writer finds most surprising - the number of apparent rays increases, they sharpen up into very distinct beams, and without any necessary intervention of refraction we get a series of focal points.

The effects are not only predictable analytically but are observed in computer plots, in moiré fringe models, in laboratory acoustic models and in full-scale experiments at sea. The chosen illustration in figure 4 is a computer plot of transmission loss contours produced at AUWE (by DG Gleaves), using the parabolic equation approach, and based closely on a plot published by Jensen and Kuperman (29).

Since figure 4 assumes isovelocity water the beams follow straight-line courses. They leave the near-bottom source at angles occupying points of symmetry relative to the modes, ie at approximate intervals of $\lambda/4H$ corresponding not only to the mode angles but also to the angles halfway between. And yet another set of beams leaves from the complementary depth to the source, an unexpected or ghost source indeed! Both sets are explicable in terms of the interferences among the modes, following generally Weston (26). Note that the beam visibility in figure 4 has been enhanced by using a close set of contour lines. In addition for this particular geometry we are actually seeing a deep median valley in a broad ridge for all the beams in the first set, and for half the beams in the complementary set.

Figure 4 shows a near-surface focal point at 7 km, ie at range $4H^2/\lambda$. Such focussing raises the possibility of image transmission in underwater acoustic and other media. Weston (11) mentioned this focussing range, Weston (26) and Rivlin and Shul'dyaev (30) discussed it for imaging in fibre-optic applications, and an up-to-date account of the effect for fibre optics appears in Kuester and Chang (31).

There is a great deal more that can be said, including the obedience of the beams to Snell in stratified water, and the occurrence of related patterns in the stage 3 infinite condition. The flux approach is not relevant.

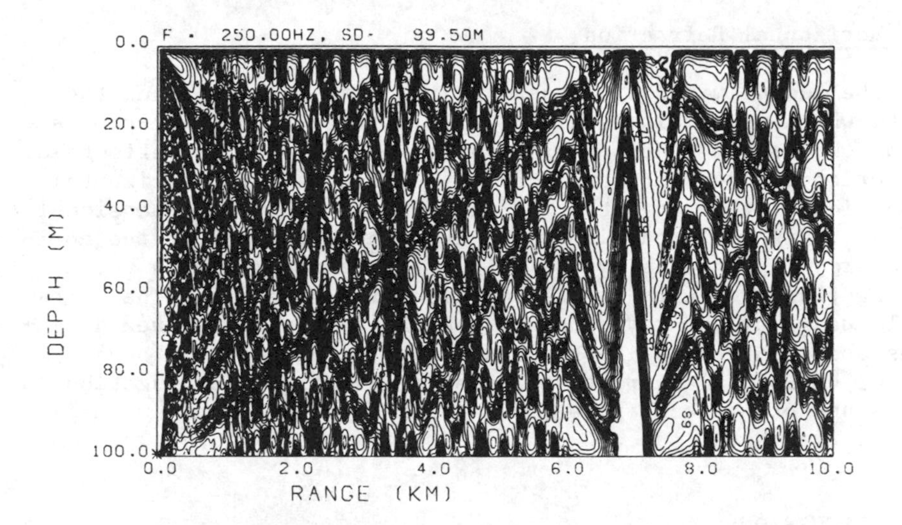

Fig 4. Interference Pattern Showing Beams and One Focal Point, 100m Isovelocity Water, 250Hz

3.13 Range-Dependent Environment

Propagation in a range-dependent environment will be discussed as basically a stage 3 or infinite-distance concept, with changes in both depth and sound velocity profile that are slow. The author has shown that there are certain properties which are invariant with range, and in his original paper (1) demonstrated this with the full trinity of approaches.

Thus considerations of ray reflection and refraction lead to a characteristic time or ray invariant, given in its simplest form by

$$T = (1/c) \int \theta \, dz. \tag{11}$$

This is equivalent to an adiabatic transfer of the energy in a given mode at a given position to the corresponding mode at another position having a different profile.

A flux approach using the reciprocity relation gives the most direct derivation of the invariance properties.

There are many questions arising, eg how slow is slow, which are still receiving attention. Note also that the flux transmission formula in section 3.11 still works in a slowly varying medium, as already pointed out, since it was originally derived to cope with this very problem.

3.14 Horizontal Refraction

The medium may vary across the track as well as along the track, which may again be modelled as a stage 3 infinite process. Weston (32) shows that for slow variation the combined effects of reflection and refraction produce a curvature in the horizontal projection of the ray path. He then shows for a two-dimensional or flat horizontal modelling that the same curvature of the horizontal ray is produced by the refraction associated with the varying phase velocity of the corresponding mode. Note here the double usage of the ray concept: for the true ray in three dimensions and then for the "horizontal" ray. Many workers have applied horizontal refraction ideas to investigate propagation in wedges and other geometries (33). Flux ideas have only a small part to play here.

4. DISCUSSION

In the course of this examination a number of symmetries or similarities have appeared. For example it is important that there are two sorts of displacement, one based on phase delay and the other on rate of change of phase delay. One appeared in the formula for phase velocity (and for eigenvalues), and the other in the formula for group velocity (and for cycle distance); and in fact they play a vital part in many other aspects of guided propagation.

Our main point concerns the three approaches, which do not always enter as mutual transforms, nor in any other tidy fashion. In particular we have met a great variety of different types of "ray". Out of the 13 applications rays and modes appear in all, though sometimes only just creeping in. But flux in its special sense does miss on 6 (sections 3.2, 3.4, 3.9, 3.10, 3.12 and 3.14). I strongly believe that most concepts or applications benefit by being looked at from more than one viewpoint. A mixing of methods can be dangerous but rewarding - try it, you'll like it!

5. ACKNOWLEDGEMENTS

I wish to acknowledge useful discussions with R Levers.

REFERENCES

1. Weston, D E. Guided Propogation in a Slowly Varying Medium. Proc Phys Soc 73 (1959) 365-384.
2. Weston, D E. Acoustic Flux Methods for Oceanic Guided Waves. J Acoust Soc Am 68 (1980) 287-296.
3. Brekhovskikh, L M. The Average Field in an Underwater Sound Channel. Sov Phys - Acoust 11 (1965) 126-134.
4. Smith, P W. Averaged Sound Transmission in Range - Dependent Channels. J Acoust Soc Am 55 (1974) 1197-1204.
5. Koopman, B O. New Ray Methods in Propagation. A D Little Report (1975).
6. Spofford, C W and L S Blumen. The ASTRAL Model, Volumes 1 and 2. Science Applications Inc Report (1978).
7. Laval, R and Y Labasque. Computation of Averaged Sound - Propagation Losses and Frequency/Space Coherence Functions in Shallow Waters. Bottom-Interacting Ocean Acoustics (NY, Plenum, 1980, Eds W A Kuperman and F B Jensen) 399-415.
8. Gloge, D and E A J Marcatili. Multimode Theory of Graded-Core Fibres. Bell System Technical Journal 52 (1973) 1563-1578.
9. Brillouin, L. Propagation d'Ondes Électromagnetique dans un Tuyau. Rev Gen Elect 40 (1936) 227.
10. Page, L and N I Adams. Electromagnetic Waves in Conducting Tubes. Phys Rev 52 (1937) 647.
11. Weston, D E. A Moiré Range Analog of Sound Propagation in Shallow Water. J Acoust Soc Am 32 (1960) 647-654.
12. Brekhovskikh, L M. Waves in Layered Media (NY, Adademic, 1960).
13. Felsen, L B. Hybrid Ray-Mode Fields in Inhomogeous Wave-guides and Ducts. J Acoust Soc Am 69 (1981) 352-361.
14. Weston, D E. Intensity-Range Relations in Oceanographic Acoustics. J Sound Vib 18 (1971) 271-287.
15. Weston, D E. Cycle Distance in Guided Propagation, Bottom-Interacting Ocean Acoustics (NY, Plenum, 1980, Eds W A Kuperman and F B Jensen) 393-398.
16. Tolstoy, I and C S Clay. Ocean Acoustics: Theory and Experiment in Underwater Sound (NY, McGraw-Hill, 1966).
17. Murphy, E L. Improvements of the Ray-Mode Analogy for Estimating the Frequency of Minimum Attenuation for Modes in Shallow Water. Sound Propagation in Shallow Water (SACLANTCEN Conference Proceedings No 14, 1974, Eds O F Hastrup and O V Olesen) Vol 2, pp 2-13.
18. Weston, D E. Influence of Bottom Profile on Ocean Acoustic Propagation. Acoustics and the Sea-Bed (UK, Bath University Press, 1983, Ed N G Pace) 1-8.
19. Tindle, C T and G E J Bold. Improved Ray Calculations in Shallow Water. J Acoust Soc Am 70 (1981) 813-819.
20. Weston, D E and C T Tindle. Reflection Loss and Mode Attenuation in a Pekeris Model. J Acoust Soc Am 66 (1979) 872-879.

21. Tindle, C T and D E Weston. Connection of Acoustic Beam Displacement, Cycle Distances, and Attenuations for Rays and Normal Modes. J Acoust Soc Am 67 (1980) 1614-1622.
22. Strutt, J W (Lord Rayleigh). On Progressive Waves. Proc London Mathematical Society 9 (1877) 21. (Included as an Appendix to the Theory of Sound Vol 1).
23. Porter, R B. Transmission and Reception of Transient Signals in a SOFAR channel. J Acoust Soc Am 54 (1973) 1081-1091.
24. Porter, R B and H D Leslie. Energy Evaluation of Wide-Band SOFAR Transmission. J Acoust Soc Am 58, 812-822.
25. Weston, D E. Shallow-Water Sound Propagation. IEEE EASCON Record IEEE Pub 78 CH 1345-4AES (1978) 252-255.
26. Weston, D E. Sound Focusing and Beaming in the Interference Field due to Several Shallow-Water Modes. J Acoust Soc Am 44 (1968) 1706-1712.
27. Tindle, C T and K M Guthrie. Rays as Interfering Modes in Underwater Acoustics. J Sound Vib 34 (1974) 291-295.
28. Kamel, A and L B Felsen. On the Ray Equivalent of a Group of Modes. J Acoust Soc Am 71 (1982) 1445-452.
29. Jensen, F B and W A Kuperman. Consistency Tests of Acoustic Propagation Models (SACLANTCEN Memorandum SM-157, 1982).
30. Rivlin, L A and V S Shul'dyaev. Multimode Waveguides for Coherent Light. Radiophys Quantum Electron 11 (1968) 318-321.
31. Kuester, E F and D C Chang. Imaging and Propagation of Beams in Metallic or Dielectric Waveguides. NATO Advanced Research Workshop on Hybrid Formulation of Wave Propagation and Scattering (1983).
32. Weston, D E. Horizontal Refraction in a Three-Dimensional Medium of Variable Stratification. Proc Phys Soc 78 (1961) 46-52.
33. Weston, D E and P B Rowlands. Guided Acoustic Waves in the Ocean. Rep Prog Phys 42 (1979) 347-387.

RAY/MODE TRAJECTORIES AND SHADOWS IN THE HORIZONTAL PLANE BY RAY INVARIANTS

C H HARRISON

CAP Scientific, 233 High Holborn, London, England

1. INTRODUCTION

Some interesting examples of the application of ray invariants to propagation in three dimensions are discussed. The initial interest in the third dimension was prompted by a study of long range reverberation in an ocean basin. Very long delay times and clear indications of returns from the edge of the basin are commonly seen (1), and the usual explanation for this type of return is scattering from the continental slope. Another possible explanation is that multiple reflection from the same continental slope causes enough horizontal bending to produce a bistatic echo. Either explanation can be valid depending on the relative magnitude of the reflection loss and the back scattering strength. Back scattering would be favoured by rough sea beds whereas reflection would be favoured by smooth. This question is an interesting one and worthy of experimental investigation although designing a suitable experiment is quite difficult.

By using ray invariants it was shown in (2) that in some cases it is possible to calculate analytically the horizontal trajectory of the ray path given the analytical form of the sea bed and the initial vertical and horizontal ray angle. Some of the cases are discussed in Section 4. They include horizontal ducting in a trough, total trapping in a basin and deflection by a ridge and a seamount.

By using the equivalence between normal modes and rays one can use this technique to study the propagation and shadowing of normal modes in the horizontal plane (3). Interesting comparisons can be made with other studies of the same phenomenon using quite different approaches (4), (5). In fact the solutions given here can be used as test cases for more sophisticated three dimensional mode calculations, and they give a feel for what is really going on.

An example of the importance of the interchange of ideas between rays and modes is that intuitively one tends to reject sharp bending in the horizontal plane but at the same time to accept the analytical mode solution in a wedge. The latter treatment includes the former phenomenon.

Clearly, the importance of horizontal curvature effects in practice depends on the magnitude of the overall loss encountered during the multiple reflections. The loss is calculated for a wedge-shaped duct for several loss laws, namely: a critical angle cut-off, a constant loss per bounce, and a loss proportional to grazing angle.

Finally it is worth stressing that rays or modes travelling obliquely up-slope can be deflected back to deep water without ever reaching a high grazing angle. Losses are therefore not necessarily prohibitive.

2. RAY INVARIANTS

The relevant ray invariants are derived for the isovelocity case in (6). The most well known invariant is due to Weston (7,8), and for the isovelocity case reduces to $H \sin \theta$. Thus, knowing the values of elevation angle and depth at the source θ_0, H_0 the elevation angle θ at a different site with depth H is given by

$$H \sin \theta = H_0 \sin \theta_0 \qquad (1)$$

The only restriction on the use of this invariant is that the water depth must only change slowly and the bottom slope must change smoothly between one reflection point and the next.

After each bottom reflection there is a slight change in heading of the ray, and there is another invariant which determines the new ray heading given the elevation angle. Unfortunately this one is not universally applicable, but it can be derived in two versions - one for troughs or ridges of constant cross section i.e. straight and parallel depth contours; the other, for basins or seamounts with rotational symmetry i.e. concentric, circular depth contours.

For straight troughs and ridges the invariant is

$$\cos\theta \sin\phi = \cos\theta_0 \sin\phi_0 \qquad (2)$$

where ϕ is the heading of the ray relative to the up-slope direction, and ϕ_0 is the initial value.

For circular basins or seamounts the invariant is

$$r\cos\theta \sin\phi = r_0 \cos\theta_0 \sin\phi_0 \qquad (3)$$

where Θ and $\varnothing$ have the same definition as for straight troughs and ridges but r is the radial coordinate of the ray.

3. RAY PATH CALCULATION

By combining the two invariants (equations 1 and 2 or 1 and 3) we can obtain an expression for the new ray heading ϕ in terms of known quantities, and an integral expression enables calculation of the ray trajectory.

3.1 Troughs and Ridges

For troughs and ridges we have

$$\sin\phi = \sin\phi_0 \cos\phi_0 \quad H/(H^2 - H_0^2 \sin^2\theta_0)^{\frac{1}{2}} \qquad (4)$$

The trajectory in the horizontal x, y plane with x measured along the depth contours and y across them is given by

$$x = \int dx = \int dy \tan\phi = \int \frac{\sin\phi_0 \cos\theta_0 \, dy}{\left[(1 - \sin^2\phi_0 \cos^2\theta_0) - \frac{H_0^2}{H^2}\sin^2\theta_0\right]^{\frac{1}{2}}} \qquad (5)$$

Given the cross section of the trough $H(y)$ this integral can be solved, and the ray path $y(x)$ is then known. In some cases, as shown below, the integral can be solved analytically.

In addition, we can calculate the total ray length from

$$s = \int ds = \int dy \sec\theta \sec\phi = \int \frac{\tan\phi \, dy}{\cos\theta_0 \sin\phi_0} = \frac{x}{\cos\theta_0 \sin\phi_0} \qquad (6)$$

Evidently, if the depth contours are straight and parallel

the ray length is always proportional to the component of distance travelled along the trough. In other words, each ray propagates along the trough at its own constant velocity regardless of any horizontal curvature.

3.2 Basins and Seamounts

For circular basins and seamounts we have

$$\sin\phi = \sin\phi_0 \cos\theta_0 \, H \frac{r_0}{r} / (H^2 - H_0^2 \sin^2\theta_0)^{\frac{1}{2}} \tag{7}$$

The trajectory in horizontal polar coordinates r, Φ is given by

$$\Phi = \int d\Phi = \int \frac{dr}{r} \tan\phi = \int \frac{\sin\phi_0 \cos\theta_0 \, dr}{r\left[\frac{r^2}{r_0^2}\left(1 - \frac{H_0^2}{H^2}\sin^2\theta_0\right) - \sin^2\phi_0 \cos^2\theta_0\right]^{\frac{1}{2}}} \tag{8}$$

The total ray length is given by

$$s = \int ds = \int dr \sec\theta \sec\phi = \int \frac{\tan\phi \, r \, dr}{\sin\phi_0 \cos\theta_0 \, r_0} = \frac{\int r^2 \, d\Phi}{r_0 \sin\phi_0 \cos\theta_0}$$

$$= \frac{2\,(\text{Area of trajectory swept out})}{r_0 \sin\phi_0 \cos\theta_0} \tag{9}$$

The ray length in a basin obeys Kepler's second law for planetary motion and is always proportional to the area swept out by the radial coordinate of the ray as it moves around the basin. It is interesting that this is true for a circular basin or a seamount of arbitrary cross section.

4 EXAMPLES OF HORIZONTAL RAY PATH

Deviations of ray paths for eight cases are given in (2), but here we give just the solutions for each of a trough, ridge, basin and seamount.

4.1 Trough: $H(y) \propto \cos(y/R)$

The solution for the cosine trough is (10)

$$\sin(y/R) = \left(1 - \frac{\cos^2(y_0/R)\sin^2\theta_0}{(1 - \sin^2\phi_0\cos^2\theta_0)}\right)^{\frac{1}{2}} \sin\left[\frac{(x+x')(1-\sin^2\phi_0\cos^2\theta_0)}{R\sin\phi_0\cos\theta_0}\right]$$

where x' is an integration constant found by setting $x=0$ at $y=y_0$.

A family of curves varying θ_0 but fixing ϕ_0 is shown in Fig 1.

Other soluble cases include $H(y) \propto [1 + (y^2/R^2)]^{-\frac{1}{2}}$ which gives exact sine wave rays, and $H(y) \propto [1 - (y^2/R^2)]^{\frac{1}{2}}$ (an elliptical trough) which gives solutions in terms of elliptic integrals.

4.2 Ridge: $H(y) \propto [1 - (y^2/R^2)]^{-\frac{1}{2}}$ (for $|y|, |y_0| < R$)

The solution consists of sinh curves which cross the axis of the ridge and cosh curves which are deflected from the axis as shown in Fig 2.

For $|y_0| < \dfrac{R\cot\theta_0\cos\phi_0}{(1 + \cot^2\theta_0\cos^2\phi_0)^{\frac{1}{2}}}$ (11)

$$y = (\cot^2\theta_0\cos^2\phi_0(R^2 - y_0^2) - y_0^2)^{\frac{1}{2}} \sinh\left[\frac{x + x'}{(R^2 - y_0^2)^{\frac{1}{2}}\cot\theta_0\sin\phi_0}\right]$$

and for $|y_0| > \dfrac{R\cot\theta_0\cos\phi_0}{(1 + \cot^2\theta_0\cos^2\phi_0)^{\frac{1}{2}}}$ (12)

$$y = (y_0^2 - \cot^2\theta_0\cos^2\phi_0(R^2 - y_0^2))^{\frac{1}{2}} \cosh\left[\frac{x + x'}{(R^2 - y_0^2)^{\frac{1}{2}}\cot\theta_0\sin\phi_0}\right]$$

Other soluble cases include $H(y) \propto \cosh(y/R)$ which gives a solution in terms of cosh and sinh functions, and the wedge $H(y) \propto y$ which gives a hyperbolic ray path. The latter will be referred to later in the discussion of ray/mode effects.

4.3 Basin: $H(r) \propto [1 + (r^2/R^2)]^{-\frac{1}{2}}$

The ray path is an ellipse centred on the middle of the basin as shown in Fig 3. For this type of basin all ray paths converge on a point diametrically opposite the source regardless of their initial heading or elevation angle. Each ray carries on around its own ellipse indefinitely.

The solution is

$$\sin[2(\Phi + \Phi')] = \frac{A - B/r^2}{(A^2 - 2B)^{\frac{1}{2}}} \qquad (13)$$

where

$$A = (R^2 + r_0^2)\ \text{cosec}^2\ \theta_0 - R^2$$

$$B = 2 \cot^2 \theta_0 \sin^2 \phi_0\ r_0^2 (R^2 + r_0^2)$$

and Φ' is given by setting $\Phi = 0$ at $r = r_0$

4.4 Seamount: $H(r) \propto r$

The solution for a conical seamount (zero water depth at the apex) is a polar plot of a cosec curve as shown in Fig 4.

$$r = r_0(1 - \cos^2 \theta_0 \cos^2 \phi_0)^{\frac{1}{2}}\ \text{cosec}\ \left[\frac{(1 - \cos^2 \theta_0 \cos^2 \phi_0)^{\frac{1}{2}}}{\sin \phi_0 \cos \theta_0} (\Phi + \Phi')\right] \qquad (14)$$

5. RAY/MODE EQUIVALENCE

Although, so far, the treatment has been entirely ray acoustics (without any mention of frequency) we can use the well known ray/mode equivalence to make some more powerful statements about either the horizontal propagation of one-dimensional normal modes or alternatively the three dimensional mode pattern.

5.1 Mode Propagation in a Wedge-shaped Duct

To plot the propagation of a mode we note that for a given water depth the mode has an equivalent ray elevation angle. Thus, we can map the progress of a mode by plotting the family of rays that have a fixed initial elevation angle θ_0 but arbitrary initial heading ϕ_0. These are shown for the case of a wedge-shaped duct in Fig 5, and the equation of each ray is a hyperbola given by

$$y^2 \; (1 - \cos^2 \theta_0 \sin^2 \phi_0) \tag{15}$$

$$= \left[x\left(\frac{1 - \cos^2 \theta_0 \sin^2 \phi_0}{\sin \phi_0 \cos \theta_0}\right) - y_0 \cos \theta_0 \cos \phi_0\right]^2 + y_0^2 \sin^2 \theta_0$$

The envelope of these curves marked in the diagram defines the edge of a horizontal shadow zone for rays with this initial elevation angle.

By elementary calculus (differentiating with respect to ϕ_0 with x, y and θ_0 constant) it is possible to calculate the envelope. This is also a hyperbola, given by

$$y^2 = x^2 \tan^2 \theta_0 + y_0^2 \sin^2 \theta_0 \tag{16}$$

It is clear from this formula and from the diagram that rays with a steeper initial elevation angle have an even larger shadow zone.

We may insert the ray/mode equivalence as follows. Taking the sea bed and sea surface to be rigid and free respectively, the source wavelength λ, water depth H, mode number n and ray elevation angle θ are related everywhere through the adiabatic approximation in isovelocity water by

$$2H \sin \theta = (n - \tfrac{1}{2})\lambda \tag{17}$$

At the source, in particular, we have

$$\sin \theta_0 = (n - \tfrac{1}{2})\lambda/2H_0 \tag{18}$$

and writing $\sin \theta_0$ as β the envelope becomes

$$y^2 = x^2/(1/\beta^2 - 1) + \beta^2 y_0^2 \tag{19}$$

Curves of this form for different n are shown in Fig 6. These agree with the asymptotic limits that were derived in (9) for the modal shadow zones, and they are also numerically identical to Figs 2-11 in (4). The similarity is most striking when it is remembered that the three approaches to the same problem have quite different starting points.

At a fixed frequency the shadow zone for a particular mode extends from the edge of the duct up to the appropriate hyperbola in Fig 6. Increasing the source frequency squeezes the whole picture towards the edge of the duct so that the shadow zones become narrow, until at infinitely high frequency they disappear altogether.

From a practical point of view the importance of the shadow zones is that they can extend to very large distances from the edge of the duct into deep water where a normal mode treatment would not usually be considered appropriate.

5.2 Propagation Over a Ridge

Although formulae are given for many paths in Section 4 it is not always possible to calculate the envelope analytically because of the appearance of trancendental equations. Qualitatively similar results may be found from a simpler ridge composed of two opposed linear slopes joined along a contour of depth D.

$$H = D + \gamma|y| \tag{20}$$

The rays are now segments of hyperbolae as shown in Fig 7, and the envelopes are also hyperbolae as shown in Fig 8. Note that on the source side of the ridge the asymptote of the envelope is inclined at θ_0 (the initial ray elevation angle) to the ridge axis as in the case of the wedge-shaped duct. On the far side of the ridge, however, the inclination of the asymptote is greater and equal to the elevation angle at the crest of the ridge. This is because the envelope is equivalent to the ray which momentarily runs along the crest of the ridge, and the result stems from the second ray invariant.

An interesting variant of this ridge case may be made by setting the bottom slope to zero in the upper part of the figure. This could model a low frequency source on the contintental slope in front of the continental shelf. The ray paths and shadow zone boundaries on the source side are unchanged and remain as in Fig 8. On the far side, though, rays are straight and almost uniformly distributed in angle so that there are no shadow zones. Thus, the shadow zone boundary on the shallow water side is simply the edge of the shelf i.e. the line $y=0$.

5.3 Other Cases

Two other cases are treated in (2). These include a circular basin for which the shadow zone is a circle if the source is near the centre, and a conical seamount for which shadow zones are shown in Fig 9 with representative rays.

6. SOUND INTENSITY IN TROUGHS AND BASINS

So far we have only considered ray paths and not sound intensity. There are two separate loss mechanisms in force now. One is the spreading loss in the horizontal plane, and the other is the reflection loss produced by the multiple bottom reflection.

The effect of horizontal spreading on a SOFAR ray as it encounters a sloping sea bed is shown in Fig 10. From the second ray invariant there is no change to the divergence $d\phi$ as the two incoming rays are deflected back to deep water. The rays behave as if they have been specularly reflected at the zero depth line (as shown by the broken lines), and the intensity falls as $1/r$ where r is the total horizontal length of the ray. A more detailed discussion is given in (2) and (8).

The total reflection loss can be evaluated for several dependences of loss with elevation angle, namely:

1) No loss up to a critical angle.

2) Loss independent of angle.

3) Loss proportional to grazing angle.

6.1 No Loss up to a Critical Angle

The effect of a critical angle is exceptionally easy to see in all the diagrams so far. From the first ray invariant (Equation 1) a critical angle implies a critical depth H_c given by

$$H_c = H_0 \sin\theta_0/\sin\theta_c \qquad (21)$$

beyond which losses rapidly become large. Thus, in the earlier diagrams the portion of any ray which is on the shallow side of the depth contour defined by H_c should be deleted. Similarly any rays returning from the shallow side of the contour to deep should be deleted.

This has the effect that there are no rays returning from close to the up-slope direction, but there are still complete ray paths setting off more obliquely to the up-slope direction. In particular, the ray which sets off along the depth contour from the source defines the edge of the shadow zone, and this ray must still exist if the initial angle θ_0 is less than the critical angle θ_c. The shadow zone is therefore largely unaffected at large ranges, but is enlarged in the up-slope direction from the source by shifting the boundary to the depth contour H_c.

The above finding is interesting in the light of the common statement that rays cannot return to deep water without reaching 90^0 elevation. In three dimensions this is not correct, and rays may turn back to deep water never having reached the critical angle. As a corollary to this, when there is penetration into the bottom (see, for instance, (10)) the penetration is not entirely in the up-slope direction. In fact the sub-bottom rays may spread right round to the constant depth direction. This may be seen by continuing the outgoing horizontal rays as straight lines from the point where they first hit the H_c depth contour.

6.2 Loss Independent of Angle

The number of bottom reflections can be calculated from

$$N = \int \frac{\sin\theta \, ds}{2H} = \frac{H_0 \sin\theta_0}{2\sin\phi_0 \cos\theta_0} \int \frac{\tan\phi \, dy}{H^2} \tag{22}$$

For a SOFAR ray reflecting from a wedge (2) the total loss L reduces to

$$L = NR = \frac{\pi}{2\gamma} R \tag{23}$$

where R is the reflection loss in dB for a single reflection and γ is the bottom slope. Note that the number of reflections $\pi/2\gamma$ can be deduced by considering the image reflecting planes in the case of a wedge.

6.3 Loss Proportional to Grazing Angle

If we assume the loss to vary with the angle as

$$R = \alpha \sin\theta \tag{24}$$

where α is a constant we can calculate the loss from

$$L = \int R dN = \frac{H_0 \sin\theta_0 \, \alpha}{2\sin\phi_0 \cos\theta_0} \int \frac{\tan\phi \sin\theta \, dy}{H^2}$$

$$= \frac{\alpha y_0^2 \sin\theta_0}{\gamma} \int \frac{dy}{y^2\left[(\mathrm{cosec}^2\theta_0 - \cot^2\theta_0 \sin^2\phi_0)y^2 - y_0^2\right]^{\frac{1}{2}}} \tag{25}$$

This loss can be evaluated for three cases of interest by substituting different integral limits.

Case 1

We allow the ray to travel to its minimum depth

$$y_m = y_0/(\operatorname{cosec}^2 \theta_0 - \cot^2 \theta_0 \sin^2 \phi_0)^{\frac{1}{2}}$$

and then return to the source depth. The loss is given by

$$L = \frac{\alpha}{\gamma} \cos \theta_0 \cos \phi_0 \tag{26}$$

For a SOFAR ray we can put $\theta_0 = 0$, and we have

$$L = \frac{\alpha \cos \phi_0}{\gamma} = \frac{R(\pi/2 - \phi_0)}{\gamma} \tag{27}$$

The interesting point here is that the equivalent vertical specular reflector along the zero depth line, which has already been mentioned, behaves as if it were made of the same material as the bottom (but scaled by the factor $1/\gamma$) since $(\pi/2 - \phi_0)$ is the asymptotic grazing angle to the edge of the wedge.

Case 2

We allow the ray to travel to its minimum depth and then back out to infinity. The loss is given by

$$L = \frac{d}{2\gamma} [\cos \theta_0 \cos \phi_0 + (1 - \cos^2 \theta_0 \sin^2 \phi_0)^{\frac{1}{2}}] \tag{28}$$

This is shown in Fig 11. It is interesting that for rays travelling more or less up-slope, i.e. low ϕ_0, the loss is greatest for low elevation angles or low order modes since these have the most reflections. For high ϕ_0, rays travelling only slightly up-slope, the situation is reversed and the loss is greatest for the high elevation angles or high order modes. There is an intermediate region at about $\phi_0 = 60^0$ where the loss is about the same for all elevation angles.

Case 3

We allow the ray to travel to its minimum depth and then back out to the shadow zone boundary at

$$y_e = \frac{y_0 \sin \theta_0}{\cos \phi_0} (1 - \sin^2 \phi_0 \sin^2 \theta_0)^{\frac{1}{2}}$$

This essentially allows us to put a limit on the validity of the shadow zone. If the loss from the source to the shadow zone boundary is prohibitively large then the shadow boundary, as derived here, is invalid. However, the shadow zone still exists, and it is made even larger by this effect. The loss is now

$$L = \frac{\alpha}{2\gamma}\left[\cos\theta_0 \cos\phi_0 + \frac{\cos\theta_0 \sin\theta_0 \sin^2\phi_0}{(1 - \sin^2\phi_0 \sin^2\theta_0)^{\frac{1}{2}}}\right] \qquad (29)$$

and this is shown in Fig 12.

7. CONCLUSION

Some interesting examples of bottom interacting rays that curve in the horizontal plane have been discussed. The analytical form of these ray paths derived from ray invariants provides useful test cases for three-dimensional ray-tracing computer programs.

The technique combined with ray/mode equivalence allows one to trace the propagation of normal modes in the horizontal plane for various troughs, basins, ridges and seamounts. Each mode has a shadow zone extending into deep water where a normal mode treatment would usually be considered inappropriate. In the case of the wedge-shaped duct the analytical solutions agree with previous authors' work, and predictions of shadow zone shapes are made for some more complicated topographies.

The intensity of such rays has been calculated by considering the spreading loss in the horizontal plane and the reflection loss. A SOFAR ray, on encountering a sloping sea bed behaves as if it were specularly reflected from the zero depth contour so its spreading loss goes as $1/r$ where r is the total horizontal ray length. The total bottom loss has been evaluated for three cases of loss dependence on local elevation angle. These are: zero loss up to the critical angle, loss independent of angle, and loss increasing with elevation angle. Since the number of reflections is roughly the inverse of the bottom slope the total loss is of order R/γ.

The simple approach used here provides a number of test cases for much more sophisticated three dimensional normal mode calculations. It also provides an easy way of distinguishing the wood from the trees.

REFERENCES

1. Goertner, J A. "Ocean Basin Reverberation from Large Underwater Explosions. Part II: Computer Model for Reverberation", SACLANTCEN Conference Proceedings No 17, Part 5, pp 25-1 - 25-14 (1975).

2. Harrison, C H. "Three-Dimensional Ray Paths in Basins, Troughs and Near Seamounts by Use of Ray Invariants", J Acoust Soc Am 62, 1382-1388 (1977).

3. Harrison, C H. "Acoustic Shadow Zones in the Horizontal Plane", J Acoust Soc Am 65, 56-61 (1979).

4. Weinberg, H and Burridge, R. "Horizontal Ray Theory for Ocean Acoustics", J Acoust Soc Am 55, 63-79 (1974).

5. Buckingham, M J. "Acoustic Propagation in a Wedge-Shaped Ocean with Perfectly Reflecting Boundaries", NATO Advanced Research Workshop Proceedings, Hybrid Formulation of Wave Propagation and Scattering (1983), this publication.

6. Harrison, C H. "Horizontal Ray Curvature Effects in Basins, Troughs, and Near Seamounts by Use of Ray Invariants", NRL Report No 8144 (1977).

7. Weston, D E. "Guided Propagation in a Slowly Varying Medium", Proc Phys Soc London 73, 365-384 (1959).

8. Weston, D E. "Horizontal Refraction in a Three-dimensional Medium of Variable Stratification", Proc Phys Soc London 78, 46-52 (1961).

9. Bradley, D L. "The Propagation of Sound in a Wedge-Shaped Shallow Water Duct", PhD thesis, The Catholic University of America, Washington, DC (1970).

10. Jensen, F B and Kuperman, W A . "Sound Propagation in a Wedge-shaped Ocean with a Penetrable Bottom", J Acoust Soc Am 67, 1564-1566 (1980).

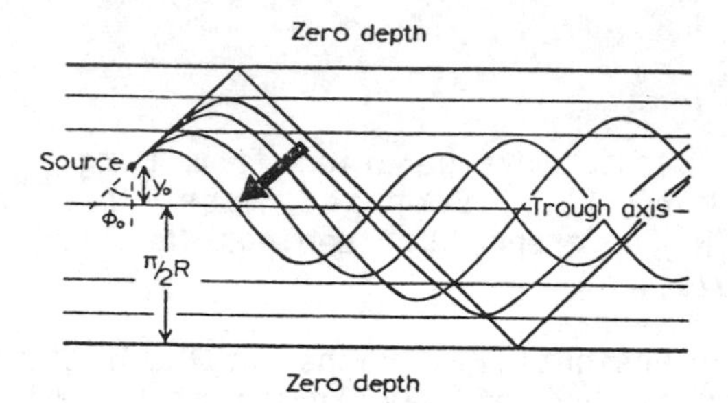

Fig 1 Horizontal projection of ray paths for a trough, for constant ϕ_0. Increasing θ_0 indicated by broad arrow.

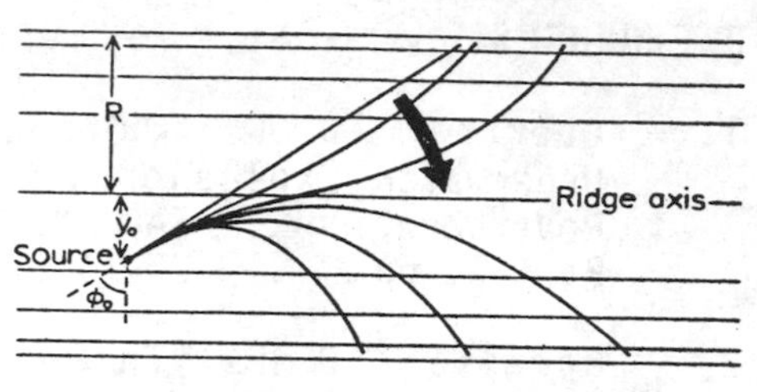

Fig 2 Horizontal projection of ray paths for a ridge, for constant ϕ_0. Increasing θ_0 indicated by broad arrow.

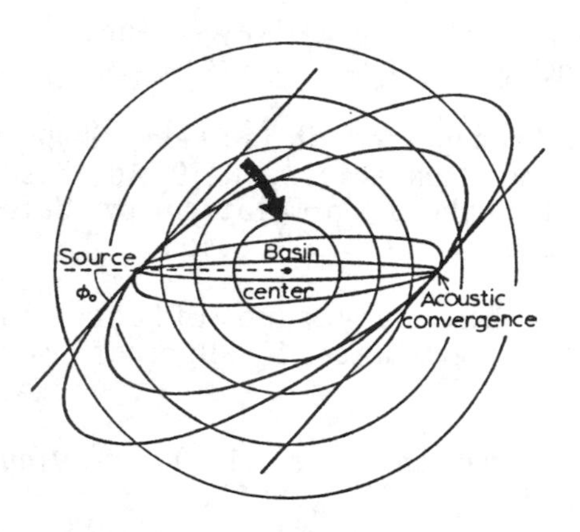

Fig 3 Horizontal projection of ray paths for a basin, for constant ϕ_0. Increasing θ_0 indicated by broad arrow.

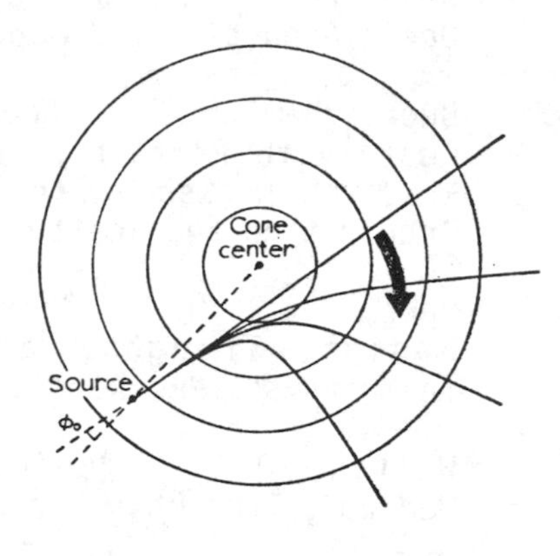

Fig 4 Horizontal projection of ray paths for a seamount, for constant ϕ_0. Increasing θ_0 indicated by broad arrow.

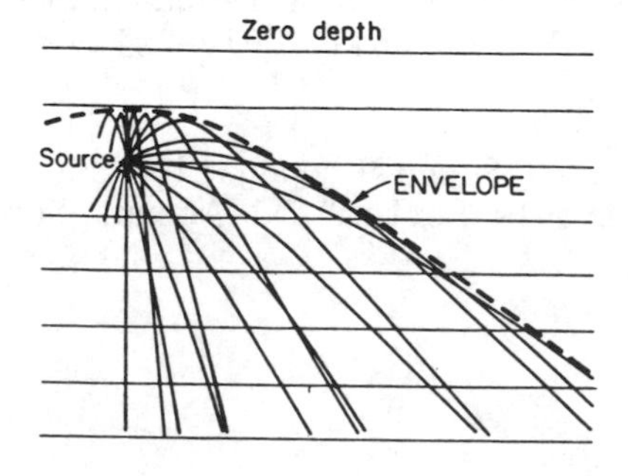

Fig 5 Horizontal ray/mode paths corresponding to n=2 in a wedge-shaped duct.

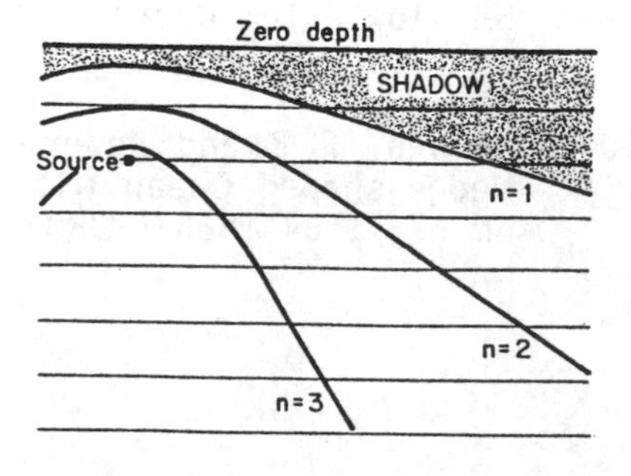

Fig 6 Shadow zone boundaries for numbered modes in a wedge-shaped duct.

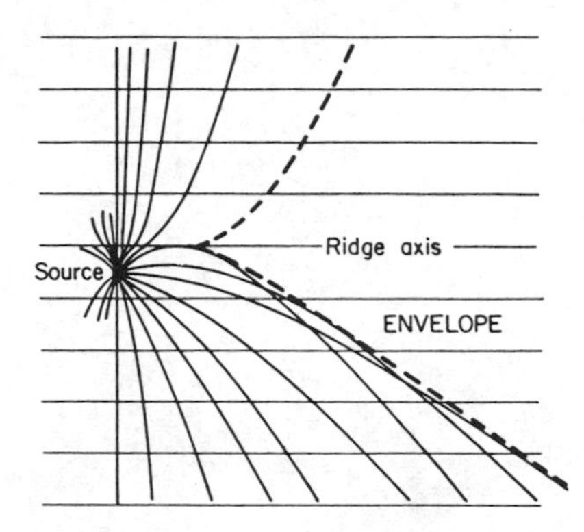

Fig 7 Horizontal ray/mode paths and shadow zone boundary construction across a ridge.

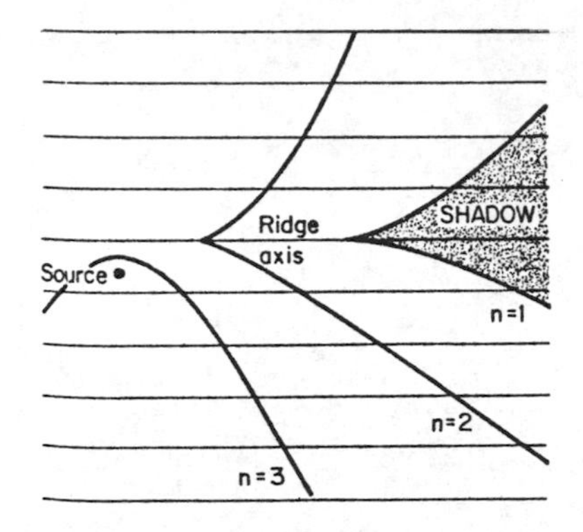

Fig 8 Shadow zone boundaries for numbered modes across a ridge.

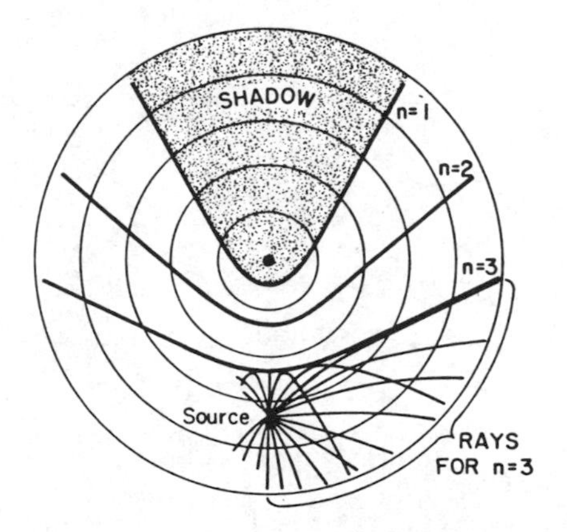

Fig 9 Shadow zone boudaries for numbered modes across a conical seamount, with some representative ray paths.

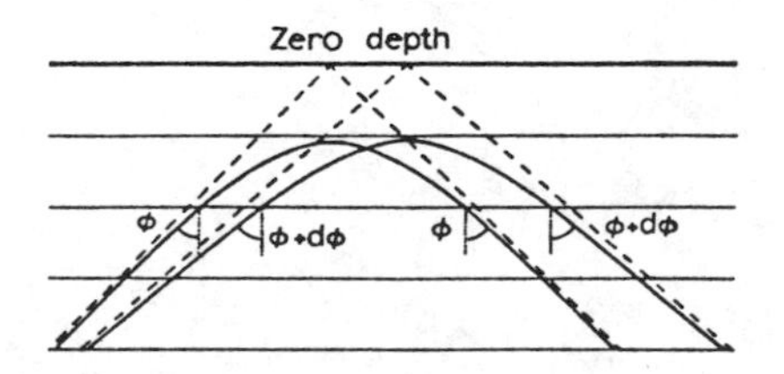

Fig 10 The horizontal divergence (dφ) of two sets of SOFAR rays is practically unchanged after encountering a straight basin edge.

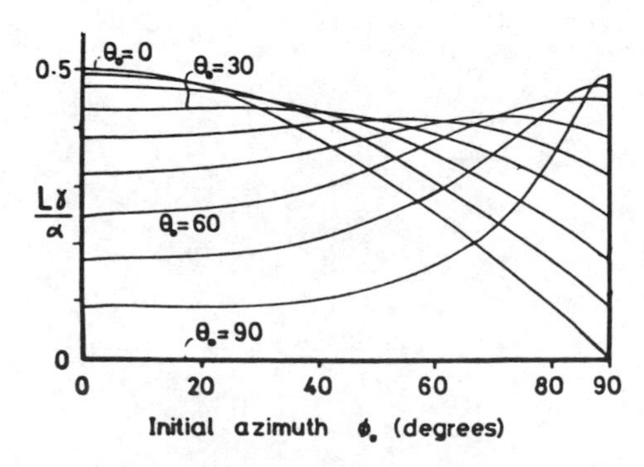

Fig 12 Dimensionless loss for Case 3 (10 degree steps in θ_0).

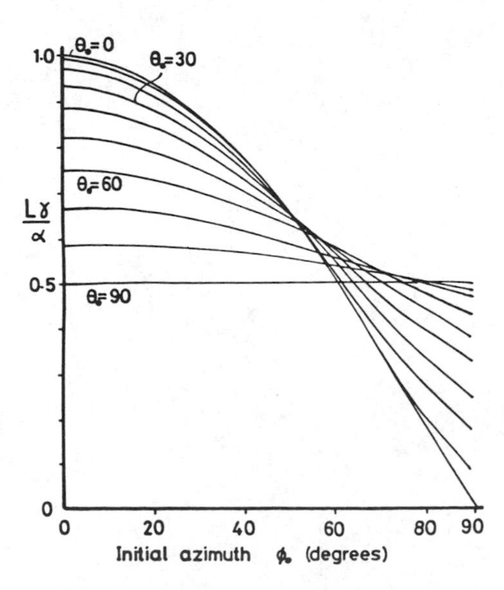

Fig 11 Dimensionless loss for Case 2 (10 degree steps in θ_0).

ACOUSTIC PROPAGATION IN A WEDGE-SHAPED OCEAN WITH PERFECTLY REFLECTING BOUNDARIES

M. J. Buckingham

Naval Research Laboratory
Washington, DC 20375
and
Royal Aircraft Establishment
Farnborough
Hants, GU14 6TD
England

SUMMARY

A new solution for the acoustic field produced by a point source in a wedge-shaped ocean channel with pressure-release boundaries is presented. The solution is in the form of a sum of normal modes, which reduces in the immediate vicinity of the source point to the free-field solution for a point source. The radiation field associated with each mode forms a well-defined beam which diverges as the energy propagates out towards deep water. Outside the beam, shadow zones are formed, where there is essentially no energy in the mode. The modal beams are nested together, with the inner and outer beams having the highest and lowest mode numbers, respectively. Thus, the spatial extent of the field in the direction parallel to the shore line is determined by the lowest order mode. The modal beams are interpreted in terms of rays, by invoking the concept of ray/mode duality. The criterion for a ray to correspond to a mode is given. Each ray path undergoes curvature in the horizontal direction, or horizontal refraction, due to the multiple acoustic interactions with the inclined boundaries of the wedge, and the rays corresponding to a given mode are found to be constrained to fall precisely within the modal beam.

INTRODUCTION

The ocean overlying the continental slope, or a sloping beach, is most simply represented as a wedge-shaped domain with perfectly reflecting (pressure release) boundaries. This model is, of course, inadequate in that it cannot account for certain phenomena, such as acoustic penetration of the bottom, which are encountered in the real ocean. On the other hand, the wave equation for the 'perfect' wedge is separable, and the analysis of the acoustic field due to a point source within the domain is tractable. By way of contrast, the wave equation for the penetrable wedge is not separable, a fact which inevitably introduces a good deal of complexity into the analysis of the acoustic field in this case. Here we shall confine our attention to the simpler case where the boundaries are perfect reflectors. Certain physical attributes of the 'perfect' wedge are discussed, most notably the fact that the radiation field is constrained to form modal beams as a result of the multiple acoustic interactions with the boundaries. This conclusion, which derives from the solution of the wave equation, can be most satisfactorily interpreted by calling on the concept of ray/mode duality. The question of the connection between rays and modes in the wedge is pursued here at some length.

In general, the radiation field in a wedge-shaped domain formed by two perfectly reflecting boundaries consists of two components: the modal component is a consequence of the contraints imposed by the boundary planes on the field, and the diffracted component arises from scattering at the apex of the wedge. It is well known (1) that when the wedge angle is a sub-multiple of π the diffracted component is identically zero; and in certain special cases, even when the wedge angle is not a sub-multiple of π, the diffracted component is again absent (2). What is more significant, however, is that in the context of ocean wedges, where the wedge angle is usually very small (typically a few degrees) and the ranges of the source and the receiver from the apex are generally many wavelengths long, the diffracted component, if not entirely absent, is negligibly small. This being so, it is ignored entirely in the following discussion, which is devoted exclusively to the modal component of the field.

An approximate solution for the modal part of the field in a 'perfect' wedge has been derived by Bradley and Hudimac (3) and examined by Graves et al (4). Unfortunately, its range of validity is extremely limited. In particular, it is unsatisfactory throughout a large proportion of the modal beams alluded to above, in just the regions which are of interest in many applications. The failure of Bradley and Hudimac's solution in these regions could, perhaps, have been anticipated, since it does not show the required form around the source point. One of the main results in the present paper is a new solution for the field in the wedge, which has a much more extensive range of validity than that of Bradley and Hudimac, and which behaves correctly in the immediate vicinity of the source.

THE FIELD DUE TO A POINT SOURCE IN THE WEDGE

The geometry for the wedge problem is shown in Fig. 1, where S and R represent a point source and a point receiver, respectively. Since we are interested in the three-dimensional field in the wedge, S and R are not necessarily in the same plane perpendicular to the apex of the wedge. The angle of the wedge is θ_o. A cylindrical coordinate system is employed for the problem, with the z-axis running down the apex of the wedge, and with the angular coordinate, θ, measured from the surface. Thus, the boundary planes are at $\theta = 0$ and $\theta = \theta_o$. The coordinates of the receiver are (r,θ,z) and those of the source, which for convenience is located in the $z = 0$ plane, are the primed quantities $(r',\theta',0)$.

The equation which must be solved for the field in the wedge is the inhomogeneous Helmholtz equation:

$$\nabla^2\phi + k^2\phi = -Q\frac{\delta(r-r')}{r}\delta(\theta-\theta')\delta(z), \tag{1}$$

where Q is the source strength, $\delta(\ )$ is the Dirac delta function, and k is the wavenumber (i.e. $k = \omega/c$, where ω is the angular frequency and c is the speed of sound in the medium, which is assumed to be independent of position). The function ϕ in equation (1) is the velocity potential excluding the time-dependent factor, in the case of an harmonic source, or it is the Fourier transform (with respect to time) of the velocity

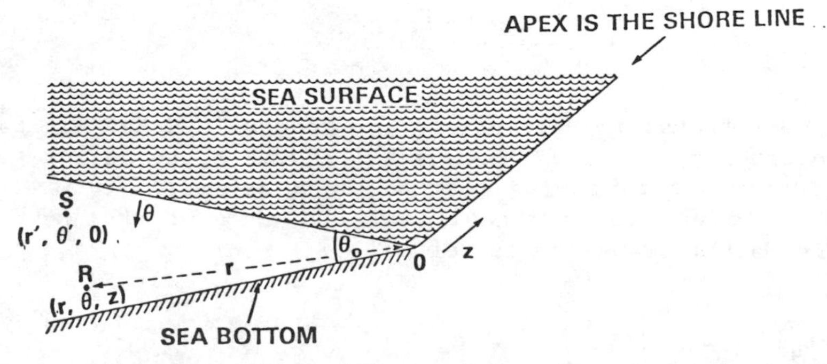

Figure 1 Cylindrical coordinate system for the wedge problem. S and R represent the source and the receiver respectively.

potential for an impulsive source. On expressing the Laplacian operator in cylindrical coordinates, equation (1) becomes

$$\frac{1}{r}\frac{\partial}{\partial r}\left(\frac{r\partial\phi}{\partial r}\right)+\frac{1}{r^2}\frac{\partial^2\phi}{\partial\theta^2}+\frac{\partial^2\phi}{\partial z^2}+k^2\phi = - Q\,\frac{\delta(r-r')\,\delta(\theta-\theta')\,\delta(z)}{r}, \tag{2}$$

Equation (2) is separable for the 'perfect' wedge, and can be solved for ϕ in several ways. The method employed here is to apply a sequence of integral transforms to both sides, and subsequently to apply the inverse transforms to arrive at the solution. Assuming pressure-release boundaries, the first of these transforms is the finite Fourier sine transform:

$$\phi_s = \int_0^{\theta_0} \phi(\theta)\ \sin\nu\theta d\theta\ , \tag{3a}$$

whose inverse is

$$\phi(\theta) = \frac{2}{\theta_0}\sum_{\nu} \phi_s(\nu)\ \sin\nu\theta, \tag{3b}$$

In these expressions, $\nu = m\pi/\theta_0$, where m is a positive integer, and the symbol $\sum_{\nu}$ means a sum over all possible values of m. It is clear from the structure of equation (3b) that the final solution for the field will be a sum of normal modes, and that m is the mode number. Note also that this formulation ensures that the pressure-release boundary conditions are satisfied. On taking the sine transform of equation (2), we obtain

$$\frac{1}{r}\frac{\partial}{\partial r}\left(r\frac{\partial\phi_s}{\partial r}\right)-\frac{\nu^2}{r^2}\phi_s+\frac{\partial^2\phi_s}{\partial z^2}+k^2\phi_s = - Q\,\frac{\delta(r-r')\,\delta(z)\sin\nu\theta'}{r}, \tag{4}$$

The next transform to apply is the generalized Hankel transform of order $\nu = m\pi/\theta_0$, which we shall assume to be an integer. This is tantamount to saying that θ_0 is a submultiple of π, so that the field due to diffraction at the apex of the wedge is zero. The Hankel transform is defined as

$$H_\nu(\phi_s) \equiv \phi_{\nu,s} = \int_0^\infty r\ \phi_s(r)\ J_\nu(pr)dr, \tag{5a}$$

and the inverse transform is

$$\phi_s(r) = \int_0^\infty p\, \phi_{\nu,s}(p) J_\nu(pr)\, dp, \tag{5b}$$

An important property of the Hankel transform is that it acts on the quantity $\Delta_\nu \phi_s$ as follows [5]:

$$H_\nu(\Delta_\nu \phi_s) = - p^2 H_\nu(\phi_s) \ , \tag{6a}$$

where Δ_ν is the differential operator

$$\Delta_\nu = \left\{ \frac{1}{r}\frac{\partial}{\partial r}\left(r\frac{\partial}{\partial r}\right) - \frac{\nu^2}{r} \right\} \ , \tag{6b}$$

Thus, when equation (4) is Hankel transformed we find that

$$\frac{\partial^2 \phi_{\nu,s}}{\partial z^2} + (k^2 - p^2)\ \phi_{\nu,s} = - Q \sin\nu\theta'\ J_\nu(pr')\,\delta(z), \tag{7}$$

The final transformation to apply is the Laplace transform, defined as

$$\phi_{\ell,\nu,s} = \int_0^\infty \phi_{\nu,s}(z) \exp{-sz} dz, \tag{8}$$

where the unilateral form has been chosen because the field must be symmetrical in z. On applying this integral transformation to equation (7), we find that

$$\phi_{\ell,\nu,s} = \frac{s\phi_{\nu,s}(o) - \frac{Q}{2} \sin\nu\theta'(J_\nu(pr')}{(s^2 + k^2 - p^2)}, \tag{9}$$

where $\phi_{\nu,s}(o)$ is the value of $\phi_{\nu,s}$ at $z = 0$. Now, the inverse Laplace transform of equation (9) is easily shown to be

$$\phi_{\nu,s} = \phi_{\nu,s}(o) \cos \eta|z| - \frac{Q}{2} \sin\nu\theta'\ J_\nu(pr')\ \frac{\sin\eta|z|}{\eta} \ , \tag{10}$$

where

$$\eta = \sqrt{k^2 - p^2} \ . \tag{11}$$

From the radiation condition, that is the requirement that the field should go to zero when $|z| \to \infty$ for $p > k$, we find that

$$\phi_{\nu,s}(o) = j\,\frac{Q}{2}\,\sin\nu\theta'\,\frac{J_\nu(pr')}{\eta}\,, \tag{12}$$

and hence

$$\phi_{\nu,s} = j\,\frac{Q}{2}\sin\nu\theta'\,\frac{J_\nu(pr')}{\eta}\,\exp j\,\eta|z|\;. \tag{13}$$

The next inversion integral to apply is the inverse Hankel transform defined in equation (5b). Using the expression in equation (13), this gives

$$\phi_s = j\,\frac{Q}{2}\sin\nu\theta'\int_0^\infty \frac{p\,\exp j\eta|z|\;J_\nu(pr)J_\nu(pr')\,dp}{\eta}\,, \tag{14}$$

and finally we obtain the solution for the field in the wedge by taking the inverse sine transform of ϕ_s defined in equation (3b):

$$\phi = \frac{Q}{\theta_o}\sum_\nu I_\nu(r,r',z)\,\sin\nu\theta\sin\nu\theta'\;. \tag{15}$$

This is a sum of normal modes whose coefficients are given by the integral

$$I_\nu(r,r',z) = j\int_0^\infty p\,\frac{\exp j\eta|z|}{\eta}\,J_\nu(pr)J_\nu(pr')\,dp. \tag{16}$$

Note that the solution as it stands in equation (15) is exact, since no approximations have yet been introduced. It is only in evaluating I_ν that some accuracy is sacrificed.

THE NORMAL MODE INTEGRAL

In order to evaluate the integral in equation (16), it is convenient to convert it to another form using the Bessel function identity (6)

$$J_\nu(pr)J_\nu(pr') = \frac{1}{\pi}\int_0^\pi J_o\left(p\sqrt{r^2 + r'^2 - 2rr'\cos\sigma}\right)\cos\nu\sigma\,d\sigma, \tag{17}$$

which is valid for ν an integer. When this expression is substituted into equation (16) we obtain the double integral

$$I_\nu = \frac{j}{\pi}\int_0^\pi \cos\nu\sigma \int_0^\infty p\,\frac{\exp j\eta|z|}{\eta} \times \tag{18}$$
$$J_0\left(p\sqrt{r^2+r'^2 - 2rr'\cos\sigma}\right)\, dp\, d\sigma .$$

Now, the inner integral here is just the field due to a point source in an infinite medium (7), and hence I_ν can be written as

$$I_\nu = \frac{1}{\pi R_0}\int_0^\pi \cos\nu\sigma\, \frac{\exp -jkR_0(1-2a\cos\sigma)^{1/2}}{(1-2a\cos\sigma)^{1/2}}\, d\sigma, \tag{19}$$

where

$$R_0 = (r^2 + r'^2 + z^2)^{1/2}, \quad a = \frac{rr'}{R_0^{\,2}} \leq 1/2. \tag{20}$$

From the definitions in equation (20) it is clear that a takes its maximum value, equal to 1/2, when r = r' and z = 0. As the source and receiver separate in the r-z plane a falls in value, approaching zero in the limit.

The form of the integral in equation (19) is more amenable to evaluation by approximation techniques than is that in equation (16). An evaluation procedure is described in Appendix I which leads to the result

$$I_\nu = \frac{1}{2}\exp j\frac{\nu\pi}{2}\left[\frac{\exp -jkR_1(1+b_1)H_\nu^{(1)}(kR_1b_1)}{R_1} + \frac{\exp -jkR_2(1-b_2)H_\nu^{(2)}(kR_2b_2)}{R_2}\right], \tag{21a}$$

which is valid provided the condition

$$kR_0 a \gg 1 \tag{21b}$$

is satisfied. In the expression for I_ν in equation (21a), $H^{(1)}(\ )$ and $H^{(2)}(\ )$ are Hankel functions of order ν of the first and second kind, respectively, and the parameters R_i and b_i, i = 1,2, are defined as follows:

$$R_1 = R_o(1-2a)^{1/2}, \; R_2 = R_o(1+2a)^{1/2},$$
$$b_1 = \frac{a}{(1-2a)}, \; b_2 = \frac{a}{(1+2a)}. \qquad (22)$$

Notice that, apart from ν, the formulation of I_ν given above depends on only two independent variables, namely kR_0 and a. Equations (15) and (21) between them specify the modal field in the wedge.

In the context of ocean wedges, where the wedge angle is usually substantially less than 10°, the inequality in equation (21b) does not represent a serious practical restriction on the validity of the solution for I_ν. As we shall demonstrate, each mode in the wedge takes the form of a beam of radiation which diverges as the range from the apex increases. Within, and to some extent beyond a modal beam, the condition in equation (21b) is always satisfied. It is only deep within those regions which are not illuminated by the source, that is to say, the acoustic shadow zones, that equation (21b) may not hold, but this is not a significant limitation because the field in these regions is essentially zero anyway. We conclude, therefore, that equation (21a) is a valid description of the modes for most scenarios appropriate to the ocean wedge.

It is perhaps worth noting in particular that equation (21a) is a valid representation of the field in the immediate vicinity of the source point (assuming that the source is many wavelengths from the apex so that equation (21b) is satisfied). Under these circumstances it can be shown that the mode sum for the field in the wedge reduces to the free-field expression for a point source (see Appendix II). This is exactly the form of limiting behaviour that is required of the solution.

It is interesting to examine equation (21a) in another limit in this case as $a \to 0$, corresponding to a large separation between the source and the receiver in the r-z plane. In this limit we have $R_1 = R_2 = R_0$ and $b_1 = b_2 = a$, which, when substituted into equation (21a) give

$$\lim_{a \to o} I_\nu = \frac{1}{R_o} \exp j \frac{\nu\pi}{2} \exp-(jkR_o) J_\nu(kR_o a), \qquad (23)$$

where the sum of the two Hankel functions has been expressed as a Bessel function of the first kind. This expression is the solution for the modal coefficients that was obtained by Bradley and Hudimac (3). Naturally, its range of validity is considerably less than that of the solution in equation (21a), since the additional constraint $a \to 0$ has now to be satisfied. This condition means that equation (23) is not a valid representation of the field in the vicinity of the source, as exemplified by the fact that it does not show the required singularity at the source point.

The implications of the condition $a \to 0$ are well illustrated by an example. If we assume that $a < 0.1$, and that the source and the receiver are in the same vertical plane perpendicular to the apex, so that $z = 0$, then a simple calculation shows that equation (23) is valid only when $r < r'/10$ or when $r > 10r'$. Taking the range of the source from the apex as 1 km. say, this means that equation (23) is invalid for all receiver ranges (measured from the apex) between 100 m. and 10 km. It is unfortunate, in view of the relatively simple and hence appealing form of equation (23), that this is precisely the range that is likely to be of interest in practical situations.

An impression of the relative accuracies of equations (21a) and (23) is conveyed by Table 1, which shows the real part, the imaginary part and the modulus of the quantity $\pi R_0 I$ for various values of a and kR_0, but chosen so that the product $kR_0 a = 100$. For comparison a numerically computed evaluation of the integral in equation (19) is included in the Table. The excellent agreement between equation (21a) and the computed values is apparent. Equation (23), on the other hand, performs rather poorly, especially as far as phase is concerned. This is true even when a takes the relatively low value of 0.1 used in the example cited above. An improvement occurs only when a is substantially less than 0.1, confirming that equation (23) has an extremely limited range of validity.

THE EIGENFUNCTIONS FOR THE WEDGE

The trigonometric functions in equation (15) contain the entire angular dependence of the modes in the 'perfect' wedge. These oscillatory functions are analogous to the eigenfunctions obtained for a shallow-water channel with pressure-release boundaries. In the shallow water case, the variable is the depth of the sensor or the source, normalized to the channel depth, whereas in the wedge it is the angular depth normalized to the wedge angle.

a	kR_0	$Re(\pi R_0 I_\nu)$	$I_m(\pi R_0 I_\nu)$	$R_0\pi\lvert I_\nu\rvert$	
10^{-4}	10^6	0.225	0.085	0.241	A
		0.225	0.085	0.241	B
		0.226	0.084	0.241	C
0.1	10^3	-0.118	-0.131	0.176	A
		-0.119	-0.132	0.178	B
		0.136	-0.199	0.241	C
0.2	500	-0.192	-0.042	0.197	A
		-0.193	-0.042	0.198	B
		-0.213	0.113	0.241	C
0.4	250	-0.025	0.244	0.245	A
		-0.025	0.245	0.246	B
		0.058	0.234	0.241	C
0.4975	201	-0.413	-0.400	0.574	A
		-0.403	-0.409	0.574	B
		0.241	0.015	0.241	C

Table 1. The modal coefficients for $\nu = 18$ (corresponding to the first mode in a 10° wedge). The symbols in the last column have the following meanings: A, numerical evaluation of the integral in equation (17); B, evaluation of the new analytical solution in equation (21a); C, evaluation of Bradley and Hudimac's solution in equation (23).

TOTAL NUMBER OF MODES IN THE FIELD

The integral for I_ν in equation (19) is essentially zero when ν exceeds a certain value. This can easily be appreciated by allowing ν to become indefinitely large, in which case the trigonometric function oscillates far more rapidly than the rest of the integrand. If the slowly varying part, that is everything except the cosine, is treated as a constant and taken outside the integral, then the integral of the cosine function is identically zero (bearing in mind that $\nu = m\pi/\theta_o$ and we are assuming that θ_o is a submultiple of π). The implication of this is that only a finite number, M, of modes contributes significantly to the field at any point in the wedge.

In order to estimate M, we set

$$\mu(\sigma) = kR_o(1-2a\ \cos\sigma)^{1/2} , \tag{24}$$

and employ the condition that the integral in equation (19) takes appreciable values only when

$$\nu = \frac{m\pi}{\theta_o} \leq \left.\frac{d\mu}{d\sigma}\right|_{max.} , \tag{25}$$

where $\left.\frac{d\mu}{d\sigma}\right|_{max}$ is the maximum "frequency" of the exponential term in the integrand. A simple calculation shows that

$$\left.\frac{d\mu}{d\sigma}\right|_{max.} = \frac{kR_o}{\sqrt{2}} \left\{1 - (1-4a^2)^{1/2}\right\}^{1/2} . \tag{26}$$

Thus, the maximum value of m from equation (25) is

$$M = E\left[\frac{kR_o\theta_o}{\pi\sqrt{2}} \left\{1 - (1-4a^2)^{1/2}\right\}^{1/2}\right] , \tag{27}$$

where the symbol E[ ] denotes "integer part of".

It is apparent from equation (27) that M, the total number of propagating modes at the receiver, depends on the positions of the source and the receiver. There are two cases in particular which are worth noting.

Case 1. Source and receiver coincident in r-z plane (a = 1/2)

In this configuration we have r = r' and z = 0, giving $R_o = \sqrt{2}\ r'$, and from equation (27)

$$M = E\left[\frac{kr'\theta_o}{\pi}\right] \tag{28}$$

Thus, the total number of modes in this case is governed by the water depth, $r'\theta_o$, at the source.

Case 2. Source and receiver well separated in range and/or z(a→0)

In this case the right hand side of equation (26) becomes equal to kR_oa, and the expression in equation (27) for M reduces to

$$M = E\left[\frac{kR_o\theta_o a}{\pi}\right] = E\left[\frac{krr'\theta_o}{\pi(r^2+r'^2+z^2)^{1/2}}\right] . \tag{29}$$

In certain specific situations, this may be simplified further, as follows

$$M = E\left[\frac{kr\theta_o}{\pi}\right] \quad \text{when } r'>>r,\ z=0, \tag{30a}$$

$$M = E\left[\frac{kr'\theta_o}{\pi}\right] \quad \text{when } r>>r',\ z=0, \tag{30b}$$

and

$$M = E\left[\frac{krr'\theta_o}{\pi|z|}\right] \quad \text{when } |z| >> r \text{ and } |z| >> r'. \tag{30c}$$

Equations (30a) and (30b) correspond to the receiver being up-slope and down-slope of the source, respectively. In both cases the total number of modes is governed by the water depth ($r\theta_o$ or $r'\theta_o$) at the shallower location. Moreover, equations (30a) and (30b) are the same results that would have been obtained from shallow water theory, had we assumed that the bottom was locally flat at r and r', respectively. (A similar statement applies to equation (28)).

SPATIAL PROPERTIES OF THE MODES

As the solution for the acoustic field has the angular dependence separated out, the behaviour of the modes in range and the z direction can be discussed independently of their angular properties. In order to examine this behaviour, we should strictly employ the expression for I_ν in equation (21a). (In certain applications, as for example in determining the response of an array of sensors in a wedge-shaped channel, where phase is of critical importance, this would indeed be essential). However, we are interested here only in establishing the broad features of the field, and for this purpose the simpler expression in equation (23) is adequate. In particular we shall focus attention on the Bessel function $J_\nu(kR_0a)$, as this contains most of the interesting properties of the field; the exponential functions merely influence the phase of the field and R_0 in the denominator is only a spreading term.

RANGE-DEPENDENCE OF THE MODES (z=0)

We begin by considering up-slope propagation, with the source much further from the apex of the wedge than the receiver (i.e. r'>> r), and with the source and the receiver in the same vertical plane perpendicular to the apex (i.e. z = 0). When

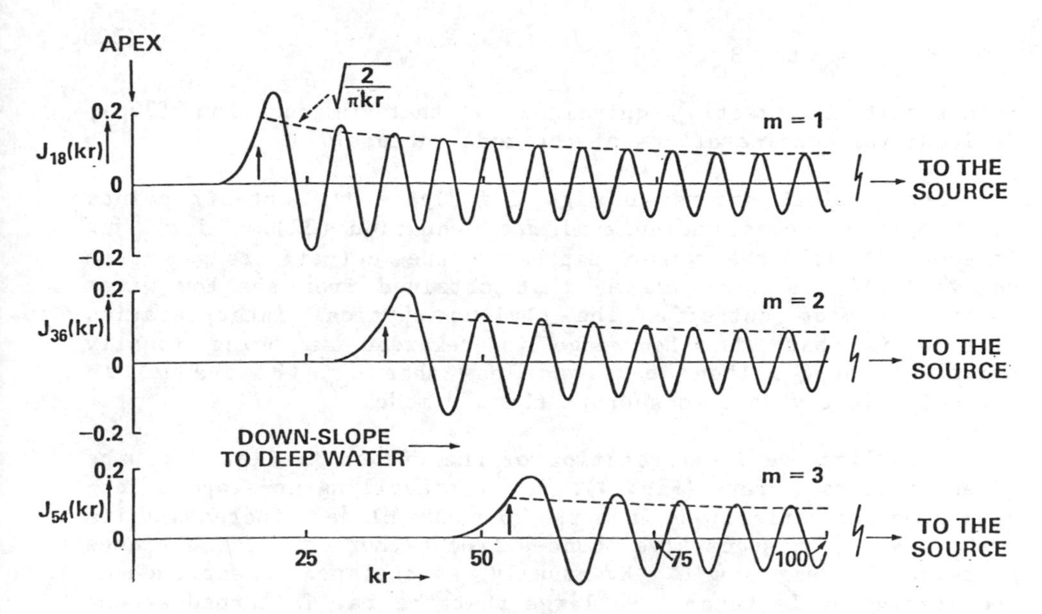

Figure 2. The mode amplitude function $J_\nu(kr)$ for $\nu = 18$, 36, and 54, corresponding to the first three modes in a 10° wedge. The vertical arrows are at the mode cut-off points calculated from shallow water theory.

these conditions prevail the Bessel function in equation (23) can be expressed simply as $J_\nu(kr)$. This function is plotted in Fig. 2 for $\nu = 18$, 36 and 54, corresponding to the first three modes in a 10° wedge. The curves are highly oscillatory beyond some cut-off value of kr, which increases with the mode number. Below cut-off (i.e. for smaller values of kr) the modal field falls rapidly to zero and remains there right up to the apex. Immediately above cut-off, the mode envelope is a maximum, and it decays away approximately as $(kr)^{-1/2}$ as kr rises.

By examining the series expansion

$$J_\nu(x) = \frac{(x/2)^\nu}{\nu!}\left[1 - \frac{(x/2)^2}{(\nu+1)} + \ldots\right] \tag{31}$$

for the Bessel function of integer order ν, it can easily be seen from the term outside the square brackets that for large ν, $J_\nu(x)$ drops extremely rapidly to zero when the argument x falls below the order ν. Taking equality between x and ν as the condition for the onset of this behaviour (i.e. when $x \le \nu$, $J_\nu(x) = 0$), we can define for the mth mode the cut-off range from the apex, r_{cm}, as

$$kr_{cm} = m\pi/\theta_o . \tag{32}$$

This result is exactly equivalent to that in equation (30a), derived from considerations of the mode integral.

The vertical arrows in Fig. 2 indicate the cut-off points for the three modes, calculated from equation (32). If $r_{cm}\,\theta_o$ is equated with the water depth at the cut-off range, then equation (32) is the same as that obtained from shallow water theory for mode cut-off. The obvious physical interpretation of this is that, if the wedge is regarded as being locally uniform in depth, then at ranges less than r_{cm} the channel is not sufficiently deep to support the mth mode.

A qualitative interpretation of the curves in Fig. 2 can be given in terms of rays (Fig. 3). A ray travelling up-slope undergoes numerous reflections from the boundary planes, increasing its grazing angle on successive bounces from either one of the planes by twice the wedge angle. Eventually, as the apex is approached, the grazing angle becomes so large that the ray is turned around and proceeds to propagate back down-slope. In the vicinity of the turn-around point, which corresponds closely with the mode cut-off range given by equation (32), the density of rays is high, which is consistent with the relatively high level of the modal envelope immediately above cut-off.

Z-DEPENDENCE OF THE MODES

The argument of the mode amplitude function, $J_\nu(kR_0a)$, depends on z, the separation of the source and receiver in the direction parallel to the apex, through the term R_0 (see definition

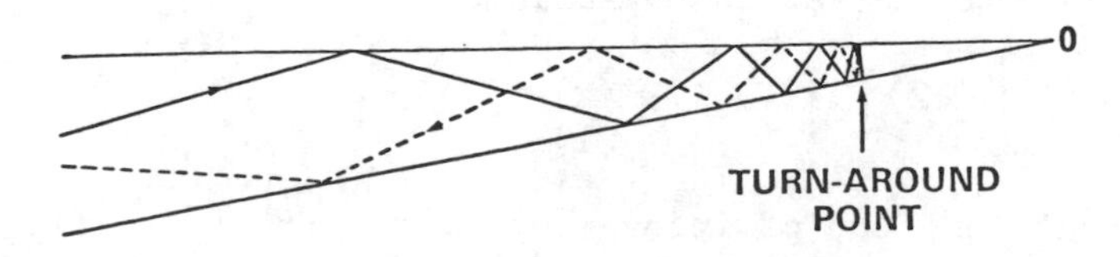

Figure 3. A ray propagation up-slope is eventually turned around and proceeds to propagate back down-slope.

in equation (20). We now examine this z-dependence down-slope of the source with the ranges of the source and the receiver held fixed. Fig. 4 shows the function

$$J_{18}\left\{\frac{krr'}{(r^2+r'^2+z^2)^{1/2}}\right\}$$

corresponding to the first mode in a 10° wedge, plotted as a function of z, assuming that r >> r'. Note that when |z| exceeds a certain critical value the field falls to zero, and where it is non-zero it is highly oscillatory suggesting strong interference. The envelope of the oscillatory function, derived from the asymptotic expansion of the Bessel function, is shown as the broken line, with the cut-off points in |z| determined as in equation (32), by equating the argument with the index. The resemblance of this outline to a butterfly with spread wings is apparent, which suggests the term 'modal butterfly'. The 'wingspan' is clearly dependent on the mode number, as illustrated in Fig. 5 for the first four modal butterflies of a 10° wedge. As the mode number increases, it is evident that the width of the non-zero field region in the z-direction becomes narrower, eventually cutting-off altogether above some value of m depending on the local conditions. The criterion governing the number of propagating modes at any given point has already been established, and is given in equation (27).

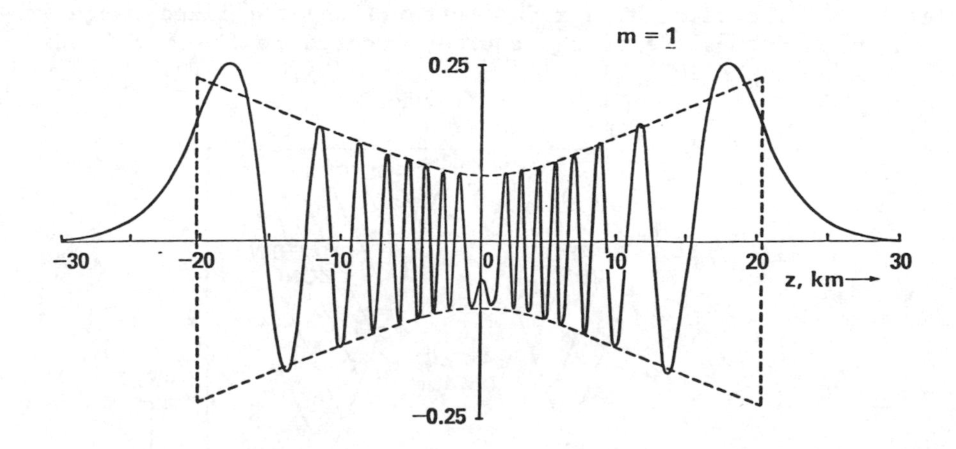

Figure 4. z-dependence of the first mode at a fixed range 5km down-slope of the source, located in the z = 0 plane (θ_o = 10°)

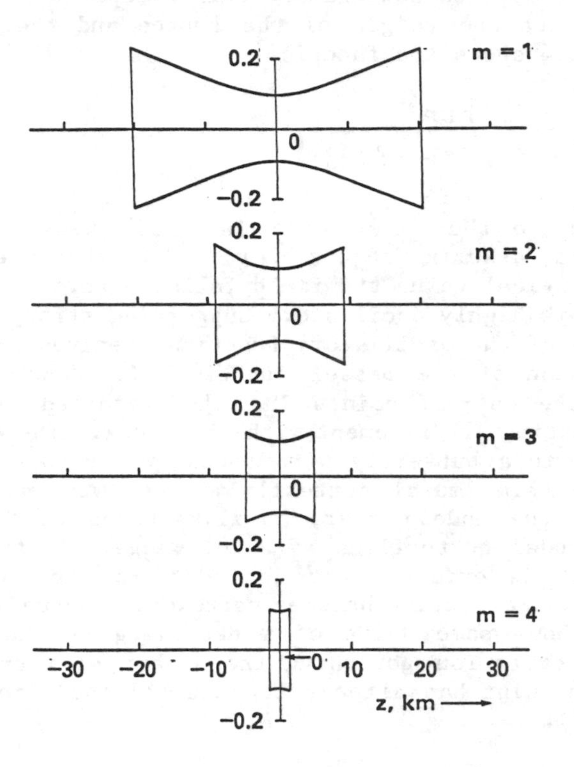

Figure 5. The first four modal butterflies at a fixed range 5km down-slope of the source, located in the z = 0 plane (θ_o = 10°).

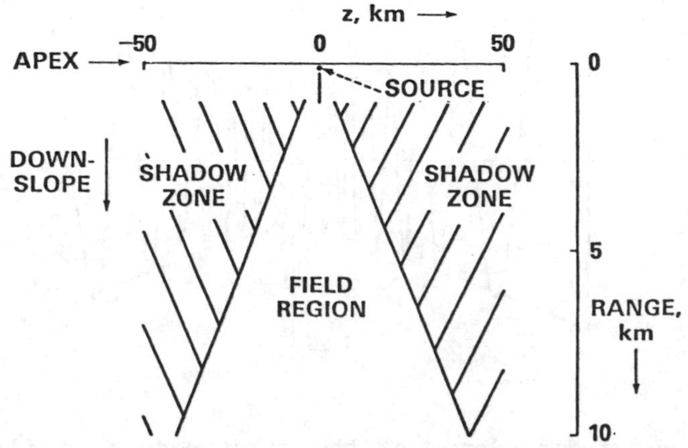

Figure 6. Range-dependence of the 'wingspan' down-slope of the source. Outside the beam the field decays rapidly to form shadow zones (θ_o = 10°).

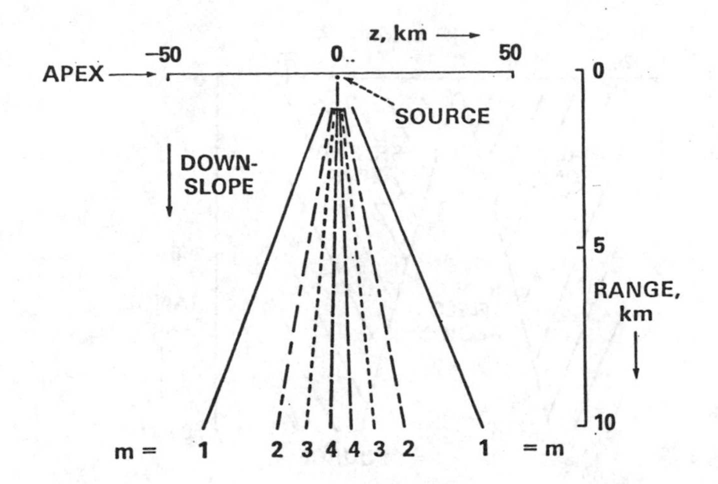

Figure 7. The first four modal beams, showing their nested structure.

The significance of the modal butterflies becomes apparent when the wingspan is examined as a function of range. This is shown for the first mode in Fig. 6. The radiation field diverges from the source to illuminate a fan-like region, beyond which there is essentially no energy, and acoustic shadow zones are formed. Thus, the interaction of the radiation with the boundaries creates a strong beaming effect, which in practical situations could be important since regions of the ocean are not being ensonified. Fig. 7 shows the same phenomenon for the first four modes, and illustrates clearly the nested structure of the modal beams: as the mode number rises, the width of the beam decreases, which of course is consistent with the argument given above for the behaviour of the wingspan with mode number.

A similar situation is encountered with the field up-slope of the source, except that in this case the field converges as the receiver recedes from the source (i.e. as the shore line is approached). This is illustrated in Fig. 8 for the first mode. The beam shape is hyperbolic (8), as is easily confirmed from the mode amplitude Bessel function. The beams corresponding to the higher-order modes again fall within those of the lower-order modes, although this is not shown in Fig. 8. As before, outside a given beam there is essentially no energy in the field associated with that particular mode. Thus, the maximum extent of the ensonified region in the $|z|$ direction is determined by the lowest order mode in the field.

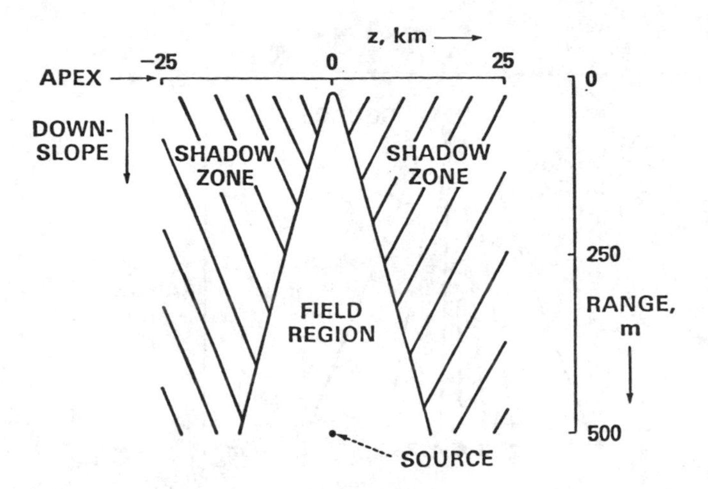

Figure 8. The field and shadow zones of the first mode up-slope of the source. The boundary curve is a hyperbola (θ_o= 10°).

RAY/MODE DUALITY

Each mode in the wedge is a superposition of two components, represented by the two terms containing the Hankel functions in equation (21a). When the arguments of these functions are very much greater than the order ν , the Hankel functions themselves may be approximated by their asymptotic forms. The range dependence of the two modal field terms is then given by the oscillatory functions exp-jkR_1 and exp-jkR_2, where from the definitions in equations (20) and (22), we see that

$$R_1 = [(r-r')^2 + z]^{1/2}, R_2 = [r+r')^2 + z^2]^{1/2}. \quad (33)$$

That is to say, in the (r,z) plane, R_1 is the range from the source to the receiver and R_2 is the range from an image source at (-r',0) to the receiver. Thus, in the asymptotic region, the two exponential functions are representative of two sets of waves, one of which is produced by the source and the other is associated with the image.

The appearance of the two field components in the modal solution can be understood by examining a ray path in the wedge, or more precisely, the horizontal projection of a ray onto the sea surface as it bounces along between the surface and the bottom.

This is a device that has been discussed by Weston (9), who has shown from geometrical arguments that the shape of the horizontally projected ray path is a hyperbola (Fig. 9). That is to say, the repeated acoustic interactions with the inclined planes forming the wedge introduce a curvature into the horizontal direction of travel of the ray, eventually leading to a turning point as the shore line is approached. (Weston refers to this phenomenon as horizontal refraction). If the asymptote associated with the branch of the hyperbola remote from the source is extended beyond the apex, it passes through the image source at (-r',0).

There are many possible hyperbolic ray paths between a source and a receiver in the wedge. Most of these do not show constructive interference, and hence do not correspond to a mode. The criterion for a ray to correspond to the mth mode is that <u>the grazing angle of the ray at the vertex of its hyperbolic path must be the same as that of the ray corresponding to the mth mode in shallow water whose depth is equal to the depth at the vertex</u>. Weston (9) has shown that the angle between the asymptotes of a hyperbolic ray path and the shore line is the same as the grazing angle at the vertex. It follows that the criterion given above for a ray to correspond to a mode fixes the directions of the arms of the modal hyperbola. This is an important factor in interpreting the modal beams and shadow zones in the wedge, as discussed below.

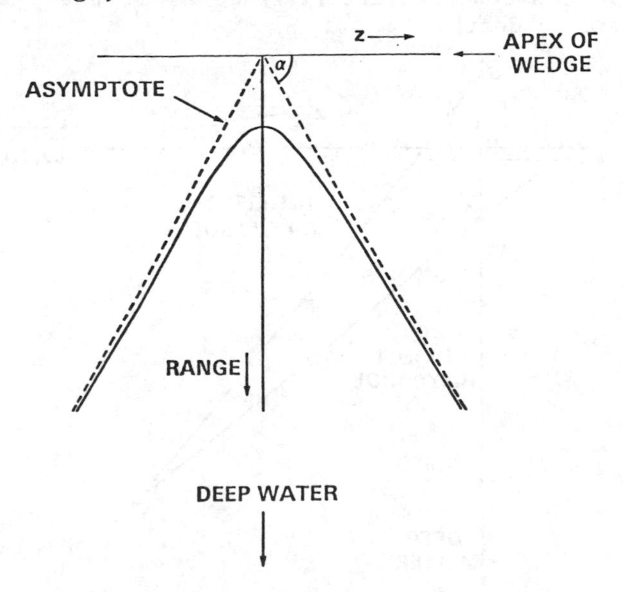

Figure 9. Horizontal projection of a ray path in the wedge. The curve is a hyperbola.

The quantity

$$T = \int_0^h \frac{\sin\alpha}{c}\, dh, \tag{34a}$$

where h is the water depth and α is the grazing angle, is a ray invariant (10). For isovelocity water (which is our case), α and c are independent of depth, and so

$$T = \frac{h\sin\alpha}{c}, \tag{34b}$$

Thus, in the wedge the quantity in equation (34b) is a constant along the hyperbolic ray path. But the condition for a ray to correspond to a mode in a shallow water channel with pressure-release boundaries is

$$h\sin\alpha = \frac{m\pi}{k} = \text{const.} \tag{35}$$

On comparing these two expressions it can be deduced that in the wedge, a hyperbolic ray corresponding to a mode behaves as though it were a mode in locally shallow water all along its track. That is to say, at each point along the hyperbolic path, the upward and downward travelling wavefronts obey the same conditions for constructive interference as if the channel were locally uniform in depth.

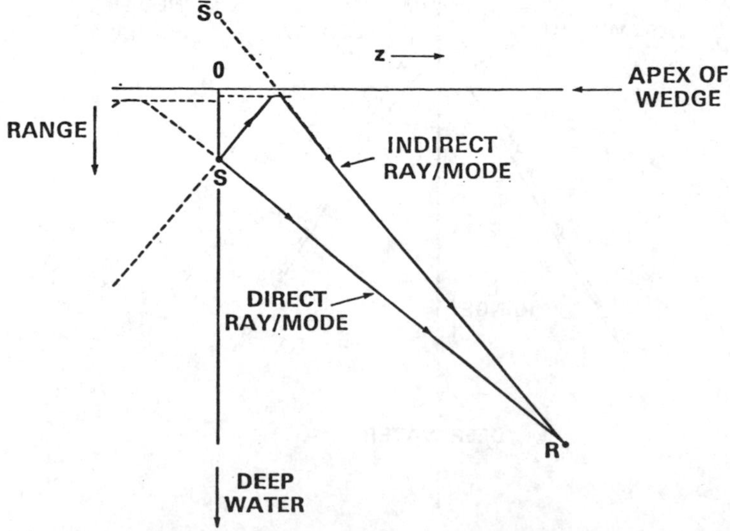

Figure 10. Horizontal projection of the direct and indirect ray paths associated with a given mode for fixed source and receiver positions.

For a fixed source/receiver configuration in the wedge, there are only two possible modal rays corresponding to the mth mode. They are illustrated in Fig. 10. One is a 'direct' ray from the source and the other is an 'indirect' ray, which has been turned around on approaching the apex, as shown in the figure. When the receiver is in the asymptotic regions of these two hyperbolic ray paths, the direct and indirect rays correspond to the field terms from the source and its image, respectively, in the modal solution (equation (21a)).

We can now use the definition of a modal ray to interpret the beaming of radiation in the wedge, as predicted by the modal solution. Fig. 11 shows up-slope propagation with the broken line representing the edge of the beam associated with the first mode. That is to say, the broken line is the same as the solid hyperbola in Fig. 8 obtained from the modal solution. The solid lines in Fig. 11 represent a few of the first-mode rays, each of which is launched from the source at a different azimuthal angle. Notice that the modal rays fall precisely within the beam. Whichever description of the field is used, either modes or rays, the energy associated with a given mode falls in the same well-defined region, beyond which are shadow zones. This beaming of the radiation can now be understood from the ray argument to be due entirely to ray curvature arising from the repeated acoustic interactions with the inclined surface and bottom planes. The curvature is such that no radiation showing constructive interference characteristic of the mth mode can extend into the region beyond the beam associated with the mth mode. Naturally, if a point receiver is placed anywhere within

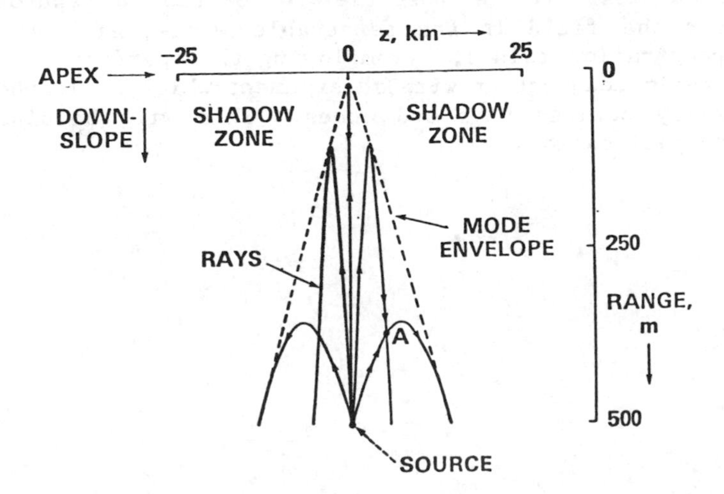

Figure 11. Up-slope propagation showing the convergent first-mode envelope and a few of the first-mode rays.

a modal beam, such as at point A in Fig. 11, only two of the modal rays will be incident upon it, as indicated by the double-headed arrows. The beam itself, of course, is comprised of very many modal rays.

CONCLUDING REMARKS

The mode and ray descriptions of the acoustic field in a wedge with perfectly reflecting boundaries are complementary. The modal solution gives a complete description of the field, including the phase, and is thus appropriate to calculations of spatial coherence, array performance, and related topics. The dual description, in terms of rays, gives a pictorial view of the field which provides a valuable intuitive understanding of propagation phenomena in the wedge. The formation of shadow zones, for example, has a simple physical interpretation in terms of modal rays.

A new solution for the field in the 'perfect' wedge has been presented here, in the form of equations (15) and (21), which is valid throughout the ensonified regions. It was possible to derive his solution because the Helmholtz equation for the problem is separable. By way of contrast, the Helmholtz equation for the penetrable wedge is not separable. This implies a degree of complexity beyond that encountered in the 'perfect' wedge problem. Indeed, this is to be expected because when one boundary is penetrable there are critical angle effects to contend with, and, even when total internal reflection occurs, there is a phase change on reflection from the boundary. Despite these difficulties, it is possible to obtain an approximate solution for the field in the penetrable wedge, at least away from the penetration points, by employing the 'perfect' solution derived here in conjunction with an argument which relies heavily on the duality between rays and modes. This will be discussed in a future publication.

REFERENCES

1. Sommerfeld, A. "Partial Differential Equations in Physics" Academic Press, New York, 1949 (§17).

2. Biot, M.A. and Tolstoy, I. J. Acoust. Soc. Am. 29, 381-391, (1957), "Formulation of wave propagation in infinite media by normal coordinates with an application to diffraction".

3. Bradley, D.L. and Hudimac, A.A. "The propagation of sound in a wedge shaped shallow duct", Naval Ord.Lab.Rep. NOLTR 70-325, November 1970 (unpublished); D.L.Bradley, Catholic Univ. Am., Washington D.C. 1970.

4. Graves, R.D., Nagl, A., Uberall, H. and Zarur, G.L. J. Acoust. Soc. Am. 58, 1171-1177, (1975), "Range dependent normal modes in underwater sound propagation: application to the wedge-shaped ocean".

5. Papoulis, A. "Systems and Transforms with Applications in Optics", McGraw-Hill, Chapter 5, (1968).

6. Lebedev, N.N. "Special Functions and their Applications", Prentice-Hall, Chapter 5, (1965).

7. Watson, G.N. "A Treatise on the Theory of Bessel Functions", Second Edition, Cambridge University Press, p.416, (1952).

8. Harrison, C. J. Acoust. Soc. Am. 65, 56-61, (1979), "Acoustic shadow zones in the horizontal plane".

9. Weston, D.E. Proc. Phys. Soc. 78, 46-52, (1961), "Horizontal refraction in a three-dimensinal medium of variable stratification".

10. Weston, D.E. Proc. Phys. Soc. 73, 365-384, (1959), "Guided propagation in a slowly varying medium".

11. Gradshteyn, I.S. and Ryzhik, I.M. "Table of Integrals, Series and Products", Academic Press, New York (1965).

12. Morse, P.M. and Feshbach, H. "Methods of Theoretical Physics," McGraw-Hill, New York, p.631, (1953).

APPENDIX 1
EVALUATION OF THE NORMAL MODE INTEGRAL

The inversion integral giving the amplitudes of the normal modes (equation 19) in the text) is

$$I_\nu = \frac{1}{\pi R_o} \int_o^\pi \frac{\cos\nu\sigma \exp -j\left\{kR_o(1-2a\cos\sigma)^{1/2}\right\} d\sigma}{(1-2a\cos\sigma)^{1/2}}, \qquad \text{(A1.1)}$$

where a and R_0 are as defined in equation (20). Now, with

$$x = \cos\sigma, \qquad \text{(A1.2)}$$

we are interested in the function

$$\rho(x) \equiv (1-2ax)^{1/2} = (1-2ax_o)^{1/2}\left[1 + \frac{a(x_o - x)}{(1-2ax_o)} + \ldots\right], \qquad \text{(A1.3)}$$

where the term on the right is the Taylor expansion of $\rho(x)$ about $x = x_0$. On splitting the range of integration into two, and setting $x_0 = 1$ in the first interval and $x_0 = -1$ in the second (these values correspond to the turning points in $\rho(x)$), the integral in equation (A1.1) can be expressed as

$$I\nu = \frac{1}{\pi R_1} \int_o^{\pi/2} \cos\nu\sigma \exp -jkR_1[1 + b_1(1-\cos\sigma) + \ldots]\, d\sigma + \frac{1}{\pi R_2} \int_{\pi/2}^{\pi} \cos\nu\sigma \exp -jkR_2[1 - b_2(1+\cos\sigma) + \ldots]\, d\sigma, \qquad \text{(A1.4)}$$

where

$$R_1 = R_o(1-2a)^{1/2}, \quad R_2 = R_o(1+2a)^{1/2} \qquad \text{(A1.5a)}$$

$$b_1 = a/(1-2a), \qquad b_2 = a/(1+2a) \qquad \text{(A1.5b)}$$

and $\rho(x)$ in the denominator has been approximated by just the first term in the expansion in equation (A1.3). The two components of I_ν can now be written in the form

$$I_\nu = \frac{\exp -jkR_1(1+b_1)\, I_1}{\pi R_1} + \frac{\exp -jkR_2(1-b_2)\, I_2}{\pi R_2} , \quad \text{(A1.6)}$$

where

$$I_1 = \int_0^{\pi/2} \cos\nu\sigma \, \exp j(kR_1 b_1 \cos\sigma)\, d\sigma \quad \text{(A1.7a)}$$

and

$$I_2 = \int_{\pi/2}^{\pi} \cos\nu\sigma \, \exp j(kR_2 b_2 \cos\sigma)\, d\sigma \quad \text{(A1.7b)}$$

In writing these expressions the terms beyond those shown in the expansions in equations (A1.4) have been neglected. This is justified because the major contributions to the integrals in equations (A1.4) come from around the turning points of ρ (x), where only those terms shown are significant.

With a little algebraic manipulation the integrals in equations (A1.7) can be expressed as standard forms which can be found in most tables of integrals (e.g. Ref. 11). For ν an integer, the results are

$$I_1 = \frac{\pi}{2} \exp j \frac{\nu\pi}{2} H_\nu^{(1)}(kR_1 b_1) \quad \text{(A1.8a)}$$

and

$$I_2 = \frac{\pi}{2} \exp j \frac{\nu\pi}{2} H_\nu^{(2)}(kR_2 b_2) , \quad \text{(A1.8b)}$$

which, when substituted into equation (A1.6) give

$$I_\nu = \frac{1}{2} \exp j \frac{\nu\pi}{2} \left[\frac{\exp -jkR_1(1+b_1) H_\nu^{(1)}(kR_1 b_1)}{R_1} + \frac{\exp -jkR_2(1-b_2) H_\nu^{(2)}(kR_2 b_2)}{R_2} \right] . \quad \text{(A1.9)}$$

This is the result in equation (21a) in the text that we set out to prove.

APPENDIX II
THE FIELD IN THE VICINITY OF THE SOURCE

The field from a point source within a bounded domain shows a singularity at the source point, and in the immediate vicinity of this point it has the same form as the free field of the source. Indeed, this fact is often employed to construct solutions for boundary value problems, in which the field is represented as a superposition of a free-field component and another component chosen to satisfy the boundary conditions.

In the case of the wedge, the solution for the field in equations (15) and (21a) has been derived using transform methods, rather than the superposition technique, but it should nevertheless be valid in the immediate vicinity of the source point. That is to say, the modal sum in this region should reduce to the free-field form. We show here that this is the case.

When the receiver is very close to the source, only the term containing the Hankel function of the first kind in equation (21a) is significant. This Hankel function has a large argument, (kR_1b_1), and a large order, ν. Bearing in mind that at the source point there is a finite number of propagating modes (see equation (28) in the text), it is always possible to choose $kR_1b_1 > \nu$. (kR_1b_1 increases as the source point is approached). On setting

$$kR_1b_1 = \nu\sec\beta, \tag{A2.1}$$

the Hankel function of the first kind can be expressed approximately as [13]

$$H_\nu^{(1)}(kR_1b_1) = H_\nu^{(1)}(\nu\sec\beta) = \sqrt{\frac{2}{\pi\nu\tan\beta}}\,\exp j[\nu(\tan\beta-\beta)-\pi/4]. \tag{A2.2}$$

From the definition of β in equation (A2.1) we have

$$\tan\beta - \beta = -\pi/2 + \frac{kR_1b_1}{\nu} + \frac{\nu}{2kR_1b_1} + \ldots, \tag{A2.3}$$

and assuming that $\nu/(kR_1b_1) \ll 1$ for all ν, so that the series may be truncated as shown, this gives

$$H_\nu^{(1)}(kR_1b_1) \simeq \tag{A2.4}$$

$$\sqrt{\frac{2}{\pi kR_1b_1}}\ \exp -j\left\{\frac{\nu\pi}{2} - kR_1b_1 - \frac{\nu^2}{2kR_1b_1} + \pi/4\right\}.$$

Thus, the expression for I_ν in equation (21a) can be approximated as

$$I_\nu =$$

$$(2\pi kR_1b_1)^{-1/2}\ \frac{\exp -jkR_1}{R_1}\ \exp -j\ \pi/4\ \exp\ j\frac{\nu^2}{2kR_1b_1}. \tag{A2.5}$$

When this is substituted into equation (15), the expression for the field becomes

$$\phi = \frac{Q}{\theta_o}\ (2\pi kR_1b_1)^{-1/2}\ \frac{\exp -jkR_1}{R_1}\ \exp -j\ \pi/4 \sum_\nu$$

$$\sin\nu\theta \sin\nu\theta' \exp\ j\frac{\nu^2}{2kR_1b_1}. \tag{A2.6}$$

In order to evaluate the summation in equation (A2.6) we first employ the trigonometric identity

$$\sin\nu\theta\sin\nu\theta' = 1/2\ [\cos\nu(\theta-\theta') - \cos\nu(\theta+\theta')], \tag{A2.7}$$

but retain only the difference term, since the remaining term makes a negligible contribution to the field when $\theta \to \theta'$. The sum in equation (A2.6) can therefore be written in the form

$$S = \frac{1}{2} \sum_\nu \cos\nu(\theta-\theta') \exp\ j\frac{\nu^2}{2kR_1b_1}. \tag{A2.8}$$

By expressing the cosine function as the sum of two exponentials, this sum can be written as

$$S = \frac{1}{4}\ (s_+ + s_-)\ , \tag{A2.9}$$

$$S_{\pm} = \sum_{m=1}^{M} \exp j \left\{ \pm \frac{m\pi}{\theta_o} (\theta-\theta') + \frac{m^2\pi^2}{2\theta_o^2 kR_1 b_1} \right\} . \quad (A2.10)$$

Here we have substituted $\nu = m\pi/\theta_o$, and the upper limit on the summation is given by equation (28). Now, provided the maximum value of the term containing m^2 in the argument of the exponential function in equation (A2.10) is greater than unity, S_+ may be approximated by the following integral:

$$S_+ = S_- = \frac{1}{\sqrt{u}} \int_o^{\infty} \exp j \left\{ y^2 + \frac{2wy}{\sqrt{u}} \right\} dy, \quad (A2.11)$$

where

$$u = \frac{\pi^2}{\theta_o^2 2kR_1 b_1} , \quad 2w = \frac{\pi}{\theta_o} (\theta-\theta') . \quad (A2.12)$$

The integral in equation (A2.11) is a standard form (11) which can be expressed exactly in terms of Fresnel integrals. However, in the limit as $\theta \to \theta'$ (which is our case) the contribution from the Fresnel integrals falls to zero and we find that

$$2S = S_+ = S_- = \frac{1}{2}\sqrt{\frac{\pi}{u}} \exp -j \frac{w^2}{u} \exp j\, \pi/4. \quad (A2.13)$$

When this expression for S is substituted into equation (A2.6), in place of the summation, the field in the vicinity of the source is found to be

$$\phi = \frac{Q}{4\pi} \frac{\exp -jkR_1}{R_1} \exp -j\frac{w^2}{u} . \quad (A2.14)$$

The argument of the second exponential function in this expression now bears examination:

$$\frac{w^2}{u} = kR_1 b_1 \frac{(\theta-\theta')^2}{2} = kR_1 b_1 \left[1 - \left\{ 1 - \frac{(\theta-\theta')^2}{2} \right\} \right]$$

$$\simeq kR_1 b_1 [1 - \cos(\theta-\theta')] , \quad (A2.15)$$

where the approximation holds because we are specifically interested in the case where $\theta \to \theta'$. Now, if R is the distance between the source and the receiver, then

$$R^2 = r^2 + r'^2 + z^2 - 2rr'\cos(\theta-\theta')$$
$$= R_1^2 + 2rr'[1 - \cos(\theta-\theta')] \qquad \text{(A2.16)}$$

so that

$$\frac{rr'}{R_1}[1 - \cos(\theta-\theta')] = \frac{R^2 - R_1^2}{2R_1} = \frac{(R-R_1)(R+R_1)}{2R_1}$$

$$\simeq (R-R_1) \qquad \text{(A2.17)}$$

where the approximation is valid because in the immediate vicinity of the source $R \simeq R_1$. Since $kR_1b_1 \equiv krr'/R_1$, it follows from equations (A2.15) and (A2.17) that

$$\frac{w^2}{u} = (R-R_1), \qquad \text{(A2.18)}$$

and hence the expression for the field in equation (A2.14) takes the form

$$\phi = \frac{Q}{4\pi}\frac{\exp{-jkR}}{R}, \qquad \text{(A2.19)}$$

where the denominator has been written as R rather than R_1. But the expression in equation (A2.19) is precisely the free-field produced by the source, which is what we set out to prove. It should be noted that Bradley and Hudimac's (3) solution for the field in the wedge (equation (23)) does not reduce to the required form in the vicinity of the source point.

ACKNOWLEDGEMENTS

I should like to acknowledge with thanks my hosts at NRL, for providing a hospitable and stimulating working environment. In particular, I am grateful to Dr. F. Ingenito for many useful discussions on the wedge problem.

PART II

RAYS AND BEAMS

A SPECTRAL EXTENDED RAY METHOD FOR EDGE DIFFRACTION

Roberto Tiberio

Dept. of Electronics Engineering, University of Florence, Florence, Italy, and The Electro-Science Lab., The Ohio State University, Columbus, Ohio, USA.

Abstract. A spectral extended ray method is applied to analyze the diffraction by two nearby edge discontinuities. Examples of edges in either perfectly conducting or impedance surfaces are considered. The basic high-frequency solution for single diffraction is given by either the uniform GTD or an analogous asymptotic approximation of the exact solution by Maliuzhinets. Uniform high-frequency solutions for the field scattered by double edge structures are obtained for the case of edge-on incidence. Results calculated in several examples are presented and found in very good agreement with those obtained by other techniques.

1. INTRODUCTION

In the last two decades, the Geometrical Theory of Diffraction (GTD) [1] has provided one of the most significant improvements to the versatility of ray methods. In applying this technique to practical problems, an approximate but yet accurate, high-frequency description of the edge diffraction process is of importance. The GTD solution to this problem, in its original formulation, suffers from some significant limitations; in particular, it fails at shadow boundaries. To overcome this difficulty two uniform formulations have been obtained, the Uniform GTD (UTD) [2] and the Uniform Asymptotic Theory of Diffraction (UAT) [3]. To the same end the Spectral Theory of Diffraction (STD) [4] can also be applied, which is not however an ordinary ray method, but can be useful to provide reference solutions for some canonical problems. Among them, the UTD has been most widely and successfully used in a large variety of engineering applications. More recently, some extensions of the UTD have also been proposed, in order to make this ray method applicable to a wider class of problems.

When the field incident at an edge is spatially, rapidly varying, as in the case of sources which have an angular pattern variation, important contributions to the diffracted field may occur which cannot be accounted for by an ordinary application of the UTD. In order to remove this limitation, a higher order term, referred to as slope diffraction term [5], can be included in the solution. This augmentation of the UTD, which makes the first derivative of the total high-frequency field continuous, is a valid high-frequency approximation provided that the incident field has a ray-optical behavior. If this is not the case, then it may be possible to decompose the incident field into ray-optical components, not only in the real space but also in the spectral domain.

The purpose of this paper is to present a survey of the application of a spectral extended ray method to the problem of calculating the high-frequency field diffracted by a pair of parallel edges. It has been recognized [6]-[9] that a satisfactory solution cannot be achieved by directly applying single-edge diffraction. In particular, the analysis is complicated when double diffraction involves the diffraction of transition region fields, and its field contribution may be of the same order as the incident field. The difficulty arises from the fact that the field incident on the second edge after diffracting from the first may exhibit a non-ray optical behavior.

The first part of the paper is concerned with the case of two parallel wedges in a perfectly conducting surface. A spectral extension of the UTD is employed which provides a uniform solution at all scattering aspects, for the case of grazing illumination [10],[11], when either the electric (TE) or magnetic (TM) field is perpendicular to the edges, including oblique incidence. Numerical calculations in several configurations of practical interest, have shown very good agreement with those obtained via other techniques [12]-[16]. Results for the examples of a strip [13] and of a square cylinder [15] are presented and compared with a moment method solution.

The analysis of the influence of material properties on edge diffraction is an interesting topic in diffraction theory and is important for many applications [17]-[26]. In the second part of the paper, some examples relevant to this subject are considered which involve diffraction at edges in impedance surfaces. A high-frequency solution to the diffraction by a wedge with two surface impedance faces illuminated by a plane wave, can be obtained by asymptotically approximating the exact solution given by Maliuzhinets. To this end, an asymptotic approximation analogous to that employed to derive the UTD formulation is used. However, in the present case, a pole singularity in the integrand of the integral representation of the field, which is related to the reflection coefficients of the two faces, may occur close to and at the saddle point. This can be conveniently accounted for in order to obtain a uniform expression for the diffracted field, which provides the correct behavior of the field for any impedance boundary condition [26].

In analysing the diffraction by an impedance strip both in free space [25] and in an impedance ground plane [26], a spectral extended ray method is used to calculate higher order, but yet significant diffracted fields. This approach is similar to that used in the case of edges in a perfectly conducting surface, illuminated by transition region fields. Surface wave contributions which may reach the observation point both directly and after diffracting at the second edge, are also included. Numerical results for the configurations mentioned above with arbitrary surface impedance, are compared with those calculated from a moment method solution and excellent agreement is found.

2. EDGES IN PERFECTLY CONDUCTING SURFACES

2.1 Formulation

In this paper the total field in the presence of a wedge is represented as the sum of a geometrical optics field and a diffracted field. The scalar field, singly-diffracted by an edge is [2]

$$u^d \sim u^i_o \; f(s') \; D(\Phi,\Phi';\beta_o;kL) \; g(s',s) \tag{1}$$

where k is the wave number; Φ, Φ', s', s are defined in Fig. 1, in which $n\pi$ denotes the exterior wedge angle; β_o is the obliquity angle between the edge and the incident ray; $u^i_o\, f(s')$ is the incident field at the point of diffraction Q on the edge, with a suppressed time dependence $\exp(j\omega t)$, and

$$f(s')=\exp(-jks') \begin{cases} 1 & \text{for plane wave illumination} \\ 1/\sqrt{ks'} & \text{for cylindrical wave illumination} \\ 1/ks' & \text{for spherical wave illumination;} \end{cases} \tag{2}$$

$$L = \frac{s's}{s'+s} \sin^2 \beta_o \tag{3}$$

is a distance parameter in which $\beta_o = \pi/2$ for cylindrical wave incidence and $s' \rightarrow \infty$ for plane wave incidence;

$$g(s',s)=\exp(jks) \begin{cases} 1/\sqrt{ks} & \text{for plane and cylindrical waves} \\ \sqrt{\dfrac{s'}{ks(s'+s)}} & \text{for spherical waves} \end{cases} \tag{4}$$

is the diffracted ray spreading factor; and D is $\sqrt{k}$ times the UTD diffraction coefficient given by eq. (25) in [2].

When the field incident on the second edge after diffracting from the first exhibits a non-ray optical behavior, the contribution from the doubly-diffracted ray is determined via a spectral extension of the uniform GTD, which is described in [10],[11]. The geometry of the configuration we are interested in is shown in Fig. 2, where O is the source and P is the observation point. However to simplify the treatment, as was done in [11], we consider the reciprocal problem where the source is located at P and the observation point at O, which lies on either the incident or reflection shadow boundary of the second wedge.

The edge diffracted field may be represented as the sum of four terms, as can be seen from the expression for the diffraction coefficient given in [2]. However, in the case of joined wedges as indicated by the dashed line in Fig. 2, the diffraction coefficient of each edge contains only two terms, since the direction of diffraction grazes the face of a wedge. For the sake of simplicity in the explanation, in the following we consider only joined wedges. Let us denote the two terms in the field incident on edge Q_1 after diffracting from edge Q_2, by u_p^d, p = 1,2 and the corresponding two terms in the diffraction coefficient by D_p. It has been shown [11] that the integrands of the integral representations of the u_p^d terms describe spatially slowly varying waves with their centers of curvature r + d from Q_1; however their angles of incidence at edge Q_1 are complex. These inhomogeneous waves are plane, cylindrical or spherical depending on the illumination of the first diffracting edge, and their diffraction can be calculated via the UTD, using the aforementioned complex angle of incidence.

In order to explain this procedure, consider the simple case of a strip, i.e., $n_1 = n_2 = 2$ in Fig. 2, illuminated by an electric line

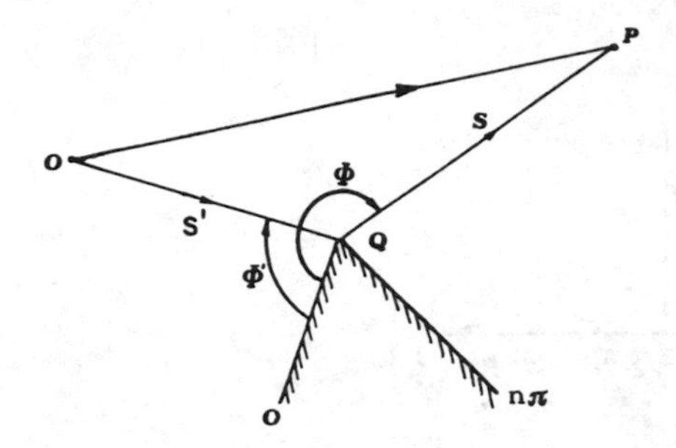

Fig. 1 - Diffraction by a wedge. The figure shows a transverse section perpendicular to the edge.

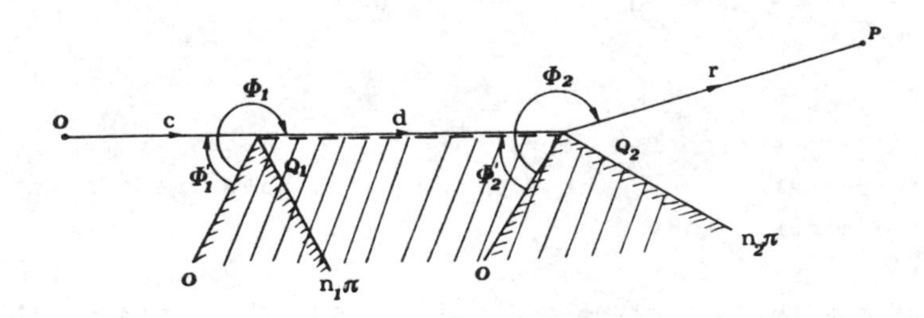

Fig. 2 - Ray geometry for the diffraction by two nearby wedges.

source. Let us use next the following high-frequency integral representation for the field $u^d(Q_1) = u_1^d + u_2^d$, impinging on Q_1 from both the faces of the strip [13]

$$u^d(Q_1) = \frac{u_o^i}{4\pi j} \int_\Gamma [A(\zeta,\phi)+A(\zeta,-\phi)] \frac{\exp[-jk\rho(\zeta)]}{\sqrt{k\rho(\zeta)}} d\zeta \tag{5}$$

where

$$\rho(\zeta) = (r^2 + d^2 - 2rd\cos\zeta)^{1/2} \tag{6}$$

$$A(\zeta,\phi) = \frac{\cos(\phi/2)}{\cos(\zeta/2)-\sin(\phi/2)} \quad . \tag{7}$$

The contour of integration Γ is the steepest descent path (SDP) depicted in Fig. 3(a), and $\zeta = \pi$ is the saddle point. The terms $A(\zeta,\phi)$ and $A(\zeta,-\phi)$ are associated with the diffracted fields propagating along the upper and lower faces of the strip, respectively. In (5) a multiplying factor of ½ has been introduced because only

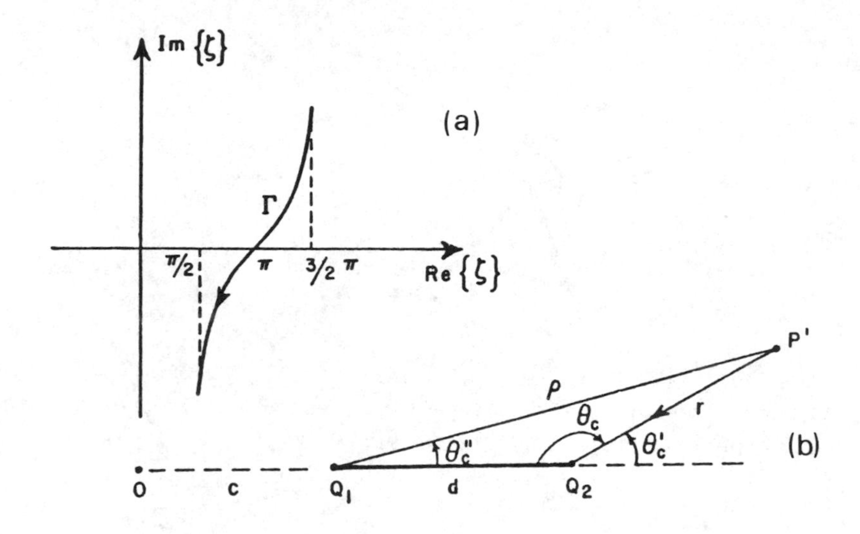

Fig. 3 - (a) The steepest descent path Γ; (b) Geometry for the interpretation of the inhomogeneous wave incident on the second edge.

one half of the total field propagating along the surface diffracts at edge Q_1, as has been pointed out in [2].

The integrand in (5) can be interpreted as a high-frequency field

$$w = \Lambda f(\rho) \tag{8}$$

incident at Q_1 from a line source at P', as shown in Fig. 3(b), where the angle between r and d is the complex angle $\theta_c = \zeta$, with ρ given in (6) and $\Lambda = [A(\zeta,\phi) + A(\zeta,-\phi)]\, d\zeta$. This field is referred to as an inhomogeneous cylindrical wave incident at Q_1. However, in order to employ the UTD diffraction coefficient the field must be expressed in terms of the local coordinate system at Q_1. To this purpose, let us introduce the following convenient transformation,

$$\cos \zeta/2 = \sqrt{j/2}\ \mu \tag{9}$$

which maps the SDP Γ onto the real axis of the complex μ plane. Along the SDP near the saddle point,

$$\zeta \simeq \pi - \sqrt{2j}\ \mu \tag{10}$$

$$\rho \simeq (r+d) - j \frac{rd}{r+d} \mu^2 . \qquad (11)$$

By employing (10) and (11) in the asymptotic approximation, it is a straightforward matter to define the appropriate UTD diffraction coefficient. Let us denote by D_q, q = 1,2 the two terms in the diffraction coefficient at edge Q_1. Each u_p^d term gives rise to two u_{pq}^d doubly-diffracted ray terms, so that the total doubly-diffracted field u_d^d at the observation point (O in Fig. 2) is given by

$$u_d^d \sim \sum_{p=1}^{2} \sum_{q=1}^{2} u_{pq}^d \quad . \qquad (12)$$

When O lies on the incident or reflection shadow boundary, the argument of the transition function of one of the D_q terms vanishes, whereas the argument of the other does not. We denote the former by q = 1 and the latter by q = 2. Those terms with q = 1 give rise to shadow boundary fields and their contributions to the diffracted field are stronger than those with q = 2.

In the TE case (electric field perpendicular to the edge) for joined wedges, it can be shown [11],[15] that the leading term in (12 is given by

$$\sum_{p=1}^{2} u_{p1}^d = \sum_{p=1}^{2} \tfrac{1}{2} u_{po}^d \quad , \qquad (13)$$

where u_{po}^d is the field at O singly-diffracted from the Q_2 edge. The factor of $\frac{1}{2}$ accounts for diffraction by Q_1, assuming the field u_p^d incident on it is a ray-optical field. The u_{p2}^d terms in (12) have the form of ordinary double-diffraction terms; however, the distance parameters involved are different. It can be seen that they allow the discontinuities which may occur at the shadow boundaries to be exactly compensated, even when the distance between the edges is not very large.

In the TM case (electric field parallel to the edge) the field u_p^d incident on the second edge vanishes, but it exhibits a rapid spatial variation with a non-ray optical behavior. It has been shown [11],[15] that (12) reduces to

$$u_d^d \sim \sum_{p=1}^{2} u_{p1c}^d \qquad (14)$$

the sum of two terms which accounts for the non-ray optical behavior of u_p^d. When Q_2 lies in the transition region of Q_1, i.e. the transition regions of the two edges overlap, u_{p1c} is of the same order as the incident field. Although an integral occurs in u_{p1c} which cannot be expressed in closed form, it can be efficiently calculated at all aspects. However when Q_1 is outside the transition region of Q_2, u^d_{p1c} reduces to the slope diffraction term, and the contribution of (14) is of order $k^{-3/2}$ with respect to the incident field.

In calculating the field diffracted by the two edges the contribution from a triply-diffracted field u^d_t may be of importance, particularly at backscatter aspects. Considering the reciprocal problem with its source at P (Fig. 2), the field u^d_t arises from a field singly diffracted at Q_1 and then doubly diffracted from Q_2 and Q_1. The field incident at Q_2 is again ½ the field singly diffracted from Q_1 and it has an effective wavefront curvature of $r_1 + d$. Then the subsequent double diffraction from Q_2 and Q_1 can be calculated by employing the result in (12) for a unit incident field at $r_1 + d$ and $\phi = \pi$ from Q_2. In the TE case, as a result of this extended ray method, the proper distance parameter is introduced so that u^d_t exactly compensates the discontinuity in the doubly-diffracted field which occurs at $\phi = \pi$ due to the shadowing by edge Q_1. In the TM case the result for u^d_t is the same as that determined by a direct application of slope diffraction.

Higher order multiply-diffracted rays may be easily introduced by successively applying the above procedure. However, at high-frequencies their contribution is negligible for most practical applications, except at minor discontinuities which may occur in the scattered field pattern in the TE case.

The preceding solutions for the diffracted fields directly apply to the cases of plane and spherical wave illumination at oblique incidence, i.e. $\beta_o \neq \pi/2$. The solution of the scalar problems can be directly related to the corresponding electromagnetic problem; thus, arbitrarily polarized, plane and spherical waves may be treated by properly superimposing solutions of the electric and magnetic type [11].

2.2 Numerical results

In order to demonstrate the accuracy of the spectral extended ray method described so far, numerical results are presented in this section, which have been calculated for the cases of a square cylinder illuminated by both an electric (TM) and a magnetic (TE) line source, and of a strip illuminated by a TM plane wave at edge on incidence.

The geometry of a square cylinder illuminated at grazing incidence by a uniform line source is sketched in Fig. 4. The filament of current is located at a distance $c = .8\lambda$ from the nearest edge and the side of the cylinder is $d = 1.6\lambda$.

In calculating the results for the TE case, the spectral approach outlined in the previous section has been applied to the field diffracted from the top face of the cylinder. All the other diffracted field contributions are evaluated via a conventional application of the UTD. Contributions from fields diffracted up to four times have been included. A pattern of the radiated power in the far zone ($r \to \infty$) is plotted by a solid line (GTD) in Fig. 5 [15], where the total field has been normalized with respect to the field of the soruce in the absence of the cylinder. The pattern is continuous at all aspects, and a very good agreement is obtained with the numerical results calculated via a moment method solution [8], that are plotted by dots (MM).

The solution for the TM case outlined in the previous section has also been applied to evaluate the field diffracted from the top thick edge of the cylinder, illuminated by an electric line source. In calculating the other diffracted field contributions, UTD slope diffraction is employed. Only the contributions from those triply-diffracted fields which may have the same strength as the doubly-diffracted fields, are incldued in the results shown in Fig. 6 as a solid line (GTD) [15]. The solid curve for the total field in the far zone, very closely compares with a moment method solution (dots, MM) [27]. Here the same normalization as in Fig. 5 is used.

The geometry for a strip illuminated by a TM plane wave at edge on incidence is shown in Fig. 7. In the far field approximation ($r \to \infty$), the integral in the expression for u^d_{p1c} in (14) can be

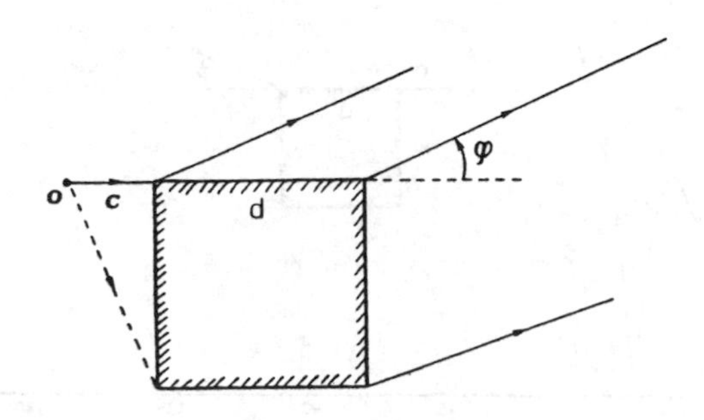

Fig. 4 - Grazing illumination of a square cylinder.

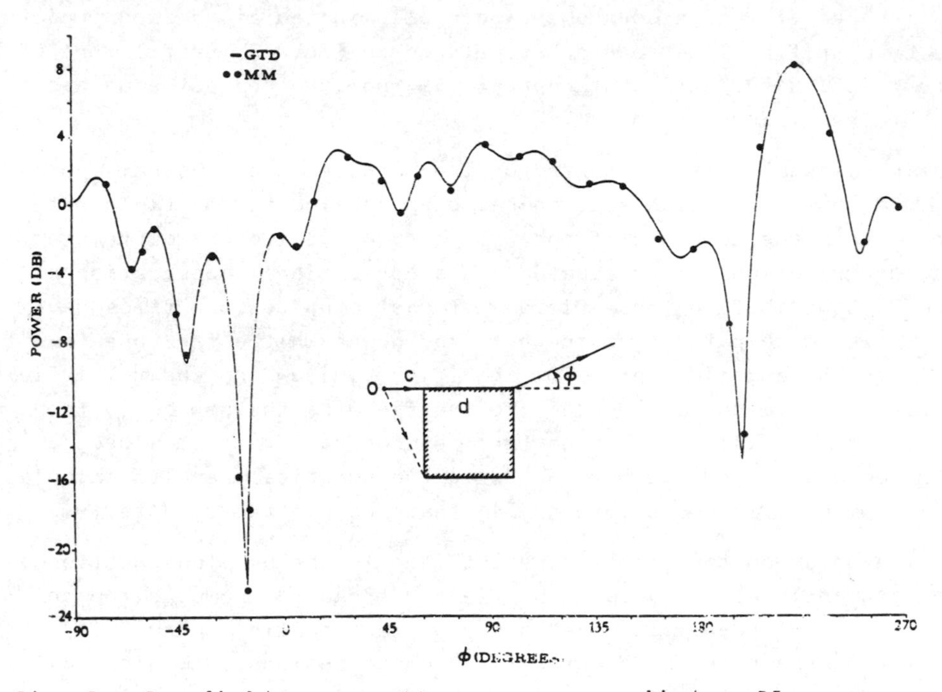

Fig. 5 - Far-field power pattern - square cylinder, TE case; $c = .8\lambda$, $d = 1.6\lambda$.

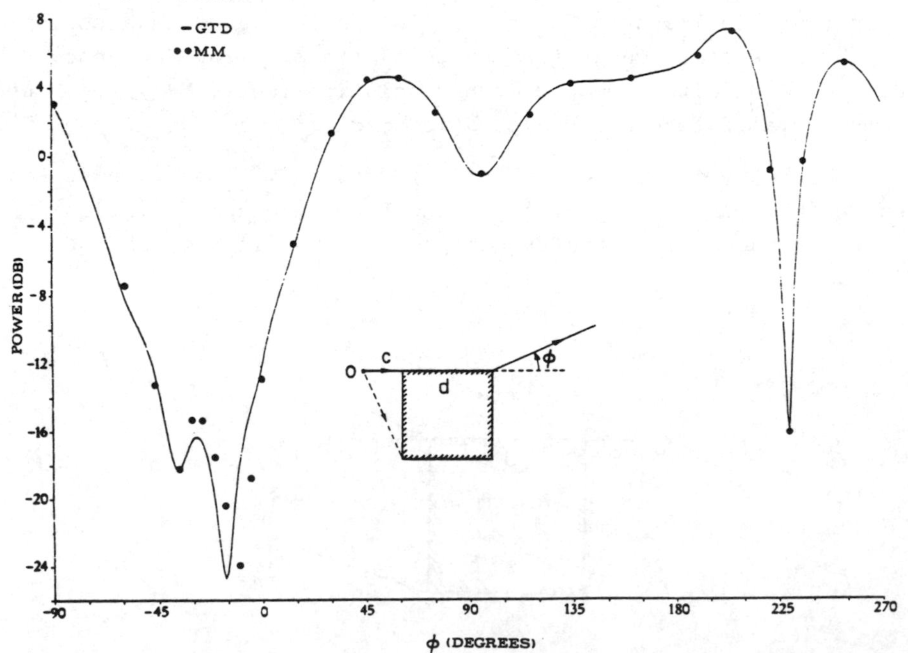

Fig. 6 - Far-field power pattern - square cylinder, TM case; $c = .8\lambda$, $d = 1.6\lambda$.

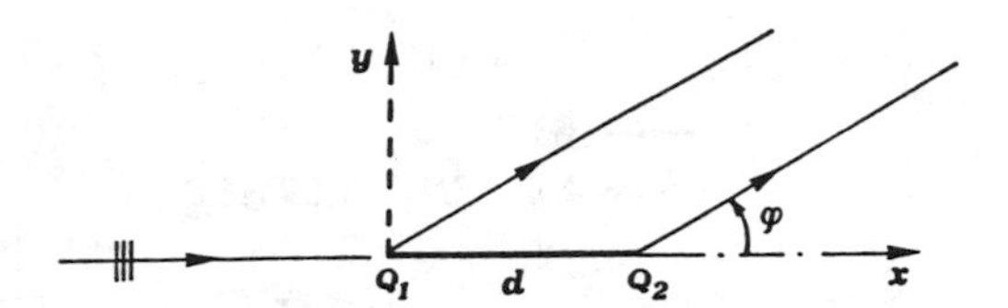

Fig. 7 - Geometry of a strip illuminated at edge on incidence.

asymptotically evaluated in closed form via the stationary phase method. Thus the expression of the total diffracted u^d field greatly simplifies in this case [13].

The scattered far field u^d is written in the form

$$u^d = f^d(\phi) \sqrt{\frac{2}{\pi}} \exp(j\pi/4) \frac{\exp(-jkr)}{\sqrt{kr}} \tag{15}$$

with the incident field at Q_1 normalized to unit amplitude and zero phase. Curves of the far-field scattered power $|f^d(\phi)|^2$ are compared with numerical results obtained from a moment method solution (MM) in Fig. 8 and Fig. 9 for kd = 20 and kd = 3, respectively [13]. The agreement is excellent in both cases, even though the narrower strip is only about one-half wavelength wide. For kd < 3 the high-frequency solution begins to fail gracefully. It is rather remarkable that a solution based on an extension of a ray method is so accurate for such a narrow strip.

3. EDGES IN IMPEDANCE SURFACES

3.1 Formulation

High-frequency solutions for the diffraction at edges in a non-perfectly conducting surface may be useful in a wide variety of engineering applications. As a part of this subject, we consider the scattering of a plane wave by an impedance strip both in free space and on an impedance ground plane where arbitrary, uniform isotropic boundary conditions may be imposed. In particular we will focus our attention to the example shown in Fig. 10. There two semi-infinite surfaces, 1 and 3, on a flat plane are separated by a strip 2, where the three different arbitrary surface impedances are denoted by

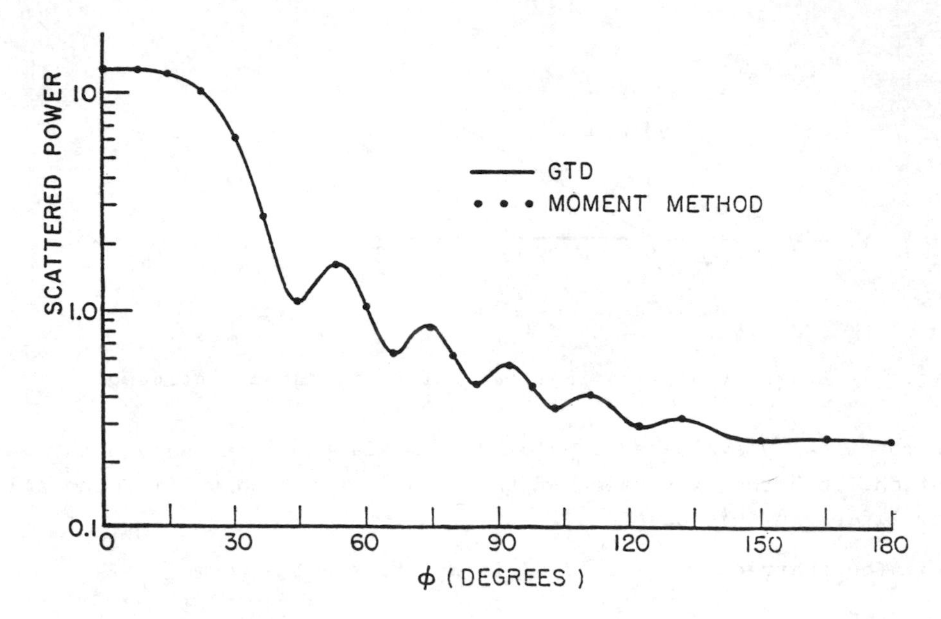

Fig. 8 - Far-field scattered power - strip; kd = 20.

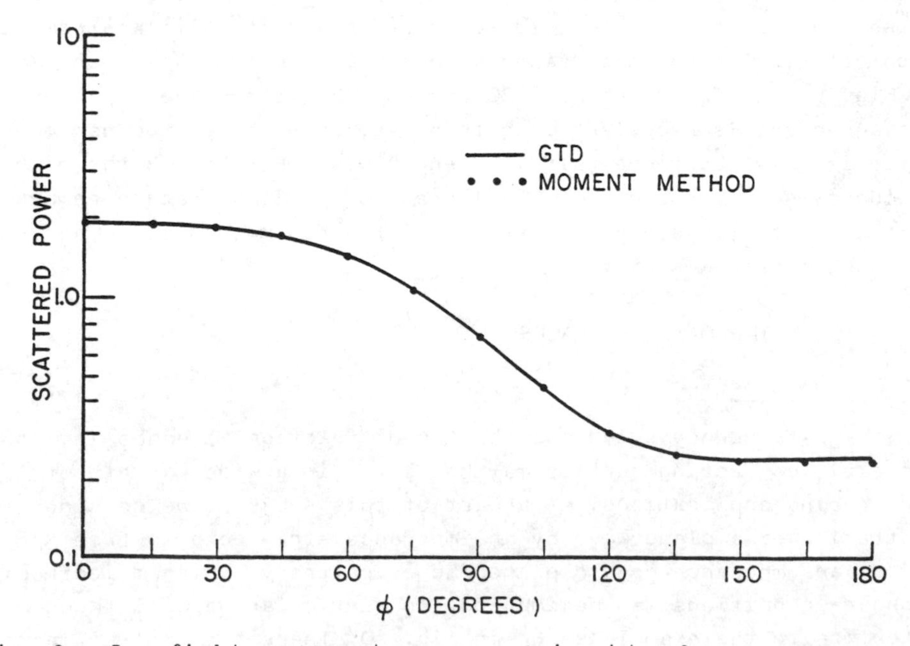

Fig. 9 - Far-field scattered power - strip; kd = 3 .

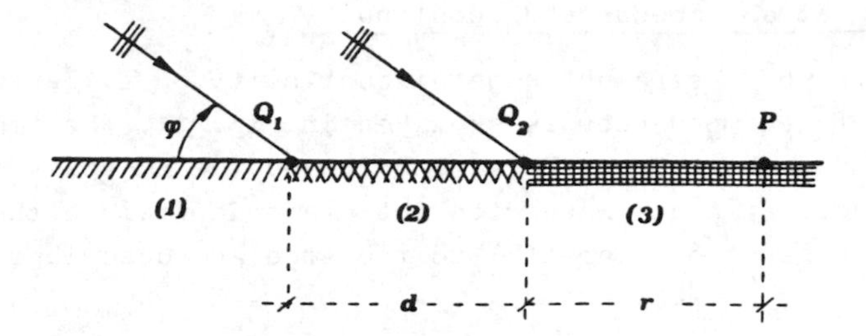

Fig. 10 - Geometry of a double impedance discontinuity.

Z_i (i = 1,2,3). Z_i is a complex constant whose real part, because of energy considerations, must be non-negative. The field scattered from the two edges Q_1 and Q_2 is received at a point P on surface 3. Both TE and TM plane wave illuminations are considered with their direction of incidence perpendicular to the edges. Referring to the reciprocal problem with its source at P and eliminating the ground plane, the same Fig. 10 describes the geometry for the scattering in the far zone by a strip in free space, illuminated at edge on incidence by a line source. In this case the impedances on the two faces of the strip may be different.

In the high-frequency solution that will be described in the sections to follow, the scattered field at P is thought of as a superposition of rays: u_2^d singly diffracted from Q_2, u_{12}^d doubly diffracted from Q_1 and Q_2, and u_{121}^d triply diffracted from Q_1, Q_2 and Q_1. Contributions from a surface wave u_2^s excited at Q_2 and from a field u_{1d}^s due to a surface wave excited at Q_1 and diffracted at Q_2 are also included when they exist.

This ray description of the scattering phenomenon is analogous to that used in the case of edges in a perfectly conducting surface. Thus, in calculating those field contributions which involve diffraction of non-ray optical fields, the same spectral extended ray method is employed. However, there the basic asymptotic solution for single diffraction of a ray optical incident field was given by the UTD. In the present case, the basic solution for single diffraction is given by a uniform asymptotic approximation of the exact solution by Maliuzhinets [28], for the diffraction of a plane wave at the edge of a wedge with surface impedance boundary conditions. This high-frequency solution for the field scattered at a straight edge, impedance discontinuity on a flat plane will be described next.

3.2 Diffraction at an impedance discontinuity

The geometry of a straight edge discontinuity on a flat plane, between two surface impedances is sketched in Fig. 11. The impedances are denoted by Z^+ at $\phi = 0$ and Z^- at $\phi = \pi$, and the boundary conditions for an incident plane wave with the electric field either parallel (TM) or perpendicular (TE) to the edge are described by either

$$\sin \theta_e^{\pm} = Z_o/Z^{\pm} \qquad \text{or} \qquad \sin \theta_h^{\pm} = Z^{\pm}/Z_o \quad , \tag{16}$$

respectively, in which Z_o is the free space impedance. Thus by employing the notation $\sin \theta^{\pm}$, the TM and TE cases are treated together, and the results presented later on apply to both cases provided the proper value of $\theta^{\pm}_{e,h}$ is used.

The total field u at P_o in Fig. 11 may be expressed as a sum of field u^i directly incident at P_o, a surface wave u^s and a diffracted field. Both u^i and u^s are given by the residues at the poles in the integral representation of u [28]. According to [28] a surface wave exists if

$$\theta = \theta_r + \cos^{-1}(1/\cos h\ \theta_i)\mathrm{sgn}(\theta_i) < 0 \tag{17}$$

with $\theta = \theta_r - j\theta_i$, and its contribution at P_o can be expressed [26] as

$$u^s = -2\ \frac{\Psi(\pi/2+\theta^-)}{\Psi(\pi/2-\phi)}\ \frac{\sin\phi\ \ \sin\theta^-}{(\cos\phi-\cos\theta^-)\cos\theta^-}\ \exp(-jk\rho\cos\theta^-)\ . \tag{18}$$

The definition of $\Psi(\alpha)$ is given in Appendix A, where an expression is also presented which is suitable for its computation.

An exact integral representation of the diffracted field u^d at P_o on the surface $\phi = \pi$ is given by [26]

$$u^d(P_o) = -j\ \frac{1}{\pi}\ \frac{\sin\phi}{\Psi(\pi/2-\phi)} \int_{\Gamma_o} \frac{\Psi(-\xi+\pi/2)}{\cos\xi - \cos\phi}\ \exp(-jk\rho\ \cos\xi)\ \cdot \left(\frac{1}{\sin\xi+\sin\theta^+} - \frac{1}{\sin\xi+\sin\theta^-}\right) \sin\xi\ d\xi \tag{19}$$

in which Γ_o is a steepest descent path through a saddle point at $\xi = 0$, as shown in Fig. 3a with $\zeta = \xi + \pi$. We note that the integrand in (19) vanishes at $\xi = 0$, except for $\sin\theta^- = 0$. In asymptotically evaluating u^d in (19), it is found that the expression which yields

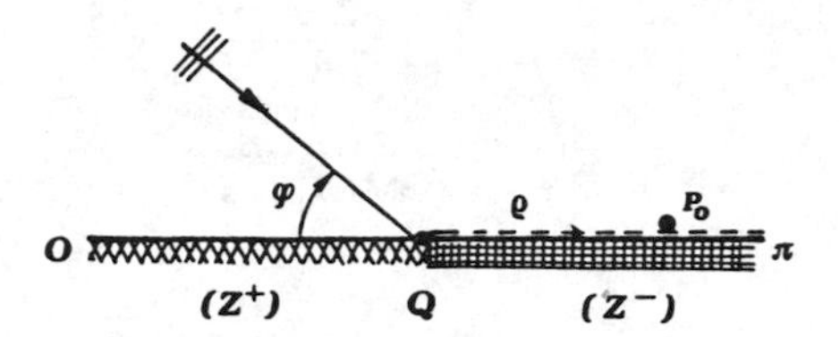

Fig. 11 - Diffraction at the edge of a surface impedance discontinuity on a flat plane.

the first non-vanishing term in the asymptotic expansion, exhibits a pole singularity at the saddle point when $\sin\theta^- = 0$. This is a consequence of the discontinuity which occurs in the reflection coefficient at grazing aspects $\xi = 0$ for $\theta^- = 0$.

By employing the modified Pauli-Clemmow method of steepest descent, the following high-frequency expression is obtained

$$u^d(P_o) \sim -\frac{2\sqrt{2}}{\sqrt{\pi}} \exp(-j\pi/4) \frac{\Psi(\pi/2)}{\Psi(\pi/2-\phi)} \sin\phi \frac{\sin\theta^+ - \sin\theta^-}{\sin\theta^+(1+\cos\theta^-)} \cdot$$
$$\cdot \frac{F[k\rho a(\phi)] - F[k\rho a(\theta^-)]}{a(\phi) - a(\theta^-)} \frac{\exp(-jk\rho)}{\sqrt{k\rho}} \tag{20}$$

In (20) $F[x]$ is the transition function given by Kouyoumjian and Pathak [2], generalized to the case of a complex argument [25] by the definition

$$\sqrt{a(\alpha)} = \sqrt{2}\ \sin(\alpha/2)\ \mathrm{sgn}\left[\alpha_1 + \mathrm{sgn}(\alpha_2)\cos^{-1}\left(\frac{1}{\cos h\ \alpha_2}\right)\right] \tag{21}$$

in which $\alpha = \alpha_1 - j\alpha_2$, so that $-3\pi/4 < \arg(\sqrt{a}) < \pi/4$.

In deriving (20) it has been found that two poles may occur close to the saddle point. These poles which will be referred to as an electric pole $a(\theta)$ and a geometric pole $a(\phi)$, are properly accounted for by the two transition functions in (20). We note that (20) provides a smooth transition between the two limiting conditions of a hard (perfectly conducting, TE case) boundary condition and a soft (perfectly conducting, TM case) boundary condition on the face $\phi = \pi$. It yields the expected behavior of the field in the various conditions, i.e., u^d is of order $k^{-1/2}$ for $a(\theta^-) = 0$ (hard b.c.), of order $k^{-3/2}$ for $a(\theta^-)$ large and $u^d = 0$ for $a(\theta^-) = \infty$ (soft b.c.). The dependence of the strength of the diffracted field on the impedance discontinuity is also clearly expressed and understood. Further-

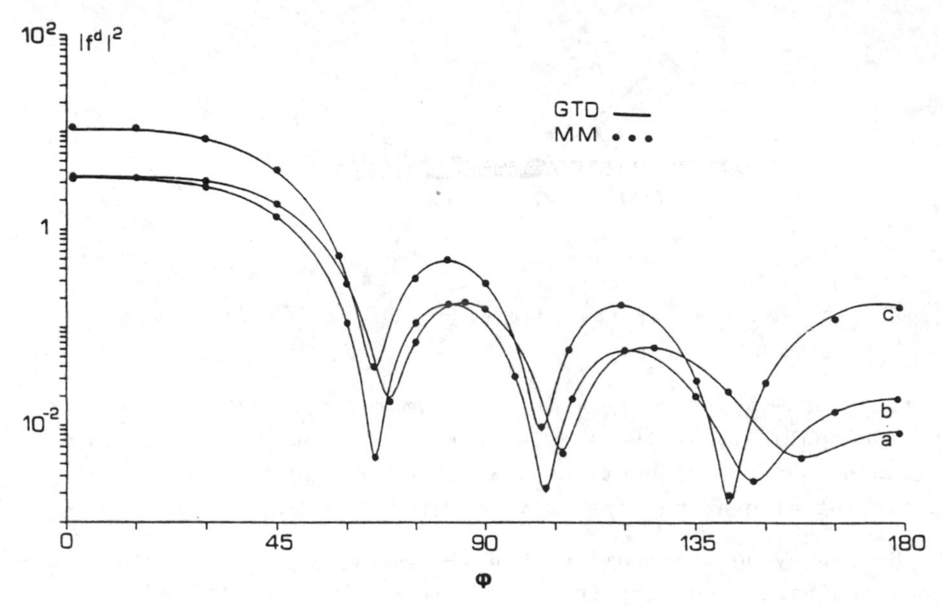

Fig. 12 - Far-field scattered power; kd = 10, $\sin\theta_1 = \sin\theta_3 = 0$, (a) $\sin\theta_2 = 0.25$, (b) $\sin\theta_2 = -j\ 0.25$, (c) $\sin\theta_2 = j\ 0.25$.

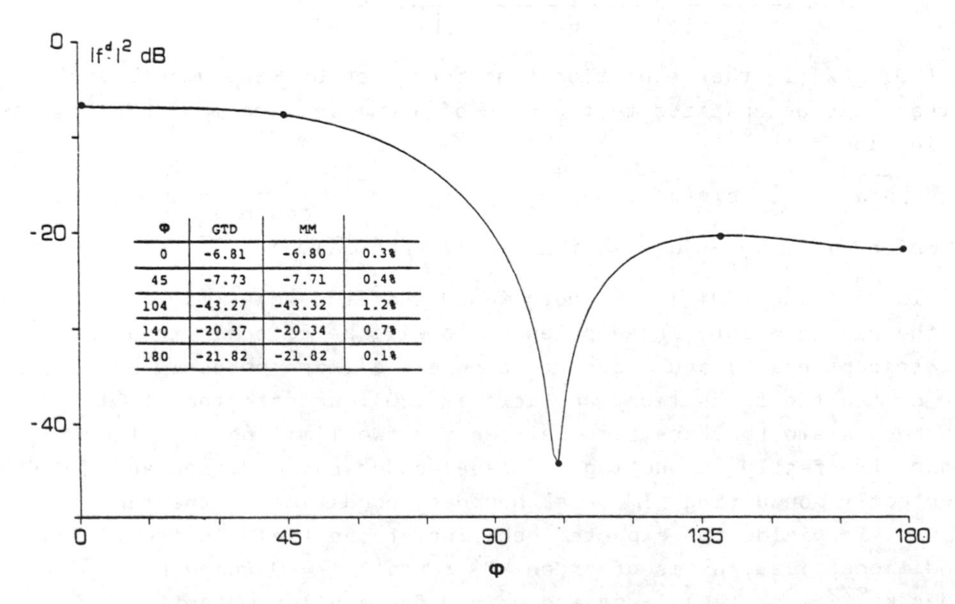

φ	GTD	MM	
0	-6.81	-6.80	0.3%
45	-7.73	-7.71	0.4%
104	-43.27	-43.32	1.2%
140	-20.37	-20.34	0.7%
180	-21.82	-21.82	0.1%

Fig. 13 - Far-field scattered power; kd = 5, $\sin\theta_1 = \sin\theta_3 = 0$, $\sin\theta_2 = 0.25$.

more, expression (20) can be easily calculated by using the formulation for $\Psi(\pi/2)/\Psi(\pi/2-\phi)$ given in Appendix A, and employing the same series expansions of F[x] for small and large arguments as in [2].

This asymptotic solution for single diffraction together with the spectral extended ray method, will be used to evaluate the diffracted field contributions in the example shown in Fig. 10.

3.3 Diffracted field contributions

As mentioned earlier, the total scattered field at P in Fig. 10 is evaluated by summing contributions from diffracted fields and surface waves when they exist. Expressions (18) and (20) are directly employed to calculate the surface wave u_2^s which may be excited at Q_2 and the field u_2^d singly diffracted from Q_2, respectively.

The doubly-diffracted field u_{12}^d arises from a field incident at Q_2 after diffracting from Q_1. As in (7) for the case of a perfectly conducting strip, the integrand in the integral representation (19) for the field impinging on edge Q_2 is interpreted as an inhomogeneous plane wave incident at Q_2 with a complex angle ξ. Its diffraction can be calculated by analytically continuing into complex space the solution (20) for single diffraction of a ray optical incident field. The integral representation given by the superposition of these contributions, is then asymptotically approximated to obtain a high-frequency expression for u_{12}^d.

The same spectral extended ray method is applied to derive a high-frequency expression for the triply-diffracted field u_{121}^d, which arises from a field singly diffracted at Q_1 and then doubly diffracted from Q_2 and Q_1. To this purpose the above mentioned expression of u_{12}^d is employed for complex angles of incidence.

It has also been found that an important contribution may arise from a diffracted field u_{1d}^s, due to a surface wave u_1^s which is excited at Q_1 and impinges on Q_2. Expression (18) for u_1^s can be interpreted as an inhomogeneous plane wave incident at Q_2 with a complex angle θ_2. Thus, its diffraction can be calculated by employing expression (20) with $\phi = \theta_2$.

Explicit expressions for u_{12}^d and u_{1d}^s are given in Appendix I and II of [26]. In u_{12}^d an integral occurs which has the same form as that encountered in evaluating the field doubly diffracted from edges in a perfectly conducting surface illuminated by a TM incident field. However it can be efficiently calculated even if its argument is complex as may occur in the case of impedance surfaces. It should be also noted that the expressions of all the field contributions considered here involve a ratio of two $\Psi(\alpha)$ functions. However, in the present case their calculation is not difficult and can be greatly simplified by using the formulation given in Appendix A. A procedure analogous to that described so far has been applied to treat the example of a strip in free space, which may have different face impedances and is illuminated at edge on incidence [25].

3.4 Numerical results

The aim of this section is to compare the numerical results obtained via this extended ray method with those calculated from an entirely different technique. To this purpose we have chosen the example of an impedance strip on a hard ground plane, $\sin\theta_1=\sin\theta_2=0$. It is apparent that in this case the scattered field, except for a multiplying factor of 2 due to specular reflection, is the same as the field scattered by an impedance strip in free space illuminated at edge on incidence. A moment method solution for this example has been derived from [21].

Curves of the far-field scattered power $|f(\phi)|^2$ due to an incident unit plane wave, are plotted on a logarithmic scale in Fig. 12 [26]. There three examples of impedance strips with kd = 10 are presented: (a) a resistive strip, $\sin\theta_2 = 0.25$; (b) a reactive strip, $\sin\theta_2 = -j\ 0.25$, where no surface wave is excited; (c) a reactive strip, $\sin\theta_2 = j\ 0.25$, where surface waves are excited. The agreement is seen to be very good in all the cases, even though the scattering phenomenon is complicated by surface wave contributions. It is rather remarkable that such a good agreement is obtained among results calculated by so different techniques.

Calculations for narrower strips have shown that when no surface wave is excited, excellent agreement is still obtained for a width $d \geqslant 3/4\ \lambda$. An example of a resistive strip with $\sin\theta_2 = 0.25$, kd = 5, is shown in Fig. 13. Few samples of the percent error of the asymptotic technique with respect to the moment method are also shown in the same Fig. 13. Then the high-frequency solution fails gracefully. However, when surface waves are excited and their attenuation is weak, this ray method solution breaks down more quickly. In this case a strong interaction occurs between the two edges of the strip which is not properly accounted for by this simple ray description.

4. CONCLUSION

A survey of the application of a spectral extended ray method to the diffraction by two nearby edge discontinuities has been presented. Examples of edges in both perfectly conducting and impedance surfaces have been considered. In the former case, the basic high-frequency solution for single diffraction is given by the uniform geometrical theory of diffraction (UTD). In the latter case, an analogous asymptotic solution is used which has been derived from the exact solution given by Maliuzhinets. In treating the diffraction of transition region fields and surface waves, a spectral extension of these solutions for single diffraction is employed. Uniform high-frequency expressions for the field scattered by double edge structures have been obtained for the case of edge on incidence. Although two integrals may occur which cannot be expressed in closed form, they can be efficiently calculated at all aspects and for any surface impedances. Numerical results obtained in several examples are compared with those calculated by a moment method solution. It is rather remarkable that such a good agreement is found between two entirely different techniques, and the results from this extended ray method are still very accurate even in examples where a high-frequency solution is not expected to work.

Acknowledgments

The author is indebted to Prof. R.G. Kouyoumjian, Dr. G. Pelosi, Dr. G. Manara and Dr. F. Bessi for their contributions in obtaining most of the results presented in this paper. Furthermore, he wishes to express his appreciation to Prof. P.H. Pathak for useful discussions and suggestions. This work was supported in part by the Associate Joint Services Electronics Program under Contract N00014-78-C-0049 between the Office of Naval Research and The Ohio State University Research Program and in part by the Italian Ministero Pubblica Istruzione.

REFERENCES

[1] Keller J.B. Geometrical Theory of Diffraction, J. Opt. Soc. Amer., vol. 52, 116-130, 1962.

[2] Kouyoumjian R.G. and P.H. Pathak. A Uniform Geometrical Theory of Diffraction for an Edge in a Perfectly Conducting Surface, Proc. IEEE, vol. 62, 1448-1461, 1974.

[3] Lee S.W. and G. A. Deschamps. A Uniform Asymptotic Theory of Electromagnetic Diffraction by a Curved Wedge, IEEE Trans. Antennas Propagat., vol. AP-24, 25-34, 1976.

[4] Rahmat-Samii Y. and R. Mittra. A Spectral Domain Interpretation of High-Frequency Phenomena, IEEE Trans. Antennas Propagat., vol. AP-25, 676-687, 1977.

[5] Kouyoumjian R.G. The Geometrical Theory of Diffraction and its Applications, in Numerical and Asymptotic Techniques in Electromagnetics, edited by R. Mittra, 165-215,Springer, New York, 1975.

[6] Jones D.S. Double Knife-Edge Diffraction and Ray Theory, Q. J. Mech. Appl. Math., vol. 26, 1-18, 1973.

[7] Lee S.W. and J. Boersma. Ray-Optical Analysis of Fields on Shadow Boundaries of Two Parallel Plates, J. Math. Phys., vol. 16, 1746-1764, 1975.

[8] Mautz J.R. and R.F. Harrington. Radiation and Scattering from Large Polygonal Cylinders, Transverse Electric Fields, IEEE Trans. Antennas Propagat., vol. AP-24, 469-477, 1976.

[9] Rahmat-Samii Y. and R. Mittra. On the Investigation of Diffracted Fields at the Shadow Boundaries of Staggered Parallel Plates. A Spectral Domain Approach, Radio Science, vol. 12, 659-670, 1977.

[10] Tiberio R. and R.G. Kouyoumjian. Application of the Uniform GTD to the Diffraction at Edges Illuminated by Transition Region Fields, USNC/URSI Annual Meeting, Stanford, Calif., 1977.

[11] Tiberio R. and R.G. Kouyoumjian. An Analysis of Diffraction at Edges Illuminated by Transition Region Fields, Radio Science, vol. 17, 323-336, 1982.

[12] Tiberio R. and R.G. Kouyoumjian. A Uniform GTD Analysis of Diffraction by Thick Edges and Strips Illuminated at Grazing Incidence, IEEE AP-S International Symp., College Park, Md., 1978.

[13] Tiberio R. and R.G. Kouyoumjian. A Uniform GTD Solution for the Diffraction by Strips Illuminated at Grazing Incidence, Radio Science, vol. 14, 933-941, 1979.

[14] Tiberio R. and R.G. Kouyoumjian. Application of the Uniform GTD to the Diffraction by an Aperture in a Thick Screen, USNC/URSI Annual Meeting, Seattle, Wash., 1979.

[15] Tiberio R. and R.G. Kouyoumjian. Calculation of High-Frequency Diffraction by Two Nearby Edges Illuminated at Grazing Incidence, IEEE Trans. Antennas Propagat. to be published.

[16] Tiberio R. and G. Pelosi. On the Accuracy of Some Extensions of the Uniform GTD, USNC/URSI National Radio Science Meeting, Albuquerque, NM, 1982.

[17] Bowman J.J. High-Frequency Backscattering from an Absorbing Infinite Strip with Arbitrary Face Impedance, Can. J. Phys., vol. 45, 2409-2430, 1967.

[18] Christiansen P.L. Comparison between Edge Diffraction Processes, Proc. IEEE, vol. 62, 1462-1468, 1974.

[19] Bucci O.M. and G. Franceschetti. Electromagnetic Scattering by a Half-Plane with Two Face Impedances, Radio Science, vol. 11, 49-59, 1976.

[20] Senior T.B.A. Some Problems Involving Imperfect Half-Planes, in Electromagnetic Scattering, edited by P.L.E. Uslenghi, 185-219, Academic Press, New York, 1978.

[21] Senior T.B.A. Backscattering from Resistive Sheets, IEEE Trans. Antennas Propagat., vol. AP-27, 808-813, 1979.

[22] Senior T.B.A. Scattering by Resistive Strips, Radio Science, vol. 14, 911-924, 1979.

[23] Pathak P.H. and R.G. Kouyoumjian. Surface Wave Diffraction by a Truncated Dielectric Slab Recessed in a Perfectly Conducting Surface, Radio Science, vol. 14, 405-417, 1979.

[24] Vaccaro V.G. Electromagnetic Diffraction from a Right-Angled Wedge with Soft Conditions on One Face, Opt. Acta, vol. 28, 293-311, 1981.

[25] Tiberio R., F. Bessi, G. Manara and G. Pelosi. Scattering by a Strip with Two Face Impedances at Edge on Incidence, Radio Science, vol. 17, 1199-1210, 1982.

[26] Tiberio R. and G. Pelosi. High-frequency Scattering from the Edges of Impedance Discontinuities on a Flat Plane, IEEE Trans. Antennas Propagat., vol. AP-31, 590-596, 1983.

[27] Richmond J.H. Scattering by an Arbitrary Array of Parallel Wires, IEEE Trans. Microwave Theory Tech., vol. MTT-13, 408-412, 1965.

[28] Maliuzhinets G.D. Excitation, Reflection and Emission of Surface Waves from a Wedge with Given Face Impedances, Sov. Phys. Dokl., Engl. Transl. 3, 752-755, 1958.

APPENDIX A

A special function $\Psi(\alpha)$ occurs in the exact solution given by Maliuzhinets [28] for the diffraction at a wedge with impedance faces illuminated by a plane wave. This is defined as

$$\Psi(\alpha) = \Psi^{+}(\alpha)\,\Psi^{-}(\alpha) \tag{1A}$$

with

$$\Psi^{\pm}(\alpha) = \Psi^{2}_{\frac{n\pi}{2}}(\pi/2)\cos\left[\frac{\alpha\mp\theta^{\pm}}{2n} \pm \frac{\pi}{4}\right] \frac{\Psi_{\frac{n\pi}{2}}\left(\alpha\pm[(n-1)\pi/2+\theta^{\pm}]\right)}{\Psi_{\frac{n\pi}{2}}\left(\alpha\pm[(n-1)\pi/2-\theta^{\pm}]\right)} \tag{2A}$$

where $n\pi$ is the exterior wedge angle, and $\Psi_{\frac{n\pi}{2}}$ is a meromorphic function which in the case n=1,2 can be expressed as

$$\Psi_{\frac{n\pi}{2}}(x) = \exp\left[\int_{0}^{x} f_{\frac{n\pi}{2}}(t)\,dt\right] \qquad (n=1,2) \tag{3A}$$

with

$$f_{\pi/2}(x) = \frac{2x-\pi\sin x}{4\pi\cos x} \tag{4A}$$

and

$$f(x) = -\frac{\pi\sin x-(2)^{3/2}\pi\sin(x/2)+2x}{8\pi\cos x} \tag{5A}$$

All the high-frequency scattered field contributions can be expressed in terms of a ratio $\Psi(\alpha_{o})/\Psi(\alpha)$, therefore the following formulation can be used which is more convenient for numerical computation [25],[26]

$$\frac{\Psi^{\pm}(\alpha_{o})}{\Psi^{\pm}(\alpha)} = \frac{\cos[(\alpha_{o}\mp\theta^{\pm})/2n\pm\pi/4]}{\cos[(\alpha\mp\theta^{\pm})/2n\pm\pi/4]}\exp\left[M_{\frac{n\pi}{2}}(\alpha-\alpha_{o},\theta^{\pm})\right] \tag{6A}$$

in which

$$M_{\frac{n\pi}{2}}(x,\theta^{\pm}) = \int_{0}^{x} f_{\frac{n\pi}{2}}\left(t+\alpha_{o}\pm[(n-1)\pi/2-\theta^{\pm}]\right) - f_{\frac{n\pi}{2}}\left(t+\alpha_{o}\pm[(n-1)\pi/2+\theta^{\pm}]\right)dt \qquad (n=1,2) \tag{7A}$$

It is worth pointing out that both α_{o} and α are real except in those expressions involving surface waves. However, in calculating a scattered field pattern the computation of $M_{\frac{n\pi}{2}}$ for $\alpha-\alpha_{o}$ complex is required only one time [26].

MODAL EDGE-DIFFRACTION COEFFICIENTS FOR CYLINDRICALLY AND SPHERICALLY CURVED CONCAVE SURFACES

Mithat Idemen

Applied Mathematics Division, Marmara Research Institute, Gebze, Kocaeli, Turkey

1 INTRODUCTION

The Geometrical Theory of Diffraction (GTD), started about thirty years ago by Keller in order to study the diffraction of very high frequency waves, relies essentially on the notion of ray. In a homogeneous medium the rays consist of consecutive segments of straight lines, along which field values can be easily transported from one end to the other. The end points of a ray are always placed on physical obstacles on which scattering phenomena take place and thereby several new space and surface rays are launched. The initial field values of these latter are modified, in a certain manner, with respect to the corresponding final values on the incident ray. If one explicitly knows all these modifications, the field values can then be evaluated at each point of the space wherever it may be. The factors showing the abovesaid modifications are called "Transfer Coefficients" and were subject to some, but insufficient, investigations [1-3].

When the rays arriving at a certain point are small in number and all the transfer coefficients are explicitly known, then the field values at this point can be evaluated by direct application of GTD. On the contrary, if the number of rays arriving at the point in consideration is very high, which causes the ray configuration to be very complicated, the direct application of the GTD becomes inadequate. Such a situation occurs especially near concave surfaces when the sources are also located near it. However, the difficulties in the application of GTD near concave surfaces can be avoided by eschewing the full ray notion in such regions. Indeed, as it was shown in [4], a hybrid representation of the field as a combination of small number of rays together with whispering gallery (WG) modes is very suitable near concave surfaces. These modes, which are always of finite number, account for the higher order reflected ray contributions. The aim of the present paper is to give some explicit expressions for the transfer coefficients, which will be useful in practical calculations as well as in theoretical discussions.

It is obvious that the abovementioned transfer coefficients depend on the curvature of the surface at the diffraction point as well as on the polarization and incidence direction of the incident field. A model problem which simultaneously has all of these properties in their most general form can not be formulated. However, by considering the fact that the curvature of a surface can be completely determined by giving its two principal curvatures, we can reduce the general problem into two, more simple, particular problems. In one of these problems the principal curvatures are equal to each other while in the second one, one principal curvature is zero. By comparing the results of these particular problems we can reveal the influence of the curvatures in different directions on the transfer coefficients. These particular problems correspond to a spherical cap and a circular cylindrical sheet, respectively. So, in what follows we consider these cases separately. As to the polarization of the field, the cases where the electric or the magnetic field is parallel to the edge will be considered separately.

In what follows we assume the time dependence $e^{-i\omega t}$ and suppose that the wave number k always has a positive imaginary part.

2 WHISPERING GALLERY MODES

Let (r,φ,z) and (r,θ,φ) denote the usual cylindrical and spherical coordinates, respectively. Then the scatterer S is supposed to be $\{r=a, \varphi_1' < \varphi < 0, z\varepsilon(-\infty,\infty)\}$ or $\{r=a, 0\leqslant\theta<\theta_1, \varphi\varepsilon[0,2\pi)\}$ (see, Fig.1). In the first case, in which S consists of a cylindical sheet, we assume that the field is independent of z while in the second, where S is a spherical cap, the field will be assumed to be independent of φ.

The simple separable solutions of Maxwell's equations in the region $0\leqslant r<a$, which also satisfy the boundary conditions, determine the WG modes. According to the type of polarization they are as follows:

2.1 WG Modes Inside A Cylindrical Sheet

When the electric field is parallel to the edge, i.e. if $\vec{E}=E\,\vec{e}_z$, the WG modes are

$$E^{(n)}(r,\varphi)=A_n\, e^{i\mu_n|\varphi-\varphi_o|} J_{\mu_n}(kr), \tag{1a}$$

where μ_n is defined by

$$J_{\mu_n}(ka)=0, \quad \mathrm{Re}\ \mu_n>0. \tag{1b}$$

The factor A_n is (1a) is yet an arbitrary constant. Later it will be fixed so as to conveniently normalize the modes. The modes given by (1a,b) propagate from the half-plane $\varphi=\varphi_o$ unperiodically in opposite directions towards $\varphi\to+\infty$ and $\varphi\to-\infty$. At the edge points M_1 and M_1' they undergo, of course, edge-diffraction phenomena whereby they

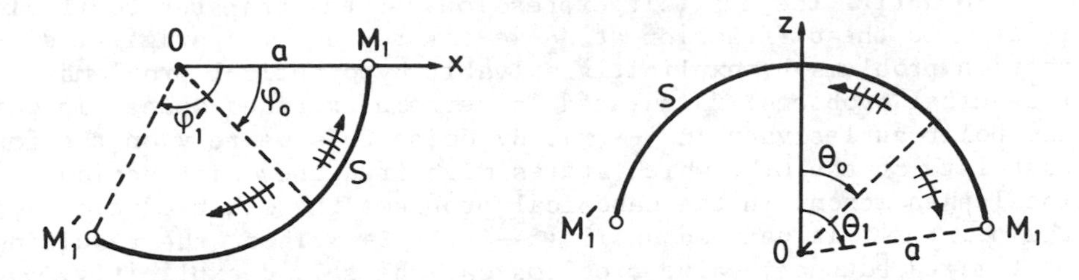

a. Cylindrical sheet b. Spherical cap

Fig.1.Geometrical parameters related to concave surfaces

excite new whispering gallery modes issuing from $\varphi=0$ and $\varphi=\varphi_1'$ as well as creeping modes on the convex side of S and space rays in different directions.

In the case where the magnetic field is parallel to the edge,i.e. $\vec{H}=H\,\vec{e}_z$, the modes are

$$H^{(n)}(r,\varphi)=A_n\, e^{i\mu_n|\varphi-\varphi_o|}\, J_{\mu_n'}(kr) \tag{2a}$$

with μ_n' defined by

$$J'_{\mu_n'}(ka)=0\ , \quad \mathrm{Re}\ \mu_n'>0\ . \tag{2b}$$

2.2. WG Modes Inside A Spherical Cap

If the surface S consists of a spherical cap shown in Fig.1b and the electric field is parallel to the edge, i.e. $\vec{E}=E\,\vec{e}_\varphi$, then the Equ.(1.a) is replaced by

$$E^{(n)}(r,\theta)=\begin{cases} A_n[J_{\mu_n}(kr)/\sqrt{kr}]\ Q^{-1}_{-\mu_n-1/2}(\cos\theta_o)Q^{-1}_{\mu_n-1/2}(\cos\theta),\ \theta>\theta_o \\ A_n[J_{\mu_n}(kr)/\sqrt{kr}]\,Q^{-1}_{-\mu_n-1/2}(\cos\theta)Q^{-1}_{\mu_n-1/2}(\cos\theta_o)\ ,\ \theta_o>\theta, \end{cases} \tag{3}$$

where μ_n is as defined in (1.b).

In the case where the magnetic field is parallel to the edge, i.e. $\vec{H}=H\,\vec{e}_\varphi$, the WG modes are as follows:

$$H^{(n)}(r,\theta)=\begin{cases} A_n[J_{\bar{\mu}_n}(kr)/\sqrt{kr}]\,Q^{-1}_{-\bar{\mu}_n-1/2}(\cos\theta_o)Q^{-1}_{\bar{\mu}_n-1/2}(\cos\theta),\theta>\theta_o \\ A_n[J_{\bar{\mu}n}(kr)/\sqrt{kr}]\,Q^{-1}_{-\bar{\mu}_n-1/2}(\cos\theta)Q^{-1}_{\bar{\mu}_n-1/2}(\cos\theta_o),\theta_o>\theta \end{cases} \tag{4a}$$

with $\bar{\mu}_n$ defined by

$$\frac{d}{dx}[\sqrt{x}\,J_{\bar{\mu}_n}(x)]=0\ , \quad \mathrm{Re}\ \bar{\mu}_n>0 \quad (x=ka). \tag{4b}$$

3 DIFFRACTION OF A WG MODE

To derive the explicit expressions of the transfer coefficients related to the diffraction at M_1 we can replace the original diffraction problems by explicitly solvable hypothetical problems (canonical problems) formulated in certain extended spaces in which the polar angles vary in $(-\infty,\infty)$. By doing this we rely on the fourth postulate of the GTD, which states high frequency diffraction as a local phenomenon. In the canonical problems the other edge of S, i.e., the point M_1', is removed until $\varphi\to-\infty$ or $\theta\to-\infty$. Then the resulting two-part mixed boundary value problems can be solved explicitly. The essentials of the method are given in [5] for the cylindrical sheet and in [3] for the spherical cap. Hence, in what follows we confine ourselves to point out briefly the fundamental steps of the method and give the results without attempting to show the computational details.

3.1. Canonical Problems For A Cylindrical Sheet

Consider first the diffraction of the WG mode given by (1a) or (2a) at the edge M_1 of the cylindrical sheet shown in Fig.1a. A canonical problem suitable to reveal this phenomenon can be formulated in the extended space, in which $r\varepsilon[0,\infty)$, $\varphi\varepsilon(-\infty,\infty)$ and $z\varepsilon(-\infty,\infty)$, so that the total axial field $E(r,\varphi)$ or $H(r,\varphi)$ can be represented as follows:

$$U(r,\varphi)=\begin{cases}\int_{-\infty}^{\infty} e^{i\nu\varphi} J_{|\nu|}(kr)A(\nu)d\nu+U^{(n)}(r,\varphi) & , r<a\\ \int_{-\infty}^{\infty} e^{i\nu\varphi} H_{\nu}^{(1)}(kr)B(\nu)d\nu & , r>a .\end{cases} \tag{6}$$

Here $U^{(n)}$ is given by (1a) if U stands for the electric field while it is given by (2a) when U is the magnetic field. The constants $A(\nu)$ and $B(\nu)$ will be determined with the aid of the (extended) boundary and edge conditions. Following the method given in [5] one gets

$$A(\nu)\sim\frac{\partial U^{(n)}(a,0)}{\partial ka}\;\frac{ika/(4\pi)}{\sqrt{\mu_n+ka}\;J_{|\nu|}(ka)}\;\frac{1}{\sqrt{\nu-ka}\;(\nu-\mu_n)} \tag{7a}$$

$$B(\nu)\sim A(\nu)J_{|\nu|}(ka)/H_{\nu}^{(1)}(ka). \tag{7b}$$

for the electric case and

$$A(\nu)\sim U^{(n)}(a,0)\;\frac{\sqrt{\mu_n'+ka}}{4\pi i\;ka\;J_{|\nu|}'(ka)}\;\frac{\sqrt{\nu-ka}}{\mu_n'-\nu} \tag{8a}$$

$$B(\nu)\sim A(\nu)J_{|\nu|}'(ka)/H_{\nu}^{(1)\prime}(ka) \tag{8b}$$

for the magnetic case.

If one puts (7a,b) or (8a,b) in (6), one gets the possibility of obtaining the leading term in the asymptotic expansion of the diffracted field E^d or H^d at any point of the space. The result is as follows (see, Fig.2):

$$E^d(P_1)\sim[\partial E^{(n)}(M_1)/\partial ka]\;T_{we}(\psi)e^{ikR_1}/\sqrt{kR_1}$$

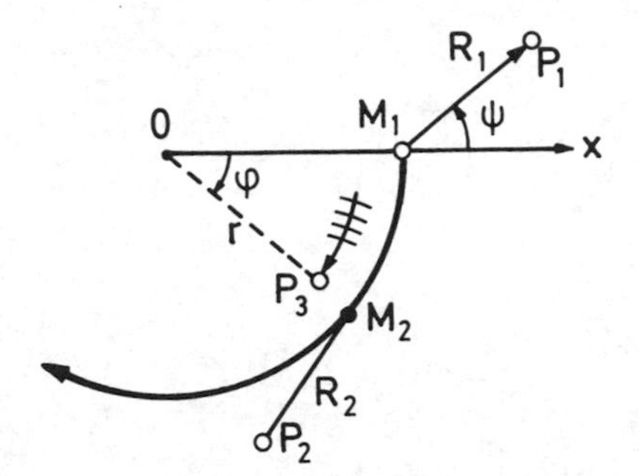

Fig.2.Diffraction at the edge of a cylindrical sheet

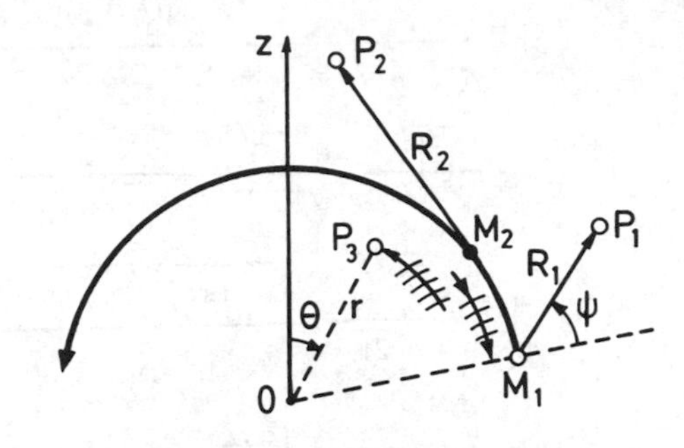

Fig.3.Diffraction at the edge of a spherical cap

$$E^d(P_2) \sim [\partial E^{(n)}(M_1)/\partial ka] T_{wc}\, e^{i\nu_1 \widehat{M_1M_2}/a} T_{cs} e^{ikR_2}/\sqrt{kR_2}$$

$$E^d(P_3) \sim \sum_m [\partial E^n(M_1)/\partial ka] T^{nm}_{ww} \Pi^m_w(r,\varphi) + (\text{a term with continuous spectrum}),$$

$$H^d(P_1) \sim H^{(n)}(M_1) T_{we}(\psi) e^{ikR_1}/\sqrt{kR_1}$$

$$H^d(P_2) \sim H^{(n)}(M_1) T_{wc}\, e^{i\nu_1' \widehat{M_1M_2}/a}\, T_{cs} e^{ikR_2}/\sqrt{kR_2}$$

$$H^d(P_3) \sim \sum_m H^{(n)}(M_1) T^{nm}_{ww} \Pi^m_w(r,\varphi) + (\text{a term with cont.spectrum}).$$

Here ν_1 and ν_1' denote the first roots of $H^{(1)}_\nu(ka)=0$ and $H'^{(1)}_{nm}(ka)=0$, respectively. The transfer coefficients T_{we}, T_{wc}, T_{cs} and T^{nm}_{ww}, and the normalized WG mode $\Pi^m_w(r,\varphi)$ are given by

$$T_{we}(\psi) = \frac{(ka)^{3/2}}{\sqrt{8\pi}} \frac{e^{-i\pi/4}}{\sqrt{\mu_n + ka}} \frac{\sqrt{1+\sin\psi}}{\mu_n - ka\sin\psi}$$

$$T_{wc} = \frac{(ka)^{3/2}}{2^{7/4}\,\pi^{1/4}} \frac{e^{-i\pi/8}}{(\nu_1+\mu_n)} \frac{1}{\sqrt{\mu_n+ka}\,\sqrt{\nu_1+ka}}$$

$$T_{cs} = \frac{2^{5/4}}{\pi^{1/4}} \frac{e^{3\pi i/8}}{\sqrt{ka}} \frac{1}{\dot{H}^{(1)}_{\nu_1}(ka)}$$

$$T^{nm}_{ww} = \frac{-e^{\pi i/4}}{4\sqrt{2\pi}} \frac{(ka)^2}{(\mu_n+\mu_m)} \frac{1}{\sqrt{\mu_n+ka}\,\sqrt{m_m+ka}}$$

$$\Pi^m_w(r,\varphi) = -\sqrt{8\pi}\, e^{\pi i/4} J_{\mu_m}(kr) e^{-i\mu_m\varphi}/[\dot{J}_{\mu_m}(ka)ka]$$

for the electric case and

$$T_{we}(\psi) = \frac{\sqrt{ka}}{\sqrt{8\pi}} \frac{\sqrt{\mu_n'+ka}}{e^{-i\pi/4}} \frac{\sqrt{1-\sin\psi}}{\mu_n' - ka\sin\psi}$$

$$T_{wc} = \frac{-e^{-i\pi/8}}{\sqrt{ka}\ 2^{7/4}\ \pi^{1/4}} \frac{\sqrt{\mu_n'+ka}\sqrt{\nu_1'+ka}}{\nu_1'+\mu_n'}$$

$$T_{cs} = \frac{2^{5/4}}{\pi^{1/4}} \frac{e^{3\pi i/8}}{\sqrt{ka}} \frac{1}{\dot{H}^{(1)\prime}_{\nu_1'}(ka)}$$

$$T^{nm}_{ww} = \frac{-e^{\pi i/4}}{4\sqrt{2\pi}} \frac{\sqrt{\mu_n'+ka}\ \sqrt{\mu_m'+ka}}{\mu_n'+\mu_m'}$$

$$\Pi^m_w(r,\varphi) = \sqrt{8\pi}\ e^{i\pi/4}\ J_{\mu_m'}(kr)\, e^{-i\mu_m'\varphi}/[\dot{J}'_{\mu_m'}(ka)ka]$$

for the magnetic case. The dot on the Bessel functions stands for the derivative with respect to the order.

The interpretation of the factors appearing in the above expressions is obvious. The expressions of $\Pi^m_w(r,\varphi)$ are so normalized that the symmetry relations $T_{we}=T_{ew}$ and $T_{wc}=T_{cw}$ are simultaneously satisfied. Here T_{ew} and T_{cw} are the transfer functions showing the modifications which occur when a space ray or the first creeping mode excites the WG mode Π^m_w [cf.2]. An essential difference between the magnetic and electric cases is that the field value at the diffraction point M_1 appears in the expressions related to the first case while its normal derivative takes place in the second case.

3.2 Canonical Problems For A Spherical Cap

Consider now the case where S consists of a spherical cap depicted in Fig.1b and the incident field is given by (3) or (4a). Then a suitable canonical problem can be formulated in the extended space $r\varepsilon[0,\infty)$, $\theta\varepsilon(-\infty,\infty)$, $\varphi\varepsilon[0,2\pi)$ so that the diffracted field $E^d(r,\theta)$ or $H^d(r,\theta)$ can be represented as follows:

$$U(r,\theta) = \begin{cases} \frac{1}{\sqrt{r}} \int_{-\infty}^{\infty} J_{|\nu|}(kr)\, Q^{-1}_{\nu-1/2}(\cos\theta) A(\nu) d\nu, & r<a \\ \frac{1}{\sqrt{r}} \int_{-\infty}^{\infty} H^{(1)}_{\nu}(kr)\, Q^{-1}_{\nu-1/2}(\cos\theta) B(\nu) d\nu, & r>a \end{cases} \quad (9)$$

where $A(\nu)$ and $B(\nu)$ are to be determined with the aid of the (extended) boundary and edge conditions. Following the method given in [3] one can easily find the leading terms in the asymptotic expansions as follows (see, Fig.3):

$$E^d(P_1) \sim [\partial E^{(n)}(M_1)/\partial ka]\, T_{we}(\psi)\, \Pi_o(M_1,P_1)$$

$$E^d(P_2) \sim [\partial E^{(n)}(M_1)/\partial ka]\ T_{wc}\Pi_c(M_1, M_2) T_{cs}\Pi_o(M_2,P_2)$$

$$E^d(P_3) \sim \sum_m [\partial E^{(n)}(M_1)/\partial ka]\, T^{nm}_{ww}\Pi^m_w(r,\theta) + (\text{a term with cont.spectrum})$$

with

$$T_{we}(\psi) = \frac{(ka)^{3/2}\ \Gamma(1-\mu_n)\, tg\mu_n\pi}{4\mu_n\Gamma(3/2-\mu_n)\sqrt{\mu_n+ka}} \frac{e^{i\mu_n\theta_1}}{\sqrt{\sin\theta_1}\ Q^{-1}_{\mu_n-1/2}(\cos\theta_1)} \frac{\sqrt{1-\sin\psi}}{\mu_n+ka\sin\psi}$$

$$T_{wc}=\frac{(ka)^{3/2}}{\sqrt{\mu_n+ka}}\;\frac{e^{\pi i/8}\pi^{1/4}2^{-9/4}}{\sqrt{\nu_1+ka}\,(\nu_1+\mu_n)}\;\frac{\Gamma(1-\mu_n)\,tg\mu_n\pi}{\mu_n\Gamma(3/2-\mu_n)}\;\frac{e^{i\mu_n\theta_1}}{\sqrt{\sin\theta_1}\;Q^{-1}_{\mu_n-1/2}(\cos\theta_1)}$$

$$T_{cs}=\frac{2^{5/4}}{\pi^{1/4}}\;\frac{e^{3\pi i/8}}{\sqrt{ka}\;\dot{H}^{(1)}_{\nu_1}(ka)}$$

$$T^{nm}_{ww}=\frac{e^{-i\pi/4}\sqrt{\pi}\,(ka)^2[(\mu_n+ka)(\mu_m+ka)]^{-1/2}\,e^{i(\mu_n+\mu_m)\theta_1}}{8\sqrt{2}\,\sin\theta_1\,Q^{-1}_{\mu_n-1/2}(\cos\theta_1)\,Q^{-1}_{\mu_m-1/2}(\cos\theta_1)\,\mu_n\mu_m\Gamma(3/2-\mu_n)}$$

$$\frac{\Gamma(1-\mu_n)\Gamma(1-\mu_m)\,tg\mu_n\pi\, tg\mu_m\pi}{\Gamma(3/2-\mu_m)(\mu_n+\mu_m)}$$

$$\Pi^m_w(r,\theta)=\frac{\mu_m(\mu_m^2-1/4)\sin\theta_1 Q^{-1}_{\mu_m-1/2}(\cos\theta_1)}{2^{-5/2}\;\sqrt{\pi ka}\;\;e^{i\pi/4}\;\;tg\mu_m\pi\;\dot{J}_{\mu_m}(ka)}\;\;\frac{J_{\mu_m}(kr)}{\sqrt{kr}}\,Q^{-1}_{-\mu_m-1/2}(\cos\theta)$$

$$\Pi_o(M,P)=\sqrt{\frac{a\,\sin\theta_M}{r_P\sin\,\theta_P}}\;\;\frac{e^{ikMP}}{\sqrt{k\,\overline{MP}}}$$

$$\Pi_c(M_1,M_2)=\sqrt{\frac{\sin\theta_1}{\sin\theta_2}}\;e^{i\nu_1(\theta_1-\theta_2)}$$

and

$$H^d(P_1)\sim H^{(n)}(M_1)T_{we}(\psi)\Pi_o(M_1,P_1)$$

$$H^d(P_2)\sim H^{(n)}(M_1)T_{wc}\Pi_c(M_1,M_2)T_{cs}\Pi_o(M_2,P_2)$$

$$H^d(P_3)\sim\sum_m H^{(n)}(M_1)T^{nm}_{ww}\Pi^m_w(r,\theta)+(\text{a term with cont.spectrum})$$

with

$$T_{we}(\psi)=\frac{i\sqrt{ka}}{4\bar{\mu}_n}\;\frac{\Gamma(1-\bar{\mu}_n)\,tg\bar{\mu}_n\pi\sqrt{\bar{\mu}_n+ka}\;e^{i\bar{\mu}_n\theta_1}\;\;\sqrt{1+\sin\psi}}{\Gamma(3/2-\bar{\mu}_n)\;\sqrt{\sin\theta_1}\,Q^{-1}_{\bar{\mu}_n-1/2}(\cos\theta_1)\;(\bar{\mu}_n+ka\;\sin\psi)}$$

$$T_{wc}=\frac{-e^{i\pi/8}}{2^{9/4}}\,\frac{\pi^{1/4}}{\sqrt{ka}}\;\;\frac{\Gamma(1-\bar{\mu}_n)\,tg\bar{\mu}_n\pi}{\bar{\mu}_n\Gamma(3/2-\bar{\mu}_n)}\;\;\frac{\sqrt{\bar{\nu}_1+ka}\;\sqrt{\bar{\mu}_n+ka}}{(\bar{\nu}_1+\bar{\mu}_n)\;\sqrt{\sin\theta_1}}\;\frac{e^{i\bar{\mu}_n\theta_1}}{Q^{-1}_{\bar{\mu}_n-1/2}(\cos\theta_1)}$$

$$T_{cs}=\frac{e^{3\pi i/8}}{\pi^{1/4}}\;\frac{2^{5/4}}{d[\sqrt{ka}\;\dot{H}^{(1)}_{\bar{\nu}_1}(ka)]/d\;ka}$$

$$T_{ww}^{nm}=\frac{e^{-i\pi/4}}{8\sqrt{2}}\ \frac{\sqrt{\pi}\sqrt{\bar{\mu}_n+ka}\ \sqrt{\bar{\mu}_m+ka}\quad e^{i(\bar{\mu}_n+\bar{\mu}_m)\theta_1}}{\sin\theta_1 Q^{-1}_{\bar{\mu}_n-1/2}(\cos\theta_1)Q^{-1}_{\bar{\mu}_m-1/2}(\cos\theta_1)\quad \bar{\mu}_n\bar{\mu}_m}$$

$$\frac{\Gamma(1-\bar{\mu}_n)\Gamma(1-\bar{\mu}_m)\,tg\bar{\mu}_n\pi\, tg\bar{\mu}_m\pi}{\Gamma(3/2-\bar{\mu}_n)\Gamma(3/2-\bar{\mu}_m)(\bar{\mu}_n+\bar{\mu}_m)}$$

$$\Pi_w^m(r,\theta)=\frac{\bar{\mu}_m(\bar{\mu}_m^2-1/4)\sin\theta_1\ Q^{-1}_{\bar{\mu}_m-1/2}(\cos\theta_1)\,2^{5/2}}{\sqrt{\pi}\,e^{5\pi i/4}\,tg\bar{\mu}_m\pi\ \frac{d}{dka}[\sqrt{ka}\ \dot{J}_{\bar{\mu}_m}(ka)]}\ \frac{J_{\bar{\mu}_m}(kr)}{\sqrt{kr}}\ Q^{-1}_{-\bar{\mu}_m-1/2}(\cos\theta)$$

$$\Pi_o(M,P)=\sqrt{\frac{a\ \sin\theta_M}{r_p\ \sin\theta_p}}\ \frac{e^{ikMP}}{\sqrt{k\overline{MP}}}$$

$$\Pi_c(M_1,M_2)=\sqrt{\frac{\sin\theta_1}{\sin\theta_2}}\ e^{i\bar{\nu}_1(\theta_1-\theta_2)}$$

Here $\bar{\nu}_1$ stands for the first root of

$$\frac{d}{dx}[\sqrt{x}\ H_\nu^{(1)}(x)]=0,\quad (x=ka).$$

The expressions of $\Pi_w^m(r,\theta)$ are so normalized that the symmetry relations $T_{we}=T_{ew}$ and $T_{wc}=T_{cw}$ are satisfied.

References

1. J.B.Keller, Diffraction By A Convex Cylinder, IRE Trans. Antennas and Propagat. Vol.AP-4, pp.312-321, 1956.
2. M.Idemen and G.Uzgören,Some Diffraction Coefficients Related To Diffractions At The Edges Of Curved Reflectors, Proc.12th European Microwave Conf., Finlandia Hall, Helsinki, Finland 1982.
3. M.Idemen and E.Erdoğan, Diffraction Of The Creeping Waves Generated On A Perfectly Conducting Spherical Scatterer By A Ring Source, IEEE Trans.Antennas and Propagat.Vol.AP-31, pp.776-784, 1983.
4. T.Ishihara, L.B.Felsen and A.Green, High Frequency Fields Excited By A Line Source Located On A Perfectly Conducting Concave Cylindrical Surface, IEEE Trans.Antennas and Propagat., Vol.AP-27,pp.757-767, 1978.
5. M.Idemen, and L.B.Felsen, Diffraction of A Whispering Gallery Mode By The Edge of A Thin Concave Cylindrically Curved Surface IEEE Trans.Antennas and Propagat., Vol.AP-29, pp.571-579,1981.

RAYS IN TAPERED WAVEGUIDES

A.M.Scheggi, R.Falciai, P.Cavigli, S.Mezza

Istituto di Ricerca sulle Onde Elettromagnetiche, C.N.R.
Florence, Italy

The study of propagation in optical dielectric tapered waveguides is reported. A numerical method based on ray optics and associated energy flux is applied in order to evaluate radiation characteristics of different tapered geometries.

1 INTRODUCTION

The study of propagation in longitudinally dependent dielectric waveguides and in particular in tapered waveguides is of interest in a variety of applications that include natural as well as man-made guiding environments.

Examples of the former are tropospheric and ionospheric ducts, underwater acoustic ducts in the ocean, seismic channels in the earth's crust, etc.

Examples of the latter are tapered dielectric antennas and optical fiber or slab waveguides. In particular these may be of interest for more efficent fiber to source coupling or for optical components as well as for high power radiation delivery systems (mechanical or medical applications) (1,2,3,4,5,6).

The present paper is to be concerned with a study of the propagation in optical dielectric tapered waveguides, by means of a numerical method based on ray optics and associated energy flux.

A number of examples are given which show how the method can be applied even with some lengthy calculations to evaluate the transverse energy distribution along the waveguide as well as the near and far field intensity patterns at the output.

2 DESCRIPTION OF THE METHOD

In order to evaluate the energy distribution across the fiber and the far field, one can solve the ray equation

$$\frac{d}{ds}\left(n\frac{d\bar{r}}{ds}\right) = \nabla n \qquad (1)$$

trace the rays, and then evaluate the flux change associated with geometrical spreading or concentration of the ray bundles.

We considered a multimode, tapered optical waveguide either with unlimited or finite cladding.

By referring the fiber to a cylindrical coordinate system ρ, φ, z and by denoting with θ, ψ the angles of the far field into which energy is radiated (Fig.1), the incremental area on a cross section of the fiber and the incremental solid angle are

$$dA = \rho d\rho d\varphi \qquad d\boldsymbol{\theta} = \sin\theta d\theta d\psi \qquad (2)$$

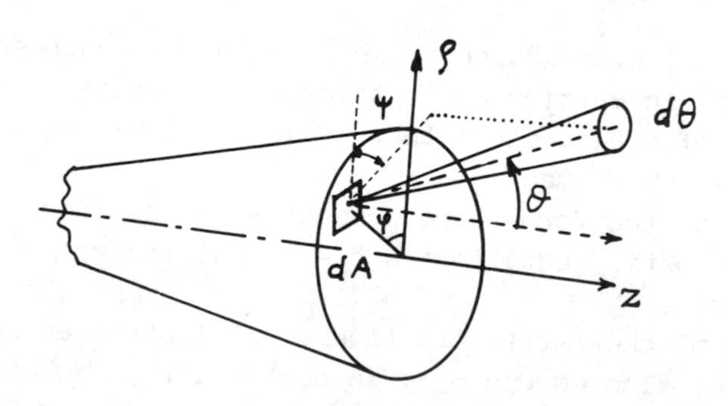

Fig.1 - Coordinate system for the near and far field distributions.

Under the hypotesis of incoherent illumination and taking into account the cylindrical symmetry of the waveguide we denote by

$I_{in}(\rho,\theta) = I_{in}(\rho)\, I_{in}(\theta)$ the intensity at the input face, where

$$I_{in}(\rho) = \frac{N_{in}(\rho)}{\Delta A} \qquad I_{in}(\theta) = \frac{M_{in}(\theta)}{\Delta \Theta} \tag{3}$$

$$\Delta A = \int_0^{2\pi} dA = 2\pi\rho\, d\rho \qquad \Delta\Theta = \int_0^{2\pi} d\Theta = 2\pi \sin\theta\, d\theta \tag{4}$$

$N(\rho)$ and $M(\theta)$ represent the spatial and angular ray distributions, respectively.

In our computations we assumed uniform spatial distribution and a Gaussian or Lambertian angular distribution depending on the case considered.

The intensity distribution $I_{\bar{z}}(\rho)$ at $z = \bar{z}$ along the fiber (or the near field intensity distribution) and the far field intensity distribution are given by

$$I_{\bar{z}}(\rho) = \frac{N_{\bar{z}}(\rho)}{\Delta A} \qquad I_{far} = \frac{M_{far}(\theta)}{\Delta\Theta} \tag{5}$$

A Fermi-Dirac distribution was assumed as index profile:

$$n(\rho,z) = n_2 + (n_1 - n_2)/(1 + \exp\left\{ C \left[\frac{\rho}{a(z)} - 1\right]\right\}) \tag{6}$$

with C = 200 corresponding to a nearly step shape. a_1 and a_2 denote the radii of the smaller and larger end faces of the fiber, $a(z)$ indicates the taper profile (Fig.3).

The method was applied to different geometrical configurations implementing step by step the computation technique according the following stages:

1 - slab tapered waveguide with infinitely extended clad.
2 - taper ended fiber with infinitely extended clad for a comparison with results of measurements performed on samples of tapered fibres fabricated at I.R.O.E.
3 - slab doubly tapered waveguide with finite clad.
4 - biconical fiber with finite clad.

The extension of the method from twodimensional to threedimensional case implies the introduction of skew rays and hence a much longer computing time.

The assumption of a finite clad is fulfilled by suitably modifying the index profile, still utilizing expression (6) (Fig.2).

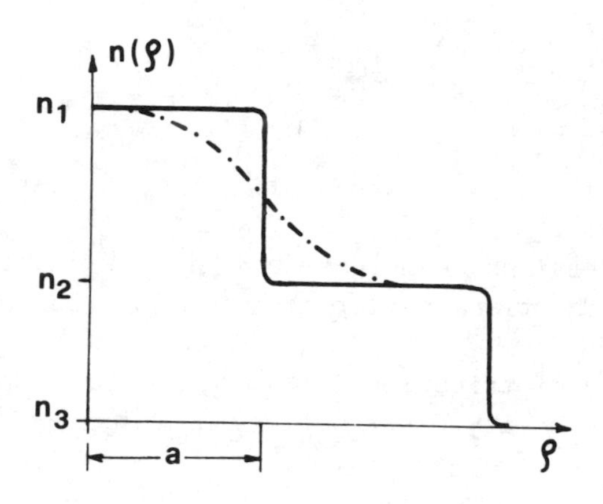

Fig.2 - Index distributions for finite clad fiber.

3 TAPERED SLAB WAVEGUIDE

As a first application we considered a slab tapered waveguide (Fig.3) with a nearly step index distribution ($C = 200$, $n_1 = 1.50$) and linear and cosinusoidal taper profiles. A uniform spatial density and a Gaussian angular density distribution were assumed so that expressions (3) become

$$I_{in}(\rho) = \frac{N_o}{\Delta\rho} \qquad I_{in}(\theta) = e^{-(\frac{\theta}{W})^2} \frac{M_o}{\Delta\theta} \tag{7}$$

choosing for W a value such that the half intensity beam width is equal to $n_1\theta_c/2$.

Figure 4 shows the near and far field intensity patterns in the cases $a_1/a_2 = 0.75$, $L = 1$ mm ($\Omega = 17'$), and $a_1/a_2 = 0.5$, $L = 5$ mm ($\Omega = 28'$) compared with those from a 5 mm long uniform waveguide. No differences are observed in the corresponding patterns for cosinusoidal taper.

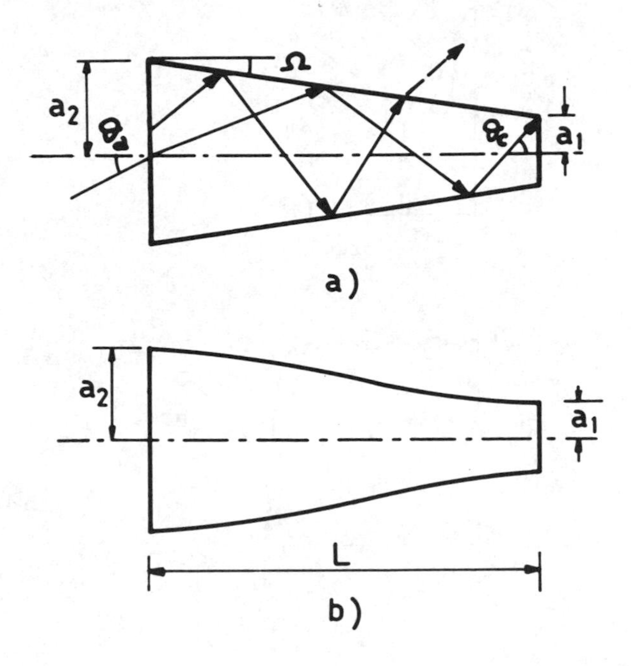

Fig.3 - Ray paths in a tapered slab waveguide.

The near-field patterns remain pratically uniform, whereas the far-field patterns present half-intensity widths that increase inversely with the taper ratio.

When the taper angle increases, the number of rays that undergo multiple reflections and are accepted at the taper end decreases, whereas the number of rays that do not impinge on the taper walls increases.

The far field patterns are modified by the presence of side lobes, which are due to the contribution of those rays that undergo only one reflection and can emerge at very wide angles (larger than the critical angle θ_c).

This effect is displayed in Fig.5 in which some far-field intensity patterns are shown corresponding to increasing values of Ω compared with the input angular density distribution and is particulary evident for the largest taper angle ($\Omega = 8^o36'$).

The pattern corresponding to $\Omega = 1^o26'$ (L = 1 mm) is also shown for comparison with that corresponding to the same taper ratio ($a_1/a_2 = 0.5$, L = 5 mm) but smaller taper angle $\Omega = 17'$.

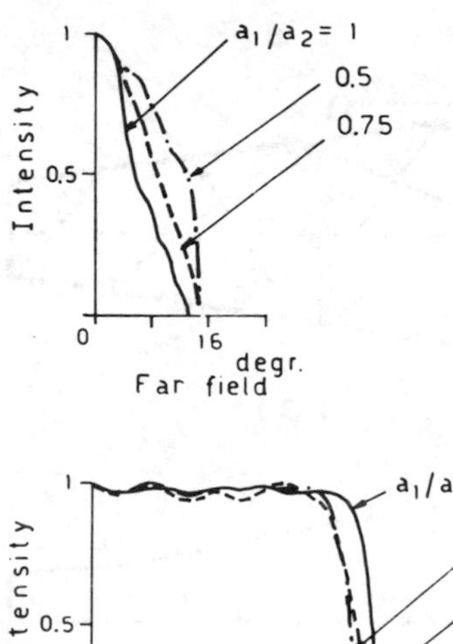

Fig.4 - Near and far field intensity patterns from two slab tapered waveguides (a_1/a_2 = 0.5, 0.75) compared with those from a uniform waveguide (a_1/a_2 = 1).

The examples examined indicate that the near-field configurations remain practically unchanged with respect to the input spatial intensity distribution for each combination of the parameters a_1/a_2 and Ω.

When the taper angle remains very small, the only parameter playing a role in the modification of the far-field intensity patterns is the taper ratio a_1/a_2. When Ω increases, the patterns become more and more modified and depend on both parameters a_1/a_2 and Ω.

4 TAPER ENDED FIBERS

As a second step we evaluated the radiation characteristics of taper ended fibres by extending the above method to the three-dimensional case.

Again a nearly step index distribution was assumed (C = 200,

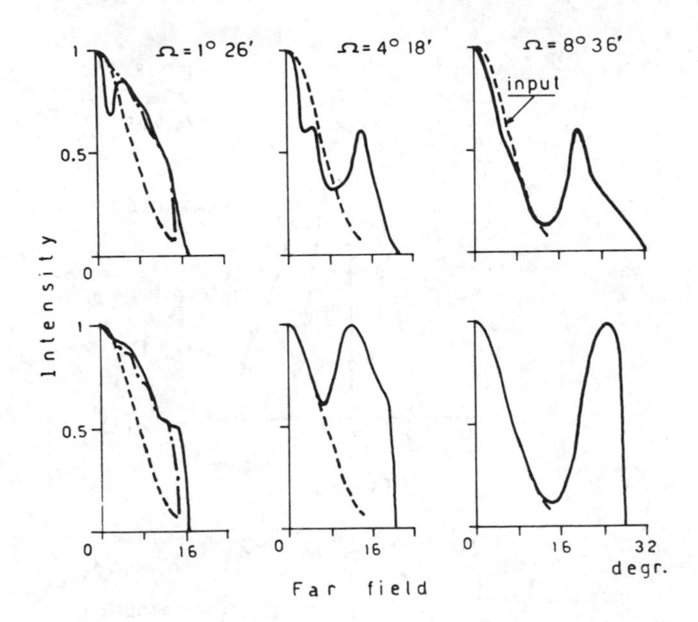

Fig.5 - Far field intensity patterns from three tapered waveguides for linear (upper) and cosinusoidal (lower) taper profile having increasing taper angle (Ω) compared with the pattern from the uniform waveguide (dashed line). The pattern for $\Omega = 1^{o}26'$ is also compared with that for $\Omega = 17'$ (dotted--dashed line) corresponding to a larger taper with same taper-ratio.

n = 1.457, n = 1.430) with linear taper profile and infinitely extended clad. The input ray distributions vary with the same laws as in the twodimensional case.

Tapered conical ends in optical fibers apart from finding use as optical components can be of interest for either mechanical or surgical applications. In fact they can provide a more efficient coupling with large-spot high-power lasers and when used at the exit of the fibre present good collimation properties which are particulary useful for endoscopic applications.

In the framework of a C.N.R. oriented mission program on Laser Application to Medicine we fabricated samples of taper ended fibres with different taper parameters (core diameter 50÷500 μm, taper length 2÷100 cm, taper ratio $a_1/a_2 \geq 0.2$).

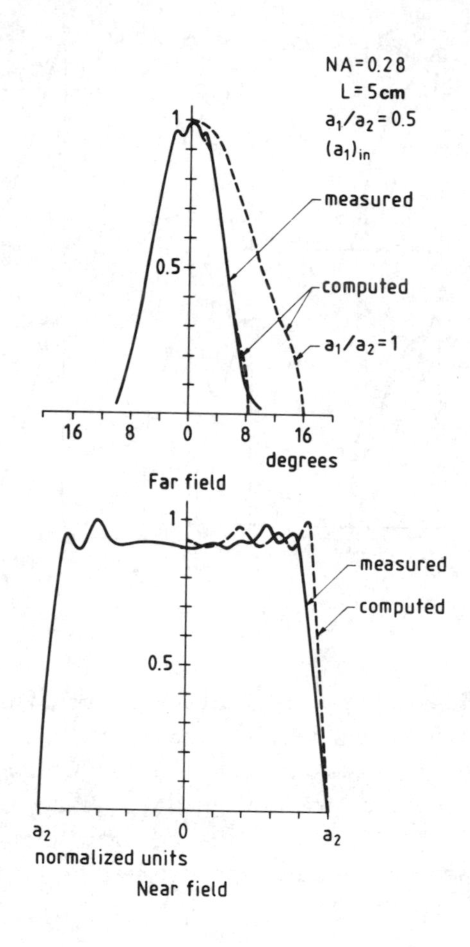

Fig.6 - Comparison between measured and computed near and far field intensity patterns for enlarged taper ended fiber. The computed far field intensity pattern for the uniform fiber is also shown.

Fig.6 shows a comparison between the measured and computed radiation patterns from an enlarged taper fiber with $a_1/a_2 = 0.5$, L = 5 cm, (a_1 = 300 μm).

The beam turns out to be collimated by a factor a_1/a_2 with respect to that from the uniform fiber while the near field distributions remain substantially unchanged.

When the fiber is illuminated at the enlarged end the radiation

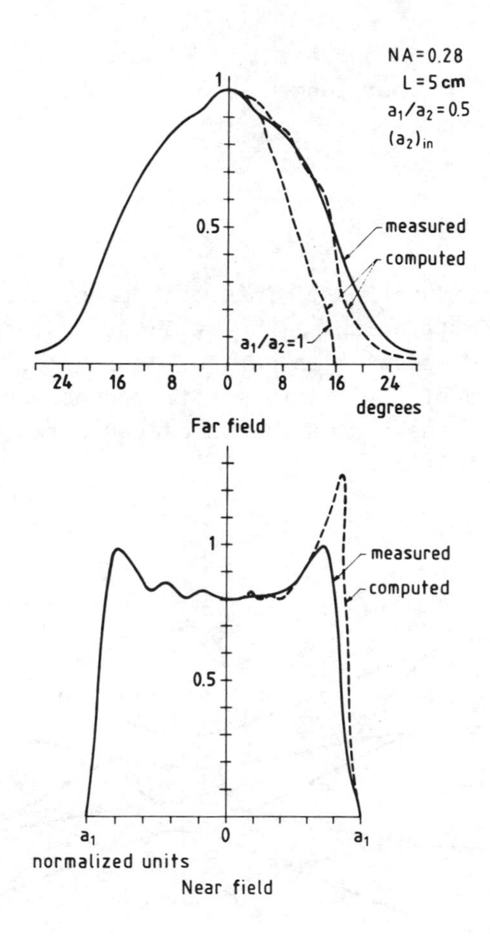

Fig.7 - Comparison between measured and computed intensity patterns for a decreasing taper ended fiber. The computed far field intensity pattern for the uniform fiber is also shown.

pattern (Fig.7) results broadened; in fact as the rays proceed on their zig-zag paths their propagation angles increase from reflection to reflection and eventually exceed the cut off angle, thus giving rise to leaky-tunneling rays. This is confirmed by the typical increasing behavior of the near field distribution at the fiber edge.

The agreement between experimental and theoretical curves is quite satisfactory even if the contribution of the leaky rays in the measured patterns appears lower because the measurement was performed after a fiber length (although short) along which the leaky rays attenuate.

5 BICONICAL WAVEGUIDE

Another configuration of interest for practical applications is the biconical fiber which can be utilized as modestripper for eliminating the clad modes or as part of an optical fiber coupler (7).

For the sake of simplicity we first considered the twodimensional case of a doubly tapered waveguide placed between two uniform waveguide sections (Fig.8).

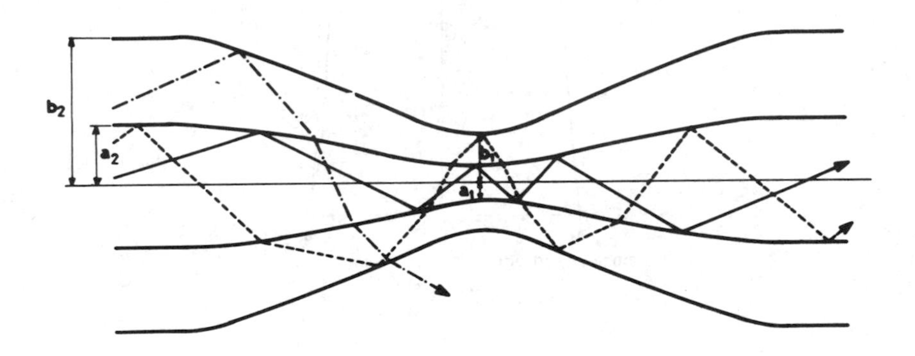

Fig. 8 - Doubly tapered "bitapered" waveguide

In this case a cosinusoidal taper profile was choosen in order to get a better matching between the two tapers, while a nearly step index profile (C = 200, n_1 = 1.473, n_2 = 1.457, n_3 = 1.430) with finite clad was assumed.

Under the assumption of uniform spatial and Lambertian angular distribution at the input of the uniform fiber joined to the taper, the rays propagating in the double taper can be divided into three groups:

1. rays which remain trapped within the core of the bitaper.

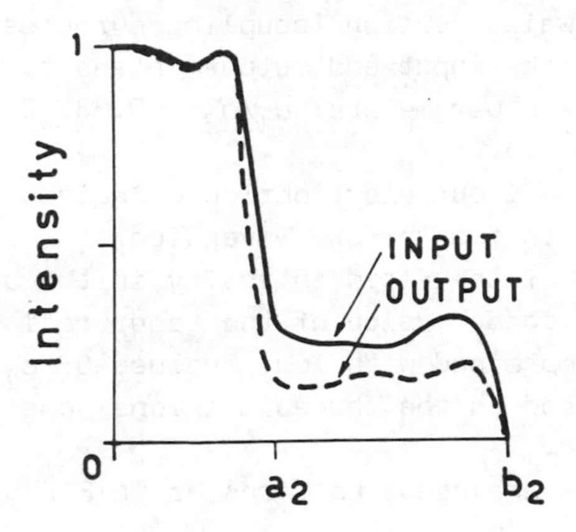

Fig.9 - Input and output intensity patterns for a bitapered slab waveguide.

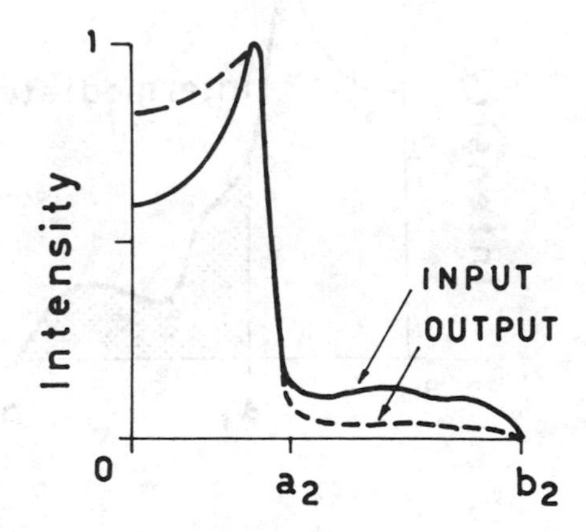

Fig.10 - Input and output intensity patterns for a biconical fiber.

2. rays radiated into the clad along the first taper, reflected back at clad-external medium interface and trapped again within the core of the second taper.
3. rays propagating in the clad which are radiated to the outside along the first taper.

For the anticipated application what is of interest is the field intensity distribution at the exit face of the bitaper (mode stripper) or at the waist section (coupling purposes).

Fig.9 shows the input and output intensity patterns of a bitaper with the following parameters: $a_1/a_2 = 0.33$, $2L = 4$ mm, $2a_2 = 50$ µm, $b_2/a_2 = b_1/a_1 = 2.5$.

Note that the input distribution coincides with the steady state distribution in the uniform waveguide.

A reduction of the field intensity in the clad of $\sim 40\%$ is obtained for the chosen value of the taper ratio but further reductions would be obtained with lower values of a_1/a_2.

We then passed to the threedimensional case of a biconical fiber with finite clad.

Fig.10 shows analogous patterns as in Fig.9; the reduction in this case is of $\sim 60\%$.

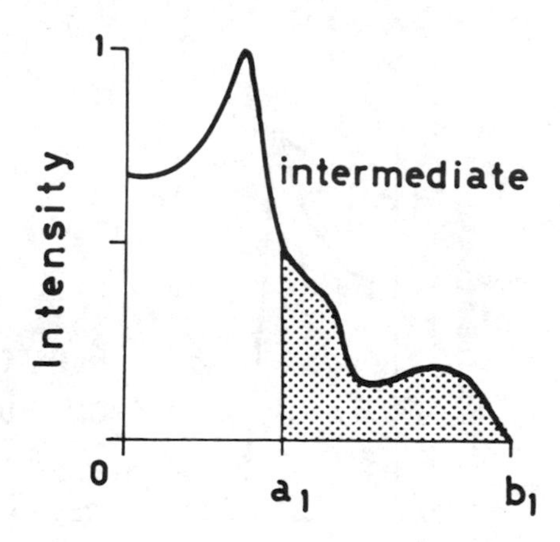

Fig.11 - Intensity pattern at the intermediate section of a biconical fiber.

Finally Fig.11 shows the intensity pattern distribution at the narrowest section of the biconical fiber, exhibiting a large mode volume in correspondence to the clad, which can be utilized for coupling this biconical fiber to an identical one (biconical taper coupler) (Fig.12).

Note that in spite of the fact that in comparing the waist region the fiber cores in such couplers are quasi elliptical (8), a good coupling is still to be expected. In fact, in elliptical fibers, the leaky ray condition varies azimuthally and is reached

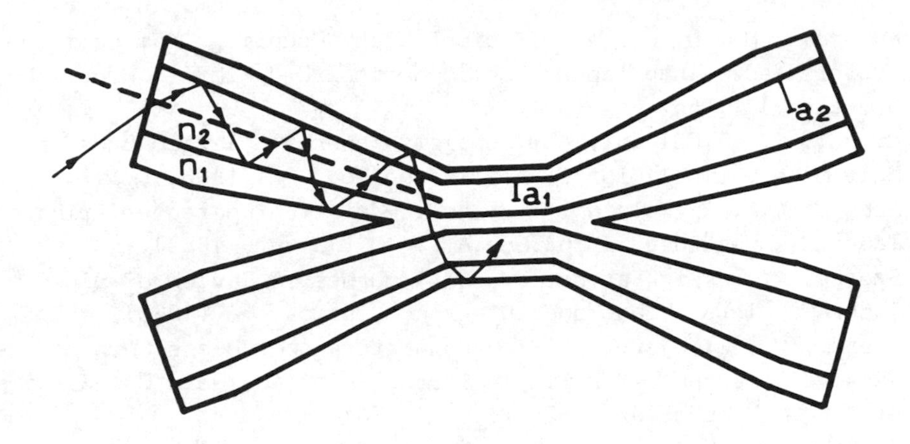

Fig.12 - Biconical taper coupler

first with respect to the circular case, at least at the points of maximum curvature (9).

In conclusion the examples considered up to now confirm that the ray method is suitable for evaluating global radiation characteristics of tapered dielectric waveguides. Three dimensional cases obviously imply relatively longer computation time due to the presence of skew rays. Next steps will be the application of the method to the graded index case.

Another problem which will be tackled will be that of considering again a tapered waveguide, but slightly overmoded and with coherent illumination.

1. Ozeki, T., T.Ito, T.Tamura. Tapered Section of Multimode Cladded Fibres as Mode Filters and Mode Analyzer, Applied Physics Letters, 26, (1975), 386.
2. Ozeki, T., B.S.Kawasaki. Optical Directional Coupler Using Tapered Sections in Multimode Fibres, Applied Physics Letters, 28 (1976), 528
3. Chang,L.T., D.C.Auth. Radiation Characteristics of a Tapered Cylindrical Optical Fiber, J.Opt.Soc.Am. 68, (1978), 1191.

4. Kuwahara, H., H.Furuta. Efficient Light Coupling from Semi-Conductor Lasers into Tapered Hemispherical-End Fibres, Proc.IEEE, 67, (1979), 1456.
5. Brenci, M., R.Falciai, A.M.Scheggi. Tapered Enlarged Ends in Multimode Optical Fibres, Applied Optics, 21, (1982), 317.
6. Scheggi A.M., R.Falciai, M.Brenci. Radiation Characteristics of Tapered Waveguides, J.Opt.Soc.Am., 73 (January 1983).
7. Szarka, F., A.Lightstone, J.Lit, R.Hughes. A Review of Biconical Taper Couplers, Fiber and Integrated Optics, 3, (1980), 285.
8. Koster, W., W.Meyer, H.Wilke. Connectors and Access Couplers as Passive Components for Single Strand Multimode Data Bus Systems, 4th ECOC, Genova 1978, 323 ff.
9. Checcacci, P.F., R.Falciai, A.M.Scheggi. Tunneling-radiating Effect in Elliptical Step Index Fibres, Opt.Lett. 4, (1980), 145.

NUMERICAL COMPUTATION OF ELASTIC WAVEFIELDS IN ANISOTROPIC ELASTIC MEDIA IN THE PRESENCE OF CAUSTICS

Andrzej Hanyga[1)2)]

1) Institute of Geophysics, PL-00973 Warsawa, Pasteura 3, Poland
2) On leave of absence at NTNF/NORSAR, Norway

1. INTRODUCTION

The objective of a forward seismological problem consists of determining the seismograms (i.e., the displacements $\underset{\sim}{u}(\underset{\sim}{x}_1,t)$) at a seismograph location $\underset{\sim}{x}_1$ for a given source $\underset{\sim}{x}_o$ and a given (often hypothetical) inhomogeneous elastic medium through which the signals propagate from $\underset{\sim}{x}_o$ to $\underset{\sim}{x}_1$. At least the first 1/3 of the seismogram is interpretable in terms of wavefront arrivals and hence can reasonably be explained in terms of geometrical optics. Although a suitable approach to the determination of wavefront arrival times and intensities would involve a Friedlander ([4]) expansion, it is a common practice to consider propagation of monochromatic high frequency signals and then to apply the (fast) Fourier transform. Although this procedure is patently inconsistent, the high-frequency part of the seismogram $\underset{\sim}{u}(\underset{\sim}{x}_1,t)$ thus obtained provides a correct description of the wavefront arrivals.

A WKBJ asymptotic expression for the wavefield $u(x;\omega)$ is the sum of terms of the form $J^{-\frac{1}{2}}g\,\exp(i\omega S)\underset{\sim}{r}$, where g is a smoothly varying amplitude, S is the eikonal, $\underset{\sim}{r}$ is the polarization vector and J is the geometric spreading factor. The WKBJ solution is made of signals propagated independently along a family of rays covering the domain of the solution. This property of WKBJ solutions is very convenient for seismological applications since it allows computation of the seismogram $\underset{\sim}{u}(\underset{\sim}{x}_1,t)$ at $\underset{\sim}{x}_1$ merely by integrating a system of ordinary differential equations, known as the dynamic ray tracing (DRT) system, along all the rays joining $\underset{\sim}{x}_o$ to $\underset{\sim}{x}_1$.

When a ray approaches a caustic (= an envelope of rays), the cross-sections of the ray tubes shrink to zero, the geometric ray spreading factor J tends to zero and the WKBJ signal blows up. Since the actual solution remains finite in the vicinity of the caustic the WKBJ expression quoted above cannot provide a uniform approximation of the actual wavefield in the caustic region. For simple caustics and caustic cusps the uniform approximation has been constructed so far in terms of Airy and Pearcey functions. This procedure has no obvious extension to all sorts of caustics.

Drawing on an idea of Maslov ⌊9⌋ we purport to show here that caustic regions and caustic shadows can be correctly accounted for within a suitably extended WKBJ method. The latter method depends on some results of symplectic geometry in quite an essential way, but we shall avoid here to bring advanced geometrical concepts into play. The reader interested in the symplectic-geometric background is referred to ⌊5⌋ and to references contained therein.

In accordance with our seismological interests we shall develop the WKBJ theory for elastic wavefields in prestressed anisotropic media. Anisotropy imposes the necessity of a "canonical" treatment involving the Hamilton-Jacobi theory. This is in turn related to the symplectic background of the Maslov ideas about which we are not allowed to speak here.

After introducing basic notions of anisotropic elasticity (Sec. 2) we shall present the DRT equations for an anisotropic elastic medium away from the caustics (following ⌊4⌋,⌊5⌋) (Sec. 3). We shall then introduce the reader to the "precanonical operator" representation of the wavefield near the caustics and derive the DRT system for the singular parts of the rays (Sec. 4). In Sec. 5 we quote the transition formulae which connect the unknown functions of the DRT systems in regular and singular regions and in Sec. 6 we make some comments on the initial data for DRT for a harmonic point source. Other kinds of sources are discussed in ⌊5⌋ and ⌊6⌋.

2. EQUATIONS OF ANISOTROPIC ELASTICITY. EIKONALS AND BICHARACTERISTICS

Harmonic waves $e^{-i\omega t}(\underset{\sim}{u}(\underset{\sim}{x};\omega)$ in a prestressed anisotropic medium satisfy the equations

$$\sum_{l,r,s=1}^{3} \frac{\partial}{\partial x^r} \lfloor B^{rs}_{kl}(\underset{\sim}{x}) \frac{\partial u_l}{\partial x_s} \rfloor + \rho\omega^2 u_k = 0 \quad \text{for } \underset{\sim}{x}\in\Omega,\ k=1,2,3 \tag{1}$$

with summation over the repeated indices l,r,s=1,2,3 ⌊4⌋. The coefficients B^{rs}_{kl} satisfy the symmetry condition

$$B^{rs}_{kl}(\underset{\sim}{x}) = B^{sr}_{lk}(\underset{\sim}{x}) \tag{2}$$

as well as the strong ellipticity condition ([8])

$$\sum_{r,s,k,l} B_{kl}^{rs}(\underset{\sim}{x})\ p_r\ p_s\ a_k\ a_l > 0 \tag{3}$$

for all non-zero vectors $\underset{\sim}{a}, \underset{\sim}{p}$.

In view of (2) and (3) the 3×3 matrix $\rho^{-1}\underset{\sim}{B}(\underset{\sim}{x},\underset{\sim}{p})$

$$\underset{\sim}{B}(\underset{\sim}{x},\underset{\sim}{p}) \equiv [B_{kl}^{rs}(\underset{\sim}{x})\ p_r\ p_s] \tag{4}$$

is positive definite for $\underset{\sim}{p} \neq 0$ and has three positive eigenvalues

$$0 < c_1(\underset{\sim}{x},\underset{\sim}{p})^2 \leqslant c_2(\underset{\sim}{x},\underset{\sim}{p})^2 \leqslant c_3(\underset{\sim}{x},\underset{\sim}{p})^2 \text{ for } \underset{\sim}{p} \neq 0 \tag{5}$$

Let $c_\sigma(\underset{\sim}{x},\underset{\sim}{p})$ be the positive square root of the eigenvalue $c_\sigma(\underset{\sim}{x},\underset{\sim}{p})^2$. It can be shown ([11]) that $c_\sigma(\underset{\sim}{x},\underset{\sim}{p})$ is a continuous function of the slowness vector $\underset{\sim}{p} \neq 0$. Since it satisfies the equation

$$\det [B_{kl}^{rs}(\underset{\sim}{x})\ p_r\ p_s - \rho(\underset{\sim}{x}) c_\sigma(\underset{\sim}{x},\underset{\sim}{p})^2 \delta_{kl}] = 0 \tag{6}$$

it is also a homogeneous function of degree one of the second argument.

We shall assume here that $c_\sigma(\underset{\sim}{x},\underset{\sim}{p})^2 \neq c_{\sigma+1}(\underset{\sim}{x},\underset{\sim}{p})^2$ unless these two eigenvalues coincide for all $\underset{\sim}{x},\underset{\sim}{p}$. This assumption excludes such phenomena as conical refraction. We shall number the eigenvalues in such a way that $c_\sigma(\underset{\sim}{x},\underset{\sim}{p})^2 < c_{\sigma+1}(\underset{\sim}{x},\underset{\sim}{p})^2$. The eigenvectors of $\underset{\sim}{B}(\underset{\sim}{x},\underset{\sim}{p})$ corresponding to an eigenvalue $c_\sigma(\underset{\sim}{x},\underset{\sim}{p})^2$ of multiplicity k_σ will be denoted by $\underset{\sim}{r}_\sigma^{(\nu)}(\underset{\sim}{x},\underset{\sim}{p})$, $\nu=1,\ldots,k_\sigma$. Applying the implicit function theorem to eq. (6) we conclude that c_σ is a continuously differententiable function provided the coefficients B_{kl}^{rs} are differentiable (which we assume henceforth).

Substituting the asymptotic series

$$\underset{\sim}{u}(\underset{\sim}{x};\omega) \sim e^{i\omega S(\underset{\sim}{x})} \sum_{n=0}^{\infty} (i\omega)^{-n}\ \underset{\sim}{u}_n(\underset{\sim}{x}) \tag{7}$$

in eq. (1) one obtains an asymptotic solution

$$\underset{\sim}{u}_{\iota\sigma}(\underset{\sim}{x};\omega) \sim \exp\{i\omega\iota S_\sigma(\underset{\sim}{x})\} \sum_{n=0}^{\infty} (i\omega)^{-n}\ \underset{\sim}{u}_n^{(\iota\sigma)}(\underset{\sim}{x}) \tag{8}$$

with $\iota = \pm 1$, $\sigma = 1,2,\ldots,$

$$\underset{\sim}{u}_0^{(\iota\sigma)}(\underset{\sim}{x}) = |J_\sigma(\underset{\sim}{x})|^{-\frac{1}{2}} \sum_{\nu=1}^{k_\sigma} g_{\iota\sigma}{}^{(\nu)}(\underset{\sim}{x})\underset{\sim}{r}_\sigma{}^{(\nu)}(\underset{\sim}{x},\nabla S_\sigma(\underset{\sim}{x})) \tag{9}$$

and S_σ satisfying the Hamilton-Jacobi equation

$$c_\sigma(\underset{\sim}{x},\nabla S_\sigma(\underset{\sim}{x})) = 1 \tag{10}$$

The equations for the geometric ray spreading factors J_σ and for the amplitudes $g_{\iota\sigma}{}^{(\nu)}$ will be discussed in the next sections. In order to satisfy the initial and boundary conditions one must consider solutions of the form $\underset{\sim}{u}(\underset{\sim}{x},\omega) = \sum_{\iota,\sigma} \underset{\sim}{u}_{\iota\sigma}(\underset{\sim}{x},\underset{\sim}{\omega})$.

The Hamilton equations associated with (10)

$$\frac{d\underset{\sim}{x}}{ds} = \frac{\partial c_\sigma}{\partial \underset{\sim}{p}}, \quad \frac{d\underset{\sim}{p}}{ds} = -\frac{\partial c_\sigma}{\partial \underset{\sim}{x}} \tag{11}$$

with appropriate initial conditions (IC)

$$\underset{\sim}{x} = \underset{\sim}{x}^o(u,v), \quad \underset{\sim}{p} = \underset{\sim}{p}^o(u,v) \quad \text{at } s = 0 \tag{12}$$

yield the bicharacteristics of (1) (= characteristic strips of (10)):

$$\underset{\sim}{x} = \bar{\underset{\sim}{x}}(s,u,v), \quad \underset{\sim}{p} = \bar{\underset{\sim}{p}}(s,u,v) \tag{13}$$

Let

$$S_\sigma(\underset{\sim}{x}^o(u,v)) = s^o(u,v) \tag{14}$$

be the initial condition for eq. (10). We then require that

$$\sum_k p_k^o(u,v)\, dx_k^o(u,v) = ds^o(u,v), \quad c_\sigma(\underset{\sim}{x}^o(u,v), \underset{\sim}{p}^o(u,v)) = 1 \tag{15}$$

If the radiating surface $\underset{\sim}{x}^o(u,v)$ and the initial phase $s^o(u,v)$ are assigned, then $\underset{\sim}{p}^o(u,v)$ has to be determined from eqs. (15). Eqs. (15) are locally uniquely solvable for $\underset{\sim}{p}^o$ provided the transversality condition

$$\text{rank}\left[\frac{\partial c_\sigma}{\partial \underset{\sim}{p}}, \frac{\partial \underset{\sim}{x}^o}{\partial u}, \frac{\partial \underset{\sim}{x}^o}{\partial v}\right] = 3 \text{ at } (\underset{\sim}{x},\underset{\sim}{p}) = (\underset{\sim}{x}^o(u,v), \underset{\sim}{p}^o(u,v)) \tag{16}$$

is satisfied. Eq. (16) means that the rays issuing from $\underset{\sim}{x}^o(u,v)$ are not tangent to it. For $\underset{\sim}{p}^o(u,v)$ in eqs. (12) we shall then substitute any solutions $\underset{\sim}{p}^o_{(r)}(u,v)$ of (15).

On account of eqs. (10) and (11)

$$\frac{d}{ds} \nabla S_\sigma(\underset{\sim}{\bar{x}}(s,u,v)) = \nabla\nabla S_\sigma(\underset{\sim}{\bar{x}}(s,u,v)) \cdot \frac{\partial c_\sigma}{\partial \underset{\sim}{p}} = - \frac{\partial c_\sigma}{\partial \underset{\sim}{x}}$$

The vector $\nabla S_\sigma(x(0,u,v))$ also satisfies (15). By the uniqueness theorem for eqs. (11) we conclude that $\underset{\sim}{\bar{p}} = \nabla S_\sigma(x)$ (for simplicity we have dropped the index "r" on $\underset{\sim}{\bar{p}}$ and S_σ).

Conversely, it can be shown that eqs. (11) and (15_1) imply existence of such a function $S(\underset{\sim}{x})$ that $\underset{\sim}{\bar{p}} = \nabla S(\underset{\sim}{\bar{x}})$. Since $c_\sigma(\underset{\sim}{x},\underset{\sim}{p})$ is an integral of motion of eqs. (11), eq. (15_2) implies that S satisfies eq. (10). Hence we can use (11) to construct solutions of eq. (10).

We also note that

$$\frac{d}{ds} S_\sigma(\underset{\sim}{\bar{x}}(s,u,v)) = \sum_{k=1}^{3} \bar{p}_k \frac{dx_k}{ds} = \sum_{k=1}^{3} \bar{p}_k \frac{\partial c_\sigma}{\partial p_k} (\underset{\sim}{\bar{x}},\underset{\sim}{\bar{p}}) = c_\sigma(\underset{\sim}{\bar{x}},\underset{\sim}{\bar{p}}) = 1$$

on account of the Euler identities for the homogeneous function $c_\sigma(\underset{\sim}{x},\cdot)$. In view of (14)

$$S_\sigma(\underset{\sim}{x}(s,u,v)) = s_o(u,v) + s \tag{17}$$

From eq. (16) it follows that for every compact subset K of the surface $\underset{\sim}{x} = \underset{\sim}{x}^o(u,v)$ there is a number s(K) such that the Jacobian

$$J_\sigma = \frac{\partial(\bar{x}_1,\bar{x}_2,\bar{x}_3)}{\partial(s,u,v)} \equiv \frac{\partial \underset{\sim}{\bar{x}}}{\partial s} \cdot \left[\frac{\partial \underset{\sim}{\bar{x}}}{\partial u} \times \frac{\partial \underset{\sim}{\bar{x}}}{\partial v} \right] \tag{18}$$

is different from zero for s<s(K). Hence <u>for s<s(K)</u> we can invert the function $\underset{\sim}{\bar{x}}(s,u,v)$ and substitute the result in (17). We thus obtain a solution $S_\sigma(\underset{\sim}{x})$ of eqs. (10) and (14), defined in a nbhd of the radiating surface.

As a by-product we have also obtained a 2-parameter family $\underset{\sim}{\bar{x}}(s,u,v)$ of <u>rays</u> associated with the <u>eikonal</u> S_σ. The coordinates (s,u,v) will be referred to as <u>ray coordinates</u>. If s^o=const, then $\underset{\sim}{x}=\underset{\sim}{x}^o(u,v)$ is an initial wavefront. In this case the surfaces s=const coincide with the wavefronts S_σ=const and the vectors $\frac{\partial \underset{\sim}{\bar{x}}}{\partial u} \times \frac{\partial \underset{\sim}{\bar{x}}}{\partial v}$ are normal to the wavefronts. Hence J_σ ds du dv is the volume of the portion of a ray tube $\lfloor u,u+du \rfloor \times \lfloor v,v+dv \rfloor$ contained between two wavefronts s,s+ds.

Condition (16) is in general too restrictive, since it excludes point and line sources. More generally, eq. (15_1) and the condition

$$\operatorname{rank}\begin{bmatrix} \frac{\partial c_\sigma}{\partial p}(\underset{\sim}{x}^o,\underset{\sim}{p}^o) & \frac{\partial \underset{\sim}{x}^o}{\partial u} & \frac{\partial \underset{\sim}{x}^o}{\partial v} \\ -\frac{\partial c_\sigma}{\partial x}(\underset{\sim}{x}^o,\underset{\sim}{p}^o) & \frac{\partial \underset{\sim}{p}^o}{\partial u} & \frac{\partial \underset{\sim}{p}^o}{\partial v} \end{bmatrix} = 3 \tag{19}$$

imply that

$$\operatorname{rank}\begin{bmatrix} \frac{\partial \bar{\underset{\sim}{x}}}{\partial s} & \frac{\partial \bar{\underset{\sim}{x}}}{\partial u} & \frac{\partial \bar{\underset{\sim}{x}}}{\partial v} \\ \frac{\partial \bar{\underset{\sim}{p}}}{\partial s} & \frac{\partial \bar{\underset{\sim}{p}}}{\partial u} & \frac{\partial \bar{\underset{\sim}{p}}}{\partial v} \end{bmatrix} = 3 \tag{20}$$

for all (s,u,v) for which the functions (13) are defined ([5]). Hence eqs. (13) define a 3-dimensional surface Λ_σ in the six-dimensional manifold $\Omega \times R^3$ of $(\underset{\sim}{x},\underset{\sim}{p})$. The ray coordinates (s,u,v) are global coordinates on Λ_σ. On account of the identity $\bar{\underset{\sim}{p}} = \nabla S_\sigma(\bar{\underset{\sim}{x}})$ any part of Λ_σ such that $J_\sigma \neq 0$ is represented by the equation $p = \nabla S_\sigma(\underset{\sim}{x})$.

It can be proved that in a nbhd U of every point $(\underset{\sim}{x},\underset{\sim}{p}) \in \Lambda_\sigma$ one of the following Jacobians

$$J_\sigma;\ J_{\sigma,1} = \frac{\partial(\bar{p}_1,\bar{x}_2,\bar{x}_3)}{\partial(s,u,v)}\ ;\quad \tilde{J}_{\sigma,2} = \frac{\partial(\bar{x}_1,\bar{p}_2,\bar{x}_3)}{\partial(s,u,v)}\ ;$$

$$\ldots,\ \tilde{J}_{\sigma,12} = \frac{\partial(\bar{x}_1,\bar{x}_2,\bar{p}_3)}{\partial(s,u,v)} \tag{21}$$

etc. does not vanish ([5]). If $\tilde{J}_{\sigma,1} \neq 0$ in U, then (p_1,x_2,x_3) can be used as local coordinates on $U \subset \Lambda_\sigma$. Moreover, there is a "<u>generating function</u>" $\tilde{S}(p_1,x_2,x_3)$ such that

$$\frac{\partial \tilde{S}}{\partial p_1} = -x_1,\quad \frac{\partial \tilde{S}}{\partial x_2} = p_2,\quad \frac{\partial \tilde{S}}{\partial x_3} = p_3 \text{ for } (\underset{\sim}{x},\underset{\sim}{p}) \in U \subset \Lambda_\sigma \tag{22}$$

Setting $\tilde{\underset{\sim}{x}} = (p_1,x_2,x_3)$, $\tilde{\underset{\sim}{p}} = (-x_1,p_2,p_3)$ we can rewrite (23) in the form

$$\tilde{\underset{\sim}{p}} = \nabla\tilde{S}(\tilde{\underset{\sim}{x}}) \tag{23}$$

Eq. (23) is a local representation of Λ_σ.

Suppose now that U,V are two intersecting simply connected open subsets of Λ_σ, $J_\sigma \neq 0$ in U, $\tilde{J}_{\sigma,1} \neq 0$ in V and $S(\underset{\sim}{x})$ is a generating function of U. We then note that in the overlap $U \cap V$ x_1 can be expressed in terms of the local coordinates (p_1,x_2,x_3), $x_1 = \tilde{x}_1(p_1,x_2,x_3)$, say. We may define

$$\tilde{S}(p_1,x_2,x_3) = S(\tilde{x}_1(p_1,x_2,x_3),x_2,x_3) - p_1\tilde{x}_1(p_1,x_2,x_3) \tag{24}$$

The function $\tilde{S}$ satisfies eqs. (22) and hence is a generating function of $U \cap V$. A reader familiar with Hamiltonian mechanics will quickly discover that $(\underset{\sim}{x},\underset{\sim}{p}) \to (\tilde{\underset{\sim}{x}},\tilde{\underset{\sim}{p}})$ is a canonical transformation of a special kind. Eq. (24) is an associated Legendre transformation.

We know that $c_\sigma(\underset{\sim}{x},\underset{\sim}{p}) = 1$ on Λ_σ. If $S(\underset{\sim}{x})$ is a generating function of a subset U of Λ_σ then $c_\sigma(\underset{\sim}{x},\nabla S(\underset{\sim}{x})) = 1$. If we define $\tilde{c}_\sigma(\tilde{\underset{\sim}{x}},\tilde{\underset{\sim}{p}}) \equiv c_\sigma(\underset{\sim}{x},\underset{\sim}{p})$, then a generating function $\tilde{S}(p_1,x_2,x_3)$ satisfies the Hamilton-Jacobi equation $\tilde{c}_\sigma(\tilde{\underset{\sim}{x}},\nabla\tilde{S}(\tilde{\underset{\sim}{x}})) = 1$.

We finally come to the notion of caustics. The <u>singular set</u> Σ_σ of Λ_σ is defined by the equation $J_\sigma = 0$. Roughly speaking Σ_σ consists of all the curves and points at which the 3D surface Λ_σ folds over Ω. The projection of Σ_σ onto Ω consists of caustics. The fold connects two sheets of Λ_σ which project onto the same subset W of Ω. The generating functions $S_1(\underset{\sim}{x})$, $S_2(\underset{\sim}{x})$, $\underset{\sim}{x} \in W$, of the two sheets may correspond to the eikonals of direct waves and waves refracted at the caustics. An exhaustive list of all the possible folds of Λ_σ is provided by the catastrophe theory ([10]). The two most common cases are a simple fold projecting onto a simple caustic and a cusp catastrophe projecting onto two caustics meeting at a cusp.

3. DYNAMIC RAY TRACING IN REGULAR REGIONS

By a <u>regular region</u> of Λ_σ we mean a simply-connected open subset U of Λ_σ which does not intersect Σ_σ. Hence $J_\sigma \neq 0$ in U and we can associate with U a generating function $S(\underset{\sim}{x})$ such that U is given by the equation $\underset{\sim}{p} = \nabla S(\underset{\sim}{x})$. The procedure outlined in Sec. 2 yields a system of ODEs for the amplitudes $g_{1\sigma}^{(\nu)}$ along a ray $\hat{\underset{\sim}{x}}(s)$. In addition we derive an ordinary differential equation for J_σ along $\hat{\underset{\sim}{x}}(s)$. The number of equations can be significantly reduced by a recourse to the ray-centered coordinate system associated with $\hat{x}(s)$, which we now define after [3].

With this in view we define the projection operators

$$\tilde{P}^i{}_j(\underset{\sim}{x},\underset{\sim}{p}) = \delta^i{}_j - c_\sigma(\underset{\sim}{x},\underset{\sim}{p})^{-1}\; p_j \frac{\partial c_\sigma}{\partial p_i}(\underset{\sim}{x},\underset{\sim}{p}) \;,\; i,j=1,2,3 \qquad (25)$$

and note that

$$\sum_j \tilde{P}^i{}_j\, \tilde{P}^j{}_k = \tilde{P}^i{}_k,\; \sum_i p_i\, \tilde{P}^i{}_j(\underset{\sim}{x},\underset{\sim}{p}) = 0,\; \sum_j \tilde{P}^i{}_j(\underset{\sim}{x},\underset{\sim}{p}) \frac{\partial c_\sigma}{\partial p_j} = 0 \qquad (26)$$

on account of Euler identities.

For definiteness we assume that

$$\frac{\partial c_\sigma}{\partial p_3} \neq 0 \text{ along an arc } s_1<s<s_2 \text{ of } \hat{\underset{\sim}{x}}(s) \qquad (27)$$

with $\underset{\sim}{p} = \hat{\underset{\sim}{p}}(s) = \nabla S(\hat{\underset{\sim}{x}}(s))$. Note that $\sum \hat{p}_k(s)(\partial c_\sigma/\partial p_k)(\hat{\underset{\sim}{x}}(s),\hat{\underset{\sim}{p}}(s)) = c_\sigma(\hat{\underset{\sim}{x}}(s),\hat{\underset{\sim}{p}}(s))=1$ excludes the possibility that $\partial c_\sigma/\partial p_i=0$ at $\hat{\underset{\sim}{x}}(s,\hat{\underset{\sim}{p}}(s)$ for all i. We then define the <u>ray-centered coordinates</u> (τ,y_1,y_2) associated with $\hat{\underset{\sim}{x}}(s)$ by the formula

$$x_k = \hat{x}_k(s) + \sum_{a=1}^{2} P^k{}_a(s) y_a,\quad P^i{}_j(s) = \tilde{P}^i{}_j(\hat{\underset{\sim}{x}}(s),\hat{\underset{\sim}{p}}(s)), \qquad (28)$$

$$j,k = 1,2,3$$

It is easily seen that τ = const is a plane tangent to the wave-front $S(\underset{\sim}{x}) = \tau$ at its intersection with $\hat{\underset{\sim}{x}}(s)$.

We now define the auxiliary unknowns

$$S_{ab}(\tau) = \sum_{i,j} \bar{S}_{ij}\, P^i{}_a\, P^j{}_b,\quad \bar{S}_{ij} = \frac{\partial^2 S}{\partial x_i\, \partial s_j}(\hat{\underset{\sim}{x}}(\tau)) \;, \qquad (29)$$

$$a,b=1,2;\quad i,j=1,2,3$$

We also need the functions G^{ab}, $w^a{}_i$, a,b=1,2; i=1,2,3, defined by the formulae

$$\frac{\partial^2 c_\sigma}{\partial p_k\, \partial p_l} = \sum_{a,b=1}^{2} G^{ab}\, p^k{}_a\, P^l{}_b,\quad \frac{\partial^2 c_\sigma}{\partial p_k\, \partial x_l} = \sum_{a=1}^{2} w^a{}_l\, P^k{}_a + \frac{\partial c_\sigma}{\partial x_l}\frac{\partial c_\sigma}{\partial p_k} \qquad (30)$$

These formulae follow from the fact that the vectors $P^i{}_1, P^i{}_2$ and $\partial c_\sigma/\partial p_i$ are linearly independent on account of (27) and from the Euler identities.

Using the identity $d \log \det A/ds = \mathrm{tr}(\dot{A} A^{-1})$ as well as eqs. (11), (30) we obtain the equation

$$\frac{d \log J_\sigma}{d\tau} = \sum_{k=1}^{3} \frac{\partial^2 c_\sigma}{\partial p_k \, \partial x_k} + \sum_{a,b=1}^{2} G^{ab}(\tau)\, S_{ab}(\tau) \tag{31}$$

Let us embed $\hat{x}(s)$ in a field of rays $\bar{x}(s,u,v)$ in such a way that $\bar{p}(s,u,v) = \nabla S(\bar{x}(s,u,v))$, $\hat{x}(s) = \bar{x}(s,0,0)$. Expanding the identity $c_\sigma(\bar{x}(s,u,v), \bar{p}(s,u,v)) = 1$ to second order in (u,v) and using eqs. (29), (30) we derive the equation

$$\frac{d S_{ab}}{d\tau} = -2 \sum_{c=1}^{2} w^c{}_{(a} S_{b)c} - \sum_{c,d=1}^{2} G^{cd} S_{ac} S_{bd} + \tag{32}$$
$$+ 2 \sum_{c,d}^{2} P_{(a} S_{b)c} G^{cd} H_d - H_{ab}, \quad a,b=1,2,$$

where

$$H_{ab} = \sum_{i,j}^{3} \frac{\partial^2 c_\sigma}{\partial x_i \, \partial x_j} P^i{}_a P^j{}_b, \quad H_a = \sum_{i}^{3} \frac{\partial c_\sigma}{\partial x_i} P^i{}_a, \tag{33}$$

and $T_{(ab)} = \frac{1}{2}(T_{ab} + T_{ba})$ generically.

Eqs. (1), (31), (32) form a closed system of 10 equations which allows to determined $\hat{x}(s), \hat{p}(s)$ and $J_\sigma(\hat{x}(s))$ provided the initial data $\hat{x}(0), \hat{p}(0)$, $J_\sigma(\hat{x}(0))$ are prescribed.

Derivation of the amplitude equations

$$\frac{d\, g_{1\sigma}{}^{(\nu)}}{d\tau} = \frac{1}{2} \left[\sum_k \frac{\partial c_\sigma}{\partial x_k} \frac{\partial c_\sigma}{\partial p_k} + \frac{\partial^2 c_\sigma}{\partial x_k \, \partial p_k} \right] g_{1\sigma}{}^{(\nu)} +$$
$$+ \sum_{\mu=1}^{k_\sigma} \left\{ \sum \frac{\partial c_\sigma}{\partial x_1} \left\langle r_\sigma^{(\nu)} \,\middle|\, \frac{\partial r_\sigma^{(\mu)}}{\partial p_1} \right\rangle - \right.$$
$$- \frac{1}{\rho} \sum_{k,1} \left[\left\langle r_\sigma^{(\nu)} \,\middle|\, B^{(k1)} p_1 \frac{\partial r_\sigma^{(\mu)}}{\partial x_k} \right\rangle - \right.$$

$$- \frac{1}{2} \langle \underset{\sim}{r}_\sigma^{(\nu)} \mid \frac{\partial \underset{\sim}{B}^{(kl)}}{\partial x_k} p_l \underset{\sim}{r}_\sigma^{(\mu)} \rangle]\} g_\sigma^{(\mu)}, \quad \nu=1,\ldots,k_\sigma \tag{34}$$

is rather lengthy ([4]). By $\langle \underset{\sim}{r} | \underset{\sim}{s} \rangle$ we denote the scalar product $\sum_k r_k, s_k$, while $\underset{\sim}{B}^{kl}$ is the 3×3 matrix $[B^{kl}_{rs}]$.

If condition (27) fails to be satisfied for $s>s_2$ or $s<s_1$, one shifts to an appropriate ray-centered coordinate system using formulae (57-59) and (29).

4. WAVEFIELD ASSOCIATED WITH A SINGULAR REGION OF Λ_σ

By a <u>singular region</u> of Λ_σ we mean simply connected open subset U of Λ_σ which intersects Σ_σ and in which one of the Jacobians (21) does not vanish.

Let U be a singular region of Λ_σ such that $\tilde{J}_{\sigma,1} \neq 0$ in U. Let V be an open simply connected subset of U such that $J_\sigma > 0$ in V and let $S(\underset{\sim}{x})$ be its generating function. Let ϕ be a smooth cut-off function such that $\phi = 1$ in a closed subset K of V and $\phi = 0$ outside V. We now apply the ω-<u>Fourier transform</u> to a WKBJ expression

$$\underset{\sim}{u}(x_1,x_2,x_3) = e^{i\omega S(\underset{\sim}{x})} g(x)\, J_\sigma(\underset{\sim}{x})^{-\frac{1}{2}}\, \underset{\sim}{r}(\underset{\sim}{x}, \nabla S(\underset{\sim}{x})), \quad \underset{\sim}{x} \in V \tag{35}$$

(cf. (9)), multiplied by ϕ:

$$\underset{\sim}{v}(k_1,x_2x_3) = \left(\frac{\omega}{2\pi}\right)^{\frac{1}{2}} \int_{-\infty}^{\infty} dx_1\, e^{-i\omega k_1 x_1} \phi(\underset{\sim}{x})\, \underset{\sim}{u}(x_1,x_2,x_3) \tag{36}$$

The fact that we transform the first coordinate x_1 only is related to the assumption that $J_{\sigma,1} \neq 0$ in U.

The inverse ω-Fourier transform yields

$$\phi(\underset{\sim}{x})\, \underset{\sim}{u}(x_1,x_2,x_3) = \left(\frac{\omega}{2\pi}\right)^{\frac{1}{2}} \int_{-\infty}^{\infty} dk_1\, e^{i\omega k_1 x_1} \underset{\sim}{v}(k_1,x_2,x_3) \equiv$$
$$\equiv (F_{\omega,1}\, \underset{\sim}{v})(\underset{\sim}{x}) \tag{37}$$

which coincides with $\underset{\sim}{u}$ in K.

We now evaluate (36) asymptotically applying the stationary phase formula ([2]). Stationary points are given by the equation

$$\frac{\partial S}{\partial x_1}(x_1,x_2,x_3) - k_1 = 0 \tag{38}$$

for each fixed value of k_1. The solutions of (38) in V are given by (22), viz. $x_1 = x_1^{(o)} = (-\partial\tilde{S}/\partial p_1)(k_1,x_2,x_3)$.

Suppose that $\underset{\sim}{x}^{(o)} = (x_1^{(o)},x_2,x_3) \in K$. On account of the identities

$$\begin{aligned} J_{\sigma,1} &= \frac{\partial(\bar{p}_1,\bar{x}_2,\bar{x}_3)}{\partial(s,u,v)} = \frac{\partial\tilde{p}_1}{\partial x_1}(x_1,x_2,x_3)\, J_\sigma = \\ &= \frac{\partial^2 S}{\partial x_1^{\,2}} J_\sigma = -\left(\frac{\partial^2\tilde{S}}{\partial p_1^{\,2}}\right)^{-1} J_\sigma \end{aligned} \tag{39}$$

and (22), (24) we then have

$$\underset{\sim}{v}(\tilde{\underset{\sim}{x}}) \sim e^{i\omega\tilde{S}(\tilde{\underset{\sim}{x}})} \mid \tilde{J}_{\sigma,1}(\tilde{\underset{\sim}{x}}) \mid^{-\frac{1}{2}} \tilde{g}(\tilde{\underset{\sim}{x}})\, \tilde{\underset{\sim}{r}}(\tilde{\underset{\sim}{x}},\nabla S(\tilde{\underset{\sim}{x}})) \tag{40}$$

with $\tilde{\underset{\sim}{x}} = (p_1,x_2,x_3)$, $\tilde{\underset{\sim}{p}} = (-x_1,p_2,p_3)$, $\tilde{\underset{\sim}{r}}(\tilde{\underset{\sim}{x}},\tilde{\underset{\sim}{p}}) \equiv r(\underset{\sim}{x},\underset{\sim}{p})$ and

$$\tilde{g}(\tilde{\underset{\sim}{x}}) = \exp\left[\frac{i\pi}{4}\,\mathrm{sgn}\left(\omega\frac{\partial^2 S}{\partial x_1^{\,2}}\right)\right] g(\underset{\sim}{x}) \text{ provided } x_1 = -\frac{\partial\tilde{S}}{\partial p_1}(\tilde{\underset{\sim}{x}}) \tag{41}$$

We now note that $\tilde{S}$, and hence $\underset{\sim}{v}(\tilde{\underset{\sim}{x}})$, is well defined and non-singular on U. In view of this it is tempting to check whether the expression

$$\tilde{\underset{\sim}{u}} = F_{\omega,1}\left[e^{i\omega\tilde{S}}\left(\underset{\sim}{v}_0 + \frac{1}{i\omega}\underset{\sim}{v}_1 + \ldots\right)\right] \tag{42}$$

with $\underset{\sim}{v}_0$ given by a sum of expressions (40) satisfies eq. (1) in an asymptotic sense. For this purpose we need the commutation formula ([9]):

$$L\, F_{\omega,1}(e^{i\omega\tilde{S}}\,\underset{\sim}{w}) = F_{\omega,1}\left[e^{i\omega\tilde{S}}\left(R_0\underset{\sim}{w} + \frac{1}{i\omega}R_1\underset{\sim}{w} + \ldots\right)\right] , \tag{43}$$

where

$$(L\underset{\sim}{w})_k = \frac{1}{\rho\omega^2}\sum_{l,r,s}\frac{\partial}{\partial x_r}\left(B_{kl}^{rs}\frac{\partial w_l}{\partial x_s}\right) + w_k , \quad k=1,2,3 , \tag{44}$$

$$\Sigma_{kl}(\underset{\sim}{x},\underset{\sim}{p};\lambda) = -\frac{1}{\rho}\sum_{r,s} B^{rs}_{kl}\, p_r p_s - \frac{\lambda}{\rho}\sum_{r,s}\frac{\partial B^{rs}_{kl}}{\partial x_r}\, p_s + \tag{45}$$
$$+ \delta_{kl} \equiv \tilde{\Sigma}_{kl}(\tilde{\underset{\sim}{x}},\tilde{\underset{\sim}{p}};\lambda)$$

is the symbol of the operator L, obtained by substituting $\underset{\sim}{p}$ for $(i\omega)^{-1}\nabla$, λ for $(i\omega)^{-1}$ in (44),

$$R_0 \underset{\sim}{w} = \tilde{\underset{\sim}{\Sigma}}(\tilde{\underset{\sim}{x}}, \nabla\tilde{S}(\tilde{\underset{\sim}{x}});0)\underset{\sim}{w} = \Sigma(-\frac{\partial\tilde{S}}{\partial p_1}, x_2x_3, p_1, \frac{\partial\tilde{S}}{\partial x_2}, \frac{\partial\tilde{S}}{\partial x_3})\underset{\sim}{w} \tag{46}$$

and R_1 is written out in full in |6|.

It follows immediately that $L\tilde{\underset{\sim}{u}} = O(\omega^{-2})$ provided

$$R_0\, \underset{\sim}{v}_0 = 0 \tag{47}$$

and

$$R_1\, \underset{\sim}{v}_0 + R_0\, \underset{\sim}{v}_1 = 0 \tag{48}$$

Eq. (47) implies that $\tilde{S} = \tilde{S}_\sigma$ satisfies the Hamilton-Jacobi equation

$$\tilde{c}_\sigma(\tilde{\underset{\sim}{x}}, \nabla\tilde{S}_\sigma(\tilde{\underset{\sim}{x}})) = 1 \tag{49}$$

(unless $\underset{\sim}{v}_0 \equiv 0$), while

$$\underset{\sim}{v}_0 = \sum_{\nu=1}^{k_r} |\tilde{J}_{\sigma,1}|^{-1}\, \tilde{g}_\sigma^{(\nu)}\, \tilde{\underset{\sim}{r}}_\sigma^{(\nu)}(\tilde{\underset{\sim}{x}}, \nabla\tilde{S}_\sigma, \tilde{\underset{\sim}{x}})) \tag{50}$$

From (48) we deduce that $\langle\tilde{\underset{\sim}{r}}_\sigma^{(\mu)}|R_1\, \underset{\sim}{v}_0\rangle = 0$, whence the amplitude equations for the functions $\tilde{g}_\sigma^{(\nu)}$ follow. The amplitude equations turn out to be merely a result of transforming (34) to the coordinates $(\tilde{\underset{\sim}{x}},\tilde{\underset{\sim}{p}})$. In particular, they coincide with eqs. (34) on V. We note that $\underset{\sim}{v}_0$ is defined only for $\underset{\sim}{x} = (k_1,x_2,x_3) \in U$, while the integral (42) extends over all real k_1. Nevertheless the values of $\underset{\sim}{v}_0$ at points distant from Σ_σ are asymptotically irrelevant for the values of $\underset{\sim}{u}$ near the caustic.

For $\tilde{J}_{\sigma,1}$ we derive as usual the equation

$$\frac{d \log \tilde{J}_{\sigma,1}}{d\tau} = \sum_{k=1}^{3} \frac{\partial^2\tilde{c}_\sigma}{\partial\tilde{x}_k\, \partial\tilde{p}_k} + \sum_{k,l} \frac{\partial^2\tilde{c}_\sigma}{\partial\tilde{p}_k\, \partial\tilde{p}_l}\, \frac{\partial^2\tilde{S}_\sigma}{\partial\tilde{x}_k\, \partial\tilde{x}_l} \tag{51}$$

which has to be supplemented by the equations

$$\frac{d\tilde{S}_{ij}}{d\tau} = - \frac{\partial^2 \tilde{c}_\sigma}{\partial \tilde{x}_i \, \partial \tilde{x}_j} - 2 \sum_k \frac{\partial^2 \tilde{c}_\sigma}{\partial \tilde{p}_k \, \partial \tilde{x}_{(i}} \tilde{S}_{j)k} - \sum_{k,l} \frac{\partial^2 \tilde{c}_\sigma}{\partial \tilde{p}_k \, \partial \tilde{p}_l} \tilde{S}_{ik} \tilde{S}_{jl} \tag{52}$$

$$i,j=1,2,3$$

for the auxiliary functions $\tilde{S}_{ij} = \partial^2\tilde{S}/\partial\tilde{x}_i \, \partial\tilde{x}_j$.

The DRT system (11), (51), (52) and (34) consists of $13+k_\sigma$ equations. It cannot be reduced by the recourse to a ray-centered coordinate system because $\tilde{c}_\sigma(\tilde{x},\tilde{p})$ is not a homogeneous function of $\tilde{p}$.

5. TRANSITION FORMULAE

Let $\tilde{u}$ be an asymptotic solution of eq. (1) given by eqs. (42) and (50), with the integration in (42) extending over such k_1 that $(k_1,x_2,x_3) \in U$. Let V_r, $r=1,\ldots,N$, be N regular regions of Λ_σ such that $V_r \cap U \neq \emptyset$ for $r=1,\ldots,N$. The regions V_r may lie on different sheets of Λ_σ over Ω. Let the generating functions of U and V_r be S and S_r, respectively.

The stationary points of the phase of (42) satisfy the equation

$$\frac{\partial\tilde{S}}{\partial p_1}(k_1,x_2,x_3) + x_1 = 0 \tag{53}$$

The solution of (53) in $U \cap V_r$ is given by the formula

$$k_1^{(r)} = \frac{\partial S_r}{\partial x_1}(x_1,x_2,x_3) \tag{54}$$

Choose U in such a way that for given x the stationary points (54) do not lie on ∂U. This requirement is met if x does not lie on the projection of ∂U onto Ω. In this case the contribution of the stationary points to $\tilde{u}$ is

$$\sum_r | J_\sigma{}^{(r)}(x)|^{-\frac{1}{2}} e^{i\omega S_r(x)} \exp\left[-\frac{i\pi}{4}\,\mathrm{sgn}\left(\omega \frac{\partial^2 S_r}{\partial x_1{}^2}\right)\right] \times \left[v_0(\tilde{x}^{(r)})+\ldots\right] \tag{55}$$

with $\tilde{x}^{(r)} = (k_1^{(r)},x_2,x_3)$, on account of the identities $S_r = \tilde{S}+k_1^{(r)}x_1$ and $J_\sigma{}^{(r)} = \tilde{J}_{\sigma,1}\left(\frac{\partial^2 S_r}{\partial x_1{}^2}\right)^{-1}$. The contribution of ∂U to $\underset{\sim}{u}$ is $O(\omega^{-\frac{1}{2}})$ (cf. [1]). Hence $\underset{\sim}{u} \sim \sum_{r=1}^{N} u_r$ with

$$\underset{\sim}{u}_r = | J_\sigma^{(r)} |^{-\frac{1}{2}} \sum_{\nu} g_{\sigma,r}^{(\nu)} e^{i\omega S_r} \underset{\sim}{r}_\sigma^{(\nu)}(\underset{\sim}{x}, \nabla S_r(\underset{\sim}{x})),$$

$$g_{\sigma,r}^{(\nu)}(\underset{\sim}{x}) = \exp\left[- \frac{i\pi}{4} \operatorname{sgn}(\omega \frac{\partial^2 S_r}{\partial x_1{}^2})\right] g_\sigma^{(\nu)}(\underset{\sim}{x}^{(r)}) \tag{56}$$

Note that N=2 for a simple fold and N=3 near the cusp of a cusp catastrophe. Phase jump at caustics is implicit in (41) and (56).

Eqs. (41), (56) and (39) provide the transition formulae for the amplitudes and for the ray spreading factors at those points of $\hat{\underset{\sim}{x}}(s)$ where the computing program shifts from eqs. (31), (32) to (51) and (52) and vice versa. For the auxiliary functions S_{ab}, $\bar{S}_{ij}$, $\tilde{S}_{ij}$ we have

$$\bar{S}_{ab} = S_{ab} + \hat{p}_a \hat{p}_b \sum_{i=1}^{3} \frac{\partial \widehat{c_\sigma}}{\partial x_i} \frac{\partial \widehat{c_\sigma}}{\partial p_i} - \frac{\partial \widehat{c_\sigma}}{\partial x_a} \hat{p}_b - \frac{\partial \widehat{c_\sigma}}{\partial x_b} \hat{p}_a, \quad a,b=1,2 \tag{57}$$

$$\bar{S}_{a3} = - \frac{\partial \widehat{c\sigma}}{\partial x_a} \hat{p}_3 - \frac{\partial \widehat{c_\sigma}}{\partial x_3} \hat{p}_a + \hat{p}_a \hat{p}_3 \sum_{i=1}^{3} \frac{\partial \widehat{c_\sigma}}{\partial x_i} \frac{\partial \widehat{c_\sigma}}{\partial p_i} - \hat{p}_3 \sum_{b,d=1}^{2} S_{ab} Q^b{}_d \frac{\partial \widehat{c_\sigma}}{\partial p_d}, \quad a=1,2 \tag{58}$$

$$\bar{S}_{33} = (\hat{p}_3)^2 \left\{ \sum_{i=1}^{2} \frac{\partial \widehat{c_\sigma}}{\partial x_i} \frac{\partial \widehat{c_\sigma}}{\partial p_i} + \sum_{c,d,a,b=1}^{2} S_{ab} Q^a{}_c Q^b{}_d \frac{\partial \widehat{c_\sigma}}{\partial p_c} \frac{\partial \widehat{c_\sigma}}{\partial p_d} \right\} - 2 \hat{p}_3 \frac{\partial \widehat{c_\sigma}}{\partial p_3} \tag{59}$$

$$Q^a{}_b = \frac{1}{\hat{p}_3 \widehat{(\partial c_\sigma / \partial p_3)}} \hat{p}_b \frac{\partial \widehat{c_\sigma}}{\partial p_a} \text{ if } a \neq b, \; a,b=1,2;$$

$$Q^1{}_1 = \frac{1}{\hat{p}_e \widehat{(\partial c_\sigma / \partial p_3)}} (1 - \hat{p}_2 \frac{\partial \widehat{c_\sigma}}{\partial p_2}), \tag{60}$$

$$\tilde{S}_{11} = -(\bar{S}_{11})^{-1},\ \tilde{S}_{1\alpha} = \bar{S}_{1\alpha}(\bar{S}_{11})^{-1},\ \tilde{S}_{\alpha\beta} = \bar{S}_{\alpha\beta} - \bar{S}_{\alpha 1}\bar{S}_{\beta 1}(\bar{S}_{11})^{-1}, \quad (61)$$

$$\bar{S}_{1\alpha} = -\tilde{S}_{1\alpha}(\tilde{S}_{11})^{-1},\ \bar{S}_{\alpha\beta} = \tilde{S}_{\alpha\beta} - \tilde{S}_{\alpha 1}\tilde{S}_{\beta 1}(\tilde{S}_{11})^{-1},\ \alpha,\beta=1,2 \quad (62)$$

([6]). The hats in (57)-(59) denote the substitution $\underset{\sim}{x}=\hat{\underset{\sim}{x}}(\tau)$, $\underset{\sim}{p}=\hat{\underset{\sim}{p}}(\tau)$.

6. INITIAL DATA FOR A POINT SOURCE AND FINAL COMMENTS

Initial data for DRT systems for various kinds of sources can be found in [6]. Here we restrict our attention to a point source corresponding to the load $\underset{\sim}{f} = e^{i\omega t}\delta(\underset{\sim}{x}-\underset{\sim}{x}_o)\underset{\sim}{s}$.

The initial conditions (IC) for eqs. (11) are $\bar{\underset{\sim}{x}}(0,\theta,\phi)=\underset{\sim}{x}_o$, $\bar{\underset{\sim}{p}}(0,\theta,\phi)=c_\sigma(\underset{\sim}{x}_o,\underset{\sim}{n}(\theta,\phi))^{-1}\underset{\sim}{n}(\theta,\phi)$, where $\underset{\sim}{n}(\theta,\phi)$ is a unit vector specified by the latitude θ and longitude ϕ. It is esy to show that condition (20) is satisfied. The point $\underset{\sim}{x}_o$ is a caustic at which Λ_σ may branch into m_σ sheets very much like the Riemann surface of $(z)^{1/m}$.

The other IC are obtained by comparison with the asymptotic solutin for the same source in a homogeneous medium with elasticities $B^{rs}_{kl}(\underset{\sim}{x}_o)$ (cf. [7]). For the latter solution we have the ray equations $\bar{\underset{\sim}{x}} = \underset{\sim}{x}_o + s(\partial c_\sigma/\partial\underset{\sim}{p})(\underset{\sim}{x}_o,\underset{\sim}{n}(\theta,\phi))$ and therefore

$$J_\sigma(\tau,\theta,\phi) = \tau^2 \det\Big[\sum_{i,j}^{3} \frac{\partial n_i}{\partial u_a}\frac{\partial^2 c_\sigma}{\partial p_i\,\partial p_j}(\underset{\sim}{x}_o,\underset{\sim}{n}(\theta,\phi))\frac{\partial n_j}{\partial u_b}\Big],$$
$$a,b=1,2,\quad u_1=\theta,\quad u_2=\phi \quad (63)$$

In terms of the notations introduced in Sec. 4 of [7] we have

$$S_\sigma^{(r)} = \frac{\underset{\sim}{n}^{(r)}(\underset{\sim}{x})\cdot(\underset{\sim}{x}-\underset{\sim}{x}_o)}{c_\sigma(\underset{\sim}{x}_o,\underset{\sim}{n}^{(r)}(\underset{\sim}{x}))},\quad r=1,\dots,m_\sigma \quad (64)$$

$$g_{\sigma,r}^{(\nu)}(0,\theta,\phi) = \frac{i}{4\pi\rho(\underset{\sim}{x}_o)}\,\frac{1}{c_\sigma(\underset{\sim}{x}_o,\underset{\sim}{n}(\theta,\phi))}\exp\!\Big(-\frac{i\pi}{4}\,\mathrm{sgn}\,B_r\Big)\ \times$$
$$\times\ \langle\underset{\sim}{r}_\sigma^{(\nu)}(\underset{\sim}{x}_o,\underset{\sim}{n}(\theta,\phi))\ |\ \underset{\sim}{s}\rangle,\quad g_{-\sigma,r}^{(\nu)}(0,\theta,\phi)=0 \quad (65)$$
$$\text{for } \sigma=1,2,\dots;\quad \nu=1,\dots,k_\sigma;\quad r=1,\dots,m_\sigma$$

In general $S_\sigma^{(r)}$ need not be differentiable at some rays (wavefronts may exhibit cusps or cuspidal edges). For any ray which does not start off in such a singular direction the initial values of $\bar{S}_{ij}$ can be derived by differentiating (64). Note that $\bar{S}_{ij}(\tau)=O(\tau)$ for $\tau\to 0$.

Boundary conditions at interfaces are discussed in [6].

For numerical applications it is important to note that the DRT equations in both regular and singular regions do not involve frequency and hence integration of the DRT yields $\underset{\sim}{u}(\underset{\sim}{x}_1;\omega)$ for all the frequencies at once. For a receiver situated at a distance from the caustics it is enough to integrate DRT equations along rays which exactly hit it. For a receiver situated near a caustic eq. (42) should be used to evaluate $\underset{\sim}{u}(\underset{\sim}{x}_1;\omega)$ and hence DRT has to be carried out along a bunch of rays covering a nbhd of $\underset{\sim}{x}_1$. Catastrophe theory yields all the 7 possible functions S up to coordinate transformations; hence 7 possible "transition functions" (42) can be derived, starting with the Airy and Pearcey functions for the simplest caustics.

REFERENCES

1. Fedoryuk, M.V. The Saddle Point Method (Moscow, Nauka, 1977).

2. Felsen, L.B. and N. Marcuvitz. Radiation and Scattering of Waves (Englewood Cliffs, Prentice Hall, 1973).

3. Hanyga, A. Dynamic Ray Tracing in Anisotropic Media. Tectonophysics 90 (1982), 243-251.

4. Hanyga, A., E. Lenartowicz, J. Pajchel. Seismic Waves in the Earth (Amsterdam, Elsevier, 1984).

5. Hanyga, A. Dynamic Ray Tracing in the Presence of Caustics, I: Basic Geometry of Lagrangian Manifolds. Acta Geophs. Polonica, to appear in 1984.

6. Hanyga, A. Dynamic Ray Tracing in the Presence of Caustics, II: Equations. Acta Geophys. Polonica, to appear in 1984.

7. Hanyga, A. Gaussian Beams in Anisotropic Elastic Media (this volume).

8. Hanyga, A. Mathematical Theory of Non-Linear Elasticity (Chichester, Ellis Horwood Ltd, 1984).

9. Maslov, V.B. Théorie des perturbations et méthodes asymptotiques (Paris, Dunod and Gauthier-Villars, 1972).

10. Saunders, P.T. An Introduction to Catastrophe Theory (Cambridge, Cambridge University Press, 1980).

11. Wilcox. D. Wave Operators and Asymptotic Solutions of Wave Propagation Problems of Classical Physics. Arch. Ratl. Mech. Anal. 22(1966), 37-78.

COMPLEX RAYS

G.Ghione, I.Montrosset, R.Orta

CESPA (CNR) - Dipartimento di Elettronica, Politecnico di Torino
Corso Duca degli Abruzzi 24, 10129 TORINO (ITALY)

1 INTRODUCTION

Since it has been noted, that a Gaussian beam can be represented in terms of a "bundle of complex rays" [1,2] (that is, as the field radiated by a point source of complex location (CSP)), complex optics and related techniques have undergone great development, with a variety of applications ranging from antenna systems [3,4,5,6] and radomes [7] to integrated optics [8]. After it was recognized that an elementary source displaced in complex space radiates a beam-like field [9], the problem arose of tracing rays from this complex source, in order to assess the behaviour of the beam thereby generated in the presence of a scatterer [10]. Although both complex geometrical optics (Complex Ray Tracing, CRT) and real optics are formally based upon the same equations, here the eikonal function, amplitude and space coordinates are allowed to take on complex values. The mathematical foundation of this method is the process of analytic continuation. Owing to the intricacies arising when one has to trace complex rays from a complex source through a system of lenses or reflectors, a new interpretation of complex optics was suggested, dealing with real quantities in real space only (Evanescent Wave Tracking,EWT) [11]. In spite of its attractive features, EWT gives rise - as a boundary value problem - to serious mathematical difficulties [3,12], which are absent from the first-order complex formulation of CRT. Practical applications of CRT to antenna problems have been concerned so far with two-dimensional or three-dimensional reflectors illuminated by a Gaussian beam, whose field is simulated by a complex line or point source . Also, problems of propagation in free space from an assigned initial field distribution have been extensively dealt with

[3,5,12]. In order to assess the behaviour of quasi-optical systems wherein diffraction effects are relevant, it is also important to extend the Geometrical Theory of Diffraction (GTD) [13] to the case where the slope of the incident ray is complex. This can be done by solving the canonical problem of a half plane illuminated by a complex point or line source [14]. In the present work we shall review some of the basic aspects related to CRT: Section 2 is devoted to complex geometrical optics, while Sections 3 and 4 are concerned with applications to free-space propagation from an assigned aperture distribution, and to the treatment of reflectors, respectively.

2 COMPLEX OPTICS

Complex Optics arises as an extension to complex coordinate space of Real Optics. We will discuss it by reviewing the steps leading from the wave equation to the ray picture of the propagation phenomenon, again limiting ourselves for simplicity to the two-dimensional case.

2.1 Eikonal and transport equations.

Complex geometrical optics deals with the asymptotic solution of the (scalar) wave equation

$$(\nabla^2 + n^2k^2)\, u(\underline{\rho}) = 0 \qquad \nabla^2 = \partial^2/\partial x^2 + \partial^2/\partial y^2 \tag{1}$$

for large wavenumber k . The position vector $\underline{\rho}=(x,y)$ is assumed to be complex. Its solution can be written in the form

$$u(\underline{\rho}) = A(\underline{\rho})\, \exp\{ikS(\underline{\rho})\}\left(1 + O(k^{-1})\right) \tag{2}$$

where $A(\underline{\rho})$ and $S(\underline{\rho})$ are complex functions and represent "amplitude" and "phase" of the wave function. On substituting this expression into the wave equation we obtain (to the lowest orders in 1/k) the eikonal equation for the phase :

$$(\nabla S)^2 = n^2 \tag{3a}$$

and the transport equation for the amplitude

$$2\, \nabla S \cdot \nabla A + A\, \nabla^2 S = 0 \tag{3b}$$

The above differential equations can be solved by the method of characteristics. We introduce a curvilinear coordinate s along the trajectory and a unit vector $\hat{s}$ tangent to it. The eikonal and trajectory equations become

$$\nabla S = n\hat{s} \qquad\qquad d/ds(n\, d\underline{\rho}/ds) = \nabla n \tag{4}$$

The phase can be now obtained by integration along the ray:

$$S(\underline{\rho}) = S(\underline{\rho}_0) + \int_{\underline{\rho}_0}^{\underline{\rho}} n\, ds \tag{5}$$

where $\underline{\rho}_0$ is the ray departure point on the initial surface. The field amplitude can be computed from the transport equation and is given by

$$A(\underline{\rho}) = A(\underline{\rho}_0)\left[\frac{n(\underline{\rho}_0)\ d\sigma(\underline{\rho})}{n(\underline{\rho})\ d\sigma(\underline{\rho}_0)}\right]^{\frac{1}{2}} = A(\underline{\rho}_0)\left[\frac{n(\underline{\rho}_0)\ J(\underline{\rho}_0)}{n(\underline{\rho})\ J(\underline{\rho})}\right]^{\frac{1}{2}} \tag{6}$$

where $d\sigma(\underline{\rho})$ is the differential cross section of the ray tube and $J(\underline{\rho})$ is the Jacobian of the transformation from rectangular to ray coordinates (see Fig.1). This equation is valid as long as $J(\underline{\rho})\neq 0$: the set of observation points C for which $J(\underline{\rho}_c)=0$ is called "caustic". The ray-optical field is singular at the caustic but also, in a neighbourhood of it, the approximation is poor and (2) should be replaced by a uniform transition function. In real optics, the simplest type of caustic (fold) is a line that separates the observation plane into two regions, where there are two or no ray contributions. Actually, the field on the "dark side of the caustic" is not zero, but has the form of an evanescent wave, and therefore can be described in terms of a complex ray. In other words, complex optics yields a closer approximation to the true field in the shadow region. Moreover, the complex caustic may have no intersections with real space: in that case the ray optical field is always finite for all (real) observation points. However, the approximation becomes poor for those regions of real space that are close to a "slightly" complex caustic.

2.2 Initial value problem.

In many practical problems, the field is known on a real initial line γ (aperture) and the field at a generic observation point P is required (see Fig.2). Let us suppose that the medium is homogeneous and the aperture field is given in the form

$$u_0(l) = A_0(l)\exp\{ikS_0(l)\} \tag{7}$$

where l is a curvilinear coordinate along the initial line γ . In order to apply complex geometrical optics, one has to analytically continue the aperture distribution to complex l values. Let Q(l) be a point of γ . The outgoing (complex) ray leaving Q has a unit vector $\hat{s}$

$$\hat{s} = \cos\beta_0 \hat{t} + \sin\beta_0 \hat{\nu} \tag{8}$$

where $\hat{t}$ and $\hat{\nu}$ are unit vectors respectively tangent and normal to γ and

$$\cos\beta_0(l) = dS_0(l)/dl = \dot{S}_0(l) \tag{9}$$

A point P on the ray is then given by

$$\underline{P} = \underline{Q}(l) + s_{QP}\, \hat{s}(l) \tag{10}$$

The complex caustic associated with the initial distribution can be obtained as follows. The distance s_{QC} between a point Q on the aperture and the corresponding point C on the caustic is given by

$$s_{QC}(l) = -(1 - \dot{S}_0^2(l))/ \ddot{S}\,(l) \tag{11}$$

and therefore the points of the caustic are determined by

$$\underline{C}(\sigma) = \underline{Q}(l) + s_{QC}(l)\, \hat{s}(l) \tag{12}$$

where σ is the curvilinear coordinate along the caustic. In terms of these results, the field computation at point P proceeds as follows: 1) determination of point Q by means of (10); 2) computation of phase and amplitude of the field according to

$$S(\underline{P}) = S_0(l) + s_{QP} \qquad A(\underline{P}) = A_0(l)\{s_{QC}/(s_{QC} - s_{QP})\}^{\frac{1}{2}} \tag{13}$$

Concerning point 1), one has that for each observation point P, there is a set of points Q_i that satisfy (9). However, not all of the corresponding rays are to be taken into account in the field computation. One is then faced with a selection problem, to which a solution will be given in Section 2.3.

2.3 Green's function approach.

The initial value problem discussed in the previous sections can be approached along a completely different route, by constructing an exact integral representation of the field at the observation point P=(x,y). Assuming for simplicity the aperture to be the y-axis, the field is given by

$$u(x,y) = \frac{ikx}{2}\int_{-\infty}^{+\infty} A_0(y_0)\, H_1^{(1)}(k\rho)\, \frac{\exp(ikS_0(y_0))}{\rho}\, dy_0 \tag{14}$$

$$\rho = \{x^2 + (y-y_0)^2\}^{\frac{1}{2}}$$

where $A(y_0) \exp[ikS_0(y_0)]$ is the field on the aperture x=0 . It is well known that evaluating the integral asymptotically by the saddle point method provides a ray picture of the wave propagation phenomenon [16]. Thus we have

$$u(x,y) = \sum_{n=1}^{N} \frac{x}{\rho^{3/2}} A_0(y_n)\{S''(x,y;y_n)\}^{-\frac{1}{2}} \exp\{ikS(x,y;y_n)\} \tag{15}$$

where

$$S(x,y;y_0) = S_0(y_0)+\rho(y_0) \quad ; \quad S''(x,y;y_0) = d^2/dy_0^2\{S(x,y;y_0)\} \tag{16}$$

The saddle points y_n are determined by

$$dS/dy_0 = dS_0(y_0)/dy_0 - \cos\beta_0(y_0) = 0 \tag{17}$$

at $y_o = y_n$. It is easy to see that the Green's function approach is equivalent to the ray approach of Section 2.2 : in fact, equation (17) coicides with (9), so that saddle points can be interpreted as the ray departure points on the complex extension of the aperture. Moreover, it may be shown that the field evaluated according to (15) is identical to the ray optical field (13).
The afore-mentioned selection problem is present also here, since not all the saddle points that satisfy (17) are to be taken into account in the field asymptotic expansion (15). The summation is extended to the relevant saddle points only, which are those lying on a deformed integration contour equivalent to the original one (i.e. the real y_o-axis). This prescription uniquely defines the saddle points to be considered but requires the study of steepest descent path (SDP) maps in order to be applied. The study of the SDP maps for complex ray systems has been shown [5] to be a useful tool for the construction of a selection rule when several complex rays are involved.

2.4 Reflection and diffraction of complex rays.

The laws of reflection and refraction of complex rays are obtained from those in real optics by analytical continuation. Also, surfaces separating media with different characteristics undergo the same process and are extended to complex space.

In order to point out the kind of problems one encounters in the application of Complex Optics, let us consider the simple geometry of Fig.3 , where a two-dimensional infinite metallic reflector of equation $x=f(y)$ is illuminated by a complex source S located at :

$$x_S = x_{OS} + ib\cos\psi \qquad\qquad y_S = y_{OS} + ib\sin\psi \tag{18}$$

x_{OS} , y_{OS} are the (real) coordinates of the center of the waist, and ψ is the angle between the beam axis and the x-axis. The observation point P has real coordinates x_P , y_P and the reflection point R has complex coordinates x , y. The point R can be found by Fermat's Principle:

$$d/dy\{d_1(y) + d_2(y)\} = 0 \tag{19a}$$

where

$$d_1(y)=\{(f(y)-y_S)^2+(y-y_S)^2\}^{\frac{1}{2}} \;;\; d_2(y)=\{(x_P-f(y))^2+(y_P-y)^2\}^{\frac{1}{2}} \tag{19b}$$

The solution of (19) is more difficult than in the case of a real source because y is complex and a two-dimensional root finding technique has to be used; moreover, the roots are distributed on a multisheeted Riemann surface: if the function $f(y)$ is a polynomial, there are four sheets, due to the presence of two square roots in (19). Since not all of these roots correspond to actual reflection points, one has to be able to reject the spurious contributions. In practical antenna problems, where the reflector is in the far field of the complex source ($k|d_1|>>1$) and the observation point is not too near the reflector, all of the contributing roots lie on the sheet characterized by $Re\{d_1\}>0$ and $Re\{d_2\}>0$. In this way outgoing field contributions are selected. Convex (for ex., hyperbolic) reflectors admit only one reflection point, whereas for concave reflectors (e.g. a parabola) the number of contributing reflection points is not always the same as the observation point changes; moreover, caustics may be present. Further problems arise when the reflector has finite size: this case will be examined in detail in Section 4.

Concerning diffraction, we consider the canonical problem of a half plane illuminated by a line or point source located in complex space. The analytic continuation of the field expression for a line source is

$$V(\underline{\rho},\underline{\rho}') = -2i\left(I^+\exp\{ik\rho^+\} \pm I^-\exp\{ik\rho^-\}\right)/\pi \tag{20a}$$

where

$$I^{\pm} = \int_{w^{\pm}}^{\infty} e^{-t^2}(t^2-2ik\rho^{\pm})^{-\frac{1}{2}}dt \qquad w^{\pm} = -e^{-i\pi/4}\{k(\rho_1-\rho^{\pm})\}^{\frac{1}{2}} \tag{20b}$$

$$\rho^{\pm} = \{(x\mp x_S)^2 + (y-y_S)^2\}^{\frac{1}{2}} \qquad \rho_1 = \rho + \rho' \tag{20c}$$

ρ^+ and ρ^- are the complex distances from the source or its image, to the observation point (Fig.4b) and the double sign in (20a) refers to the E (+) and H (-) polarizations. The $I^{\pm}$ integrals have a saddle point for $t=0$ (direct or reflected ray) and an end point contribution in $w^{\pm}$ (diffracted ray). Asymptotically the field is given as a sum of these two contributions if $Re\{w^{\pm}\}<0$ or only by the end point contribution if $Re\{w^{\pm}\}>0$ [14]. The equation $Re\{w^{\pm}\}=0$ that divides the real plane in regions with two or one ray contributions is the equivalent of equation $w^{\pm}=0$, defining shadow and reflection boundaries in the real case. In other words, for a real source the transition between regions with one or two ray contributions takes place when the SP and the end point of the field integral representation coincide, whereas for a complex source the same transition takes place when the end point crosses the saddle point Steepest Ascent Path (SAP). The expression of the diffracted component of the field is [14,17]:

$$V^d \sim \pm \frac{e^{ik\rho_1}}{k(\rho\rho')^{\frac{1}{2}}} \frac{2 \begin{matrix}\sin\\ \cos\end{matrix}\left(\frac{\phi}{2}\right) \begin{matrix}\sin\\ \cos\end{matrix}\left(\frac{\phi'}{2}\right)}{\cos\phi + \cos\phi'} \tag{21}$$

where ϕ and ϕ' are the angles defined as in Fig.4b and the upper line refers to E-polarization. The phase term $(k\rho)$ shows that the diffraction process can be interpreted in terms of a complex ray from the source to the real edge and of the real ray therefrom to the observation point. In other words, the real edge appears to be the diffraction point.

In the three-dimensional case the exact field expression at the observation point $P=(\underline{\rho},z)$ for a complex scalar point source $S=(\underline{\rho}',z')$ (Fig.4a) is [17]

$$V(\underline{\rho},z;\underline{\rho}',z') = I^+ \pm I^- \tag{22a}$$

where

$$I^{\pm} = i\int_{w^{\pm}}^{\infty} \frac{H_1^{(1)}(it^2+kR^{\pm})}{(t^2-2ikR^{\pm})^{\frac{1}{2}}}dt \qquad w^{\pm} = -e^{i\pi/4}\{k(R_1-R^{\pm})\}^{\frac{1}{2}} \tag{22b}$$

$$R^{\pm} = \{(x-x')^2+(y\mp y')^2+(z-z')^2\}^{\frac{1}{2}} \;;\; R_1 = \{(\rho+\rho')^2+(z-z')^2\}^{\frac{1}{2}} \tag{22c}$$

the double sign in (22a) refers to the soft (-) and hard (+) boundary conditions respectively. The integrals $I^{\pm}$ have the same characteristics as the previous ones (eq.(20b)) and the diffracted field V^d can be asymptotically written in the form [17]

$$V^d(\underline{\rho},z;\underline{\rho}',z) \sim \pm\left(\frac{2}{\pi kR_1}\right)^{\frac{1}{2}} \frac{e^{i(kR_1+\pi/4)}}{k(\rho\rho')^{\frac{1}{2}}} \frac{\begin{matrix}\sin\\ \cos\end{matrix}\left(\frac{\phi}{2}\right) \begin{matrix}\sin\\ \cos\end{matrix}\left(\frac{\phi'}{2}\right)}{\cos\phi + \cos\phi'} \tag{23}$$

where ϕ and ϕ' are the angles defined in Fig.4b and the upper line refers to soft boundary conditions. Examining (23), the phase term shows that R_1 can be interpreted as the sum of the complex distances d_1 from the source S to the diffraction point $T=(\underline{0},z_T)$ and d_2 from the diffraction point to the observation point P (Fig.4a) provided that the diffraction point is determined according to the generalized Fermat's Principle

$$d/dz(d_1+d_2) = (z_T-z')/d_1+(z_T-z)/d_2 = \hat{z}\cdot\hat{n}_i+\hat{z}\cdot\hat{n}_d = 0 \qquad (24)$$

In general, this equation has complex z_T solutions; z_T is real when the source location is real (ρ', z' real) or when the beam axis is normal to the edge (z' real) and $z=z'$. Therefore, in the three-dimensional case the diffraction point is generally complex.

3 ANALYSIS OF PLANAR APERTURE DISTRIBUTIONS

The complex ray analysis of the field from assigned planar aperture distributions is the simplest application of the theory described in Section 2 but has also practical interest, for instance for feed modelling purposes. In the present section CRT will be applied to tapered (Gaussian) aperture fields but also to more rapidly varying distributions; since the results have been published [5], only a summary will be given. Moreover, the equivalence between the complex source point field and the Gaussian beam will be discussed.

3.1 Complex ray analysis of tapered (gaussian) distibutions.

The complex ray analysis of planar two-dimensional aperture distributions is based upon the theory of Section 2.2 . Complex Ray Tracing can effectively describe the evolution of a tapered (gaussian) initial field distribution, since the associated system of complex rays is comparatively simple, that is, the field is given almost everywhere by one ray contribution. This kind of behaviour is customarily associated with fields wherein interference effects are irrelevant, like the Gaussian beam. However, even in this case, we have to select, among the complex rays belonging to the system described by (9) the ones giving a correct evaluation of the field. Although this selection may require a full asymptotic discussion of the field integrals by means of the saddle point method, it is often possible to obtain "a priori" rules for evaluating the ray optical field [5].

Let us consider the initial field distribution (7), where

$$A_0(y_0) = 1 \qquad S_0(y_0) = -iy_0^2/(2b) \qquad (25)$$

The associated ray system satisfies the condition (9) :

$$\cos\beta_0(y_0) = (y-y_0)\{x^2+(y-y_0)^2\}^{-\frac{1}{2}} = dS_0(y_0)/dy_0 \qquad (26)$$

As it may easily be seen, the departure angle from the initial line y_o is generally complex. When solved for y_o , (26) yields the complex departure point of each ray in the complex plane y_o , which is the analytical extention of the initial curve. It may be shown (from (26)) that two complex rays reach the observation point (x,y). The problem, how to deal with the field contributions associated with each complex ray, has been discussed in [5]; as a result, it has been found that the Gaussian beam is, apart from the near (out-of-axis) region, a one-ray field.

The asymptotic far-field behaviour of the beam having initial Gaussian taper can be easily investigated, for in this instance, (26) yields an immediate analytical solution. Taking into account the only contributing complex ray, the far-field pattern is:

$$g(\theta) = \cos\theta \exp\{-kb\sin^2(\theta/2)\} \qquad (27)$$

where θ is measured from the beam axis. Although (27) yields, near the beam axis, the same results as the well-known paraxial Gaussian beam expression, its validity extends beyond the paraxial region.

3.2 Gaussian beam and CSP field solution: a comparison.

The well-known paraxial equivalence between the field generated by an initial Gaussian taper and the sheet beam radiated by a line source having complex coordinates (CSP) [1,2] can be shown to extend over a broad angular range, so that -for all practical purposes- they represent the same field, provided that (kb) is not too small. For the CSP field, the following normalized expression will be used:

$$G(\underline{\rho},\underline{\rho}') = H_0^{(1)}(k|\underline{\rho}-\underline{\rho}'|)/H_0^{(1)}(-ikb) \quad ; \ \underline{\rho}' = ib\hat{x} \qquad (28)$$

The afore-mentioned equivalence physically follows from the fact that the Gaussian taper is very well approximated, for kb>1, by the initial field distribution associated with the CSP field, which can be easily obtained from (28), taking into account that

$$H_0^{(1)}(k\sqrt{y^2-b^2}) = (2i/\pi)K_0(k\sqrt{b^2-y^2}) + 2I_0(k\sqrt{b^2-y^2}) \qquad (29)$$

where K_0, I_0 are modified Bessel fuctions. Although the Gaussian taper has constant phase everywhere on the aperture, whereas the CSP aperture field has increasing phase for $|y|>b$ and constant phase for $|y|<b$ (apart from a small neighborhood of the point $|y|=b$, where it exhibits a logarithmic singularity), for kb>1 those discrepancies take place in regions where the field has already experienced a strong decay. In Fig.5 the aperture distribution is shown for both Gaussian beam and CSP field for different values of b/λ, with respect to the normalized distance y/b. Very good agreement is found for $b>5\lambda$ (kb>1). Assuming kb>1, the full width Δy of the aperture distribution of a CSP at the level of $-\alpha$dB is given as:

$$\Delta y/\lambda = 0.3828\sqrt{\alpha(b/\lambda)} \qquad (30)$$

where λ is the wavelength. A plot of y/λ versus b/λ for several values of α having practical interest is presented in Fig.6.

A comparison between the far-field behaviour of the CSP field and of the Gaussian beam is based on their radiation patterns. For the CSP, the radiation pattern can be obtained from (28):

$$g_{CSP}(\theta) = \exp\{kb(\cos\theta-1)\} \tag{31}$$

where θ is measured from the beam axis. The pattern of the Gaussian beam has already been reported in the last section (27); both patterns are normalized to unity on the beam axis. A comparison of the far field patterns for different values of b/λ is shown in Fig.7. The excellent agreement found between the far field patterns of the Gaussian beam and of the complex source point field also extends to the near-intermediate region [5], wherever the Gaussian field is given by only one ray contribution, as it always is for the CSP field. From (31) the total beamwidth $\Delta\theta$ at a level of $-\alpha$ dB under the maximum as a function of b/λ is found to be expressed as:

$$\Delta\theta_\alpha = 2\arccos(1-0.0183\alpha(b/\lambda)^{-1}) \tag{32}$$

Eq. (32) is plotted in Fig.6 for two values of α having practical interest.

3.3 CRT analysis of rapidly varying aperture distributions.

As already seen in the previous sections, tapered initial distributions radiate a field which is easily described in term of complex rays. In particular, fields wherein interference is negligible or absent altogether (as in the Gaussian beam) are described by a single complex ray. These attractive features are no longer present when complex ray theory is employed to describe fields radiated by a rapidly varying initial distribution (i.e., one with sharp variations of amplitude) The flattening of the central portion of the aperture field and the consequent sharp decay at the edges of the aperture leads to a radiation pattern having built-in interference phenomena (that is, secondary lobes), to be described by means of several interfering complex rays. This case has been thoroughly discussed in [5], selecting as a test problem the aperture distribution (7) with:

$$A_0(y_0) = 1 \qquad S_0(y_0) = iy_0^{2N}/(2b^{2N-1}) \tag{33}$$

For N=1 a Gaussian taper is obtained, whereas for greater N both a central flattening of the distribution and a sharper edge fall-off take place. For $N \to \infty$ a square amplitude distribution results. The ray system associated with (33) is made up of 2N complex rays; selecting among those rays the proper ones once again requires a discussion of the field integrals through saddle point method. A proper set of "a priori" selection rules has been discussed in [5].

4 ANALYSIS OF REFLECTOR ANTENNAS

The complex rays method has recently been applied to antenna problems with encouraging results. Two types of reflectors have been analyzed: a focus-fed parabolic reflector [4] and a focus-fed hyperbolic subreflector [6]. In the first case the beam was assumed to be collimated such that the diffraction contribution could be disregarded, and only the reflected field was computed. In the second case this assumption was dropped and the diffracted field was computed by means of GTD extended to complex rays, as discussed in section 2.4. When a parabolic reflector is illuminated by an off-focus source the problem is more complicated since caustics arise and the number of reflection points changes as the observation point moves.

4.1 Defocused Parabolic Reflector.

Let us refer to the geometry of Fig.3. The reflector consists of the arc of parabola between ordinates y_1 and y_2, while the feed field is modelled with a complex line source of strength I_0 located in S (see (18)); the b parameter is related to the beam divergence by

$$b = 0.0183\alpha\lambda/(1-\cos(\Delta\theta_\alpha/2)) \tag{34}$$

where $\Delta\theta_\alpha$ is the full beamwidth at $-\alpha$dB. The beam axis makes an angle ψ with the positive x axis.

The reflection point R is determined by Fermat's Principle; if the observation point P is in the far field, in the θ direction, this yields:

$$\frac{(x-x_S)dx/dy + (y-y_S)}{\{(x-x_S)^2 + (y-y_S)^2\}^{\frac{1}{2}}} - (dx/dy\cos\theta + \sin\theta) = 0 \tag{35}$$

The roots are four and lie on a two sheeted Riemann surface. The roots on the improper sheet $\mathrm{Re}\{d_1\}<0$ can be discarded as explained in section 2.4. As for the others, one has to ascertain which are the reflection points that lie "on" the limited arc of parabola that we consider. If the reflection points were real the answer would be obvious, but in the complex case there seems to be no simple geometrical rule to discard spurious contributions. Therefore, in order to develop a selection rule we have approached the problem in a different way. First, we constructed an integral representation of the scattered field and then, we evaluated it asymptotically with the method of steepest descent. For developing the selection rule, Physical Optics is adequate and the scattered field can be expressed as

$$E(\theta) = \int_{y_1}^{y_2} f(y)\exp\{ikq(y)\}dy \tag{36a}$$

where

$$f(y) = \frac{if\mu_0}{2(DR)^{\frac{1}{2}}} I_0 n_\rho (\partial\rho/\partial y)_J \qquad q(y) = D+R-(x\cos\theta+y\sin\theta) \tag{36b}$$

n_ρ is the radial component of the reflector normal, $(\partial\rho/\partial y)$ is the Jacobian of the transformation between the parabola and the y-axis, R, θ are the coordinates of the observation point and D is the complex distance between source and reflection point. The saddle point equation $q'(y_S)=0$ coincides with (35); this means that the saddle point contributions to the integral are to be interpreted as reflected ray fields. Moreover, since SDPs through saddle points extend in general to infinity, it is necessary to complete the integration contour with SDPs running from the end points to infinity. The contribution from these paths can be interpreted as a diffracted field, within the limits of Physical Optics [18]. The choice of the relevant saddle points (reflection points) is unambiguous and is connected with the possibility of travelling from one end point to the other along SDPs. The saddle point motion in the complex y-plane as function of the observation angle θ is depicted in Fig.8 for the case of a real (isotropic) source located away from the focus. One clearly sees the crossing of the two reflection boundaries (at 8.5° and 10.1°) and of a caustic (at 11.5°). If the source is now given a complex location, the picture changes as shown in Fig.9: it may be seen that the saddle points never coalesce for any real θ value. In other words, the caustic has been pushed into complex space. A simple, non uniform first order asymptotic evaluation of (36a) has been carried out and the results for the field amplitude and phase are shown in Figs.10 and 11. The solid line refers to a numerical evaluation of the radiation integral, whereas dots correspond to the asymptotic evaluation. The agreement is good, apart from the θ values where saddle point and end point strongly interact. Figs. 12a-b refer to the crossing of a reflection boundary. In the first case two end point contributions plus a saddle point contribution must be considered, whereas in the second, only the end point contributions are to be taken into account. Therefore, a reflection boundary corresponds to the θ value for which the Steepest Ascent Path (SAP) of the saddle point passes through the end point. As stated above, the relevant saddle points are defined unambiguously by the possibility of connecting the two end points by means of steepest descent paths. Alternatively, a saddle point is to be taken into account if the end points lie on different sides of its SAP. However, since a general a-priori criterion (that is, one which does not require the study of SDPs) has not yet been found, further investigation is needed.
It is to be pointed out that the approach based on the integral representation of the scattered field is used only to resolve the ambiguities connected with the choice of the relevant reflection points. The actual field computation proceeds by making use of the

GTD equations, i.e. eq. (21), or, when reflection and diffraction point are close together, of its uniform counterpart involving the complementary error function [17].

5 CONCLUSIONS

We have reviewed the methodology and some recent developments in Complex Ray Theory and related techniques, with particular emphasis on applications to free-space propagation and to reflector antenna problems. The characteristics of the field radiated by a complex line source have been reexamined and compared with the field radiated by an initial Gaussian taper; it was found that the complex source point field can be used with confidence instead of the classical Gaussian beam. Complex Ray Tracing has also been employed to describe the field evolution from different kinds of aperture distributions; the advantages but also the limitations of this theory, when dealing with initial distributions showing rapid amplitude variations, have been summarized. Finally, the treatment of metallic reflectors illuminated by beam-like fields has been discussed, with particular attention to edge diffraction, and to the distinction between relevant and spurious reflected ray contributions. These examples show that complex ray methods represent a promising tool for analysing propagation and diffraction problems due to incident beam-type fields.

ACKNOWLEDGEMENTS

We would like to thank Prof. L.B. Felsen for reviewing the manuscript and for his encouragements; many helpful discussions with Prof. R. Zich are gratefully acknowledged.

REFERENCES

1. G.A.Deschamps:"Gaussian Beam as a Bundle of Complex Rays.", Electronics Letters, vol.7, 1971, pp.684-685
2. J.B.Keller,W.Streifer:"Complex Rays with an Application to Gaussian Beams.",J.Opt.Soc.Am. , vol.61, 1971, pp.40-43
3. P.D.Einziger,L.B.Felsen: "Evanescent Waves and Complex Rays.", IEEE Trans. on Antennas and Propagation, vol.AP-30, 1982, pp.594-605.
4. F.J.V.Hasselman,L.B.Felsen:"Asymptotic Analysis of Parabolic Reflector Antennas.", IEEE Trans.on Antennas and Propagation, vol.AP-30, 1982, pp.677-685.
5. G.Ghione,I.Montrosset,L.B.Felsen:"Complex Ray Analysis of Radiation from Large Apertures with Tapered Illumination.",to be pu - blished in IEEE Trans. on Antennas Propag., July]984 Issue.

6. N.C.Albertsen,K.Pontoppidan:"Subreflector Analysis by Complex Rays.",International Symposium on Antennas and Propagation,Albuquerque (New Mexico, USA), 1982,pp.608-611.
7. X.J.Gao, L.B.Felsen: "Complex Ray Analysis of Radome-Covered Large Aperture Antennas", IEEE-AP/S and URSI Symposium, Houston, Texas, May 1983.
8. H.G.Unger: "Beam Propagation through Lenses." , this issue.
9. L.B.Felsen:"Complex Source Point Solutions of the Field Equations and their Relation to the Propagation and Scattering of Gaussian Beams.", Symposia Matematica, vol.18, 1976, Academic Press,London, pp.40-56.
10. W.D.Wang,G.A.Deschamps: "Application of Complex Ray Tracing to Scattering Problems.", Proc. IEEE, vol.62, 1974, pp.1541-1551.
11. L.B.Felsen: "Evanescent Waves.", J.Opt.Soc.Am., vol.66, 1976, pp.751-760.
12. G.Ghione,I.Montrosset: "Asymptotic Techniques for Gaussian and Gaussian-like Distributions: A Comparison.", Proceedings of the Seventh Colloquium on Microwave Communications, Budapest, 6-10 September 1982, pp.295-298.
13. J.B.Keller: "Geometrical Theory of Diffraction.",J.Opt.Soc. of Am., vol.52, 1962, pp.116-130.
14. A.C.Green,H.L.Bertoni,L.B.Felsen: "Properties of the Shadow Cast by a Half-Screen when Illuminated by a Gaussian Beam.", J.Opt.Soc.Amer., vol.69, 1979, pp.1503-1508.
15. I.Montrosset,R.Orta: "La methode des Rayons Complexes.", Annales des Telecommunications, vol.38, 1983, pp.135-144.
16. L.B.Felsen, N.Marcuvitz: Radiation and Scattering of Waves, Prentice Hall Inc., Englewood Cliffs, 1973, Sec . 4 .
17. J.J.Bowman, T.B.A.Senior, P.L.E.Uslenghi: Electromagnetic and Acoustic Scattering by Simple Shapes, North-Holland Publ. Co.,1969, Secs. 8.3,8.4 .
18. W.T.Rusch: "Physical Optics Diffraction Coefficients for a Paraboloid", Electronics Letters, vol.10, 1974, pp. 358-360.

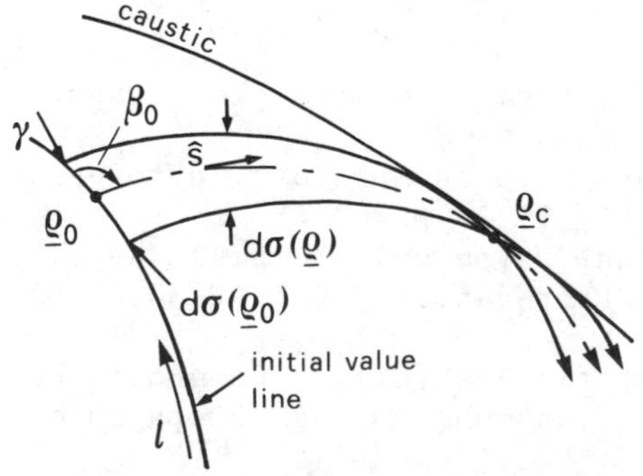

Fig.1 - Typical ray tracing in inhomogeneous media for an initial value problem.

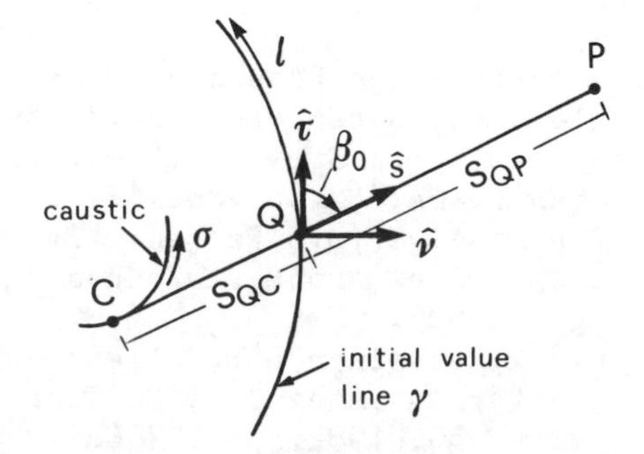

Fig.2 - Ray tracing in homogeneous media for an initial value problem.

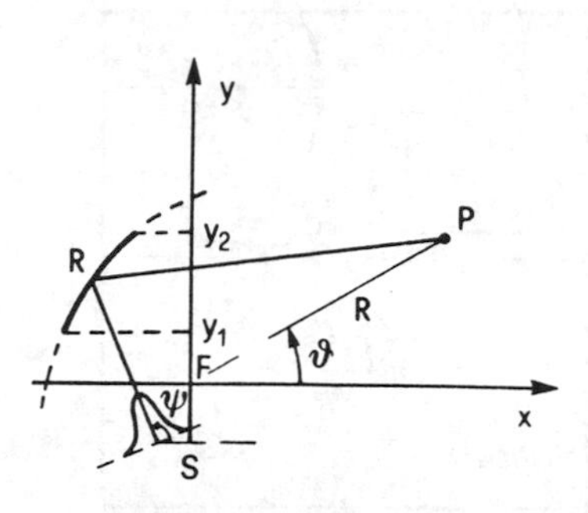

Fig.3 Reflector geometry.

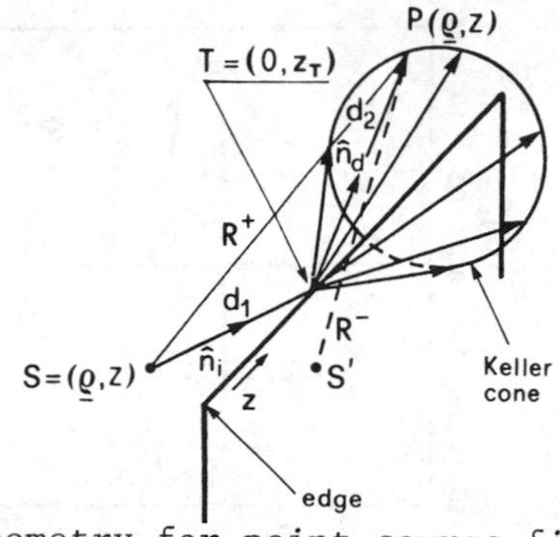

Fig.4a Geometry for point source field diffraction by a half-plane : three - dimensional ray picture .

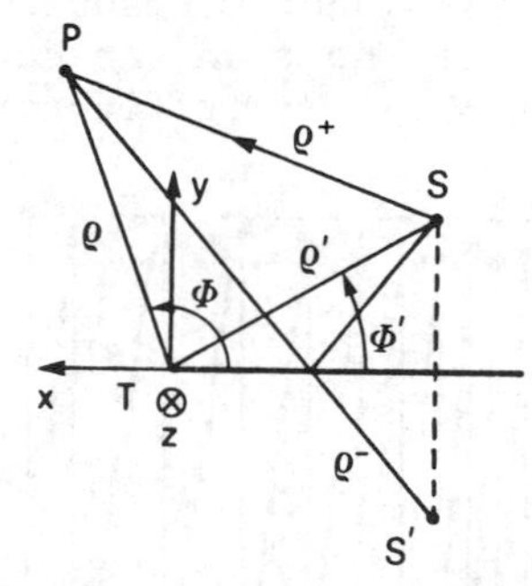

Fig. 4b Point source field diffraction by a half-plane : representation in the plane normal to the edge and geometry for the two-dimensional case.

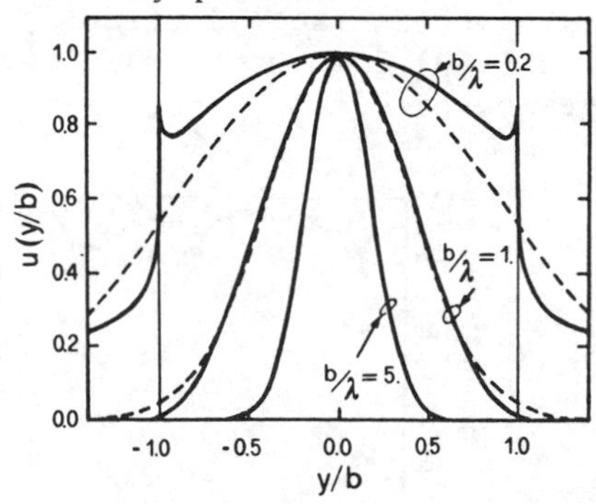

Fig.5 Aperture field amplitude for Gaussian beam (- - - -) and CSP source (———) for several values of b/λ. For |y/b|=1 the CSP field is singular .

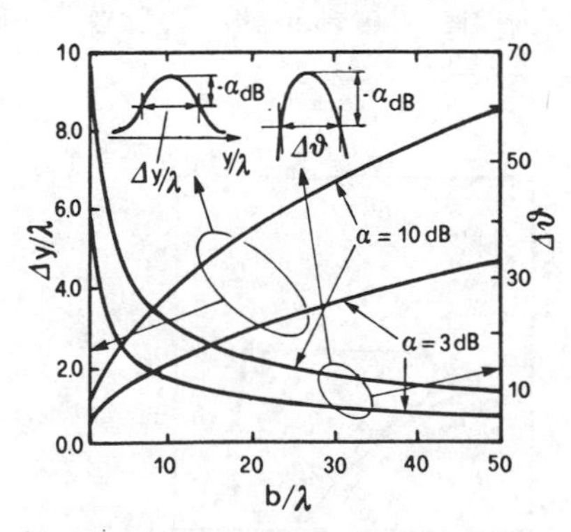

Fig.6 CSP field:normalized aperture distribution width at αdB below the maximum (left) and beamwidth of the radiated field at α dB below the maximum (right) ; α=3,10 dB .

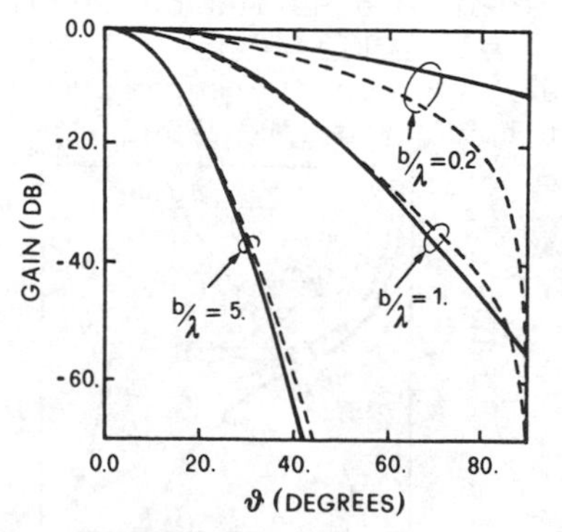

Fig.7 Far field radiation patterns for Gaussian beam(- - -) and CSP distributions (———) for several values of b/λ .

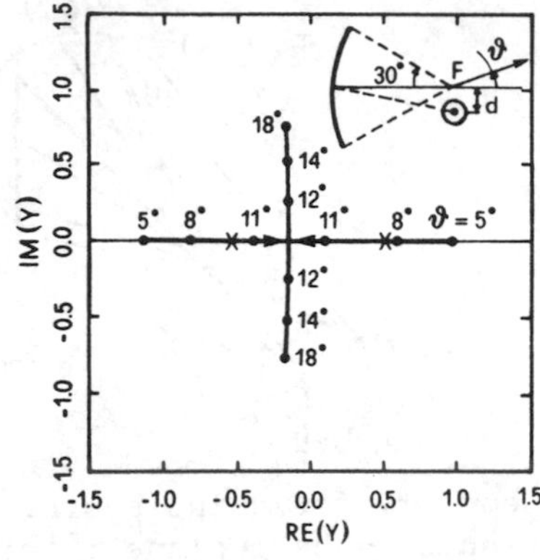

Fig.8 Saddle point locus in the complex y-plane:real(isotropic) source (b=0).Defocusing d/F=0.2, x: reflector edges.

Fig.9 Saddle point locus in the complex y-plane: CSP source b/F= 0.03,10-dB taper on the edges;defocusing d/F=0.2;x:reflector edges

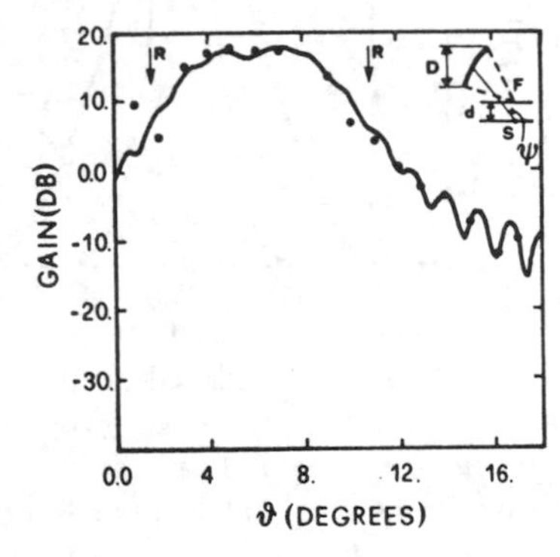

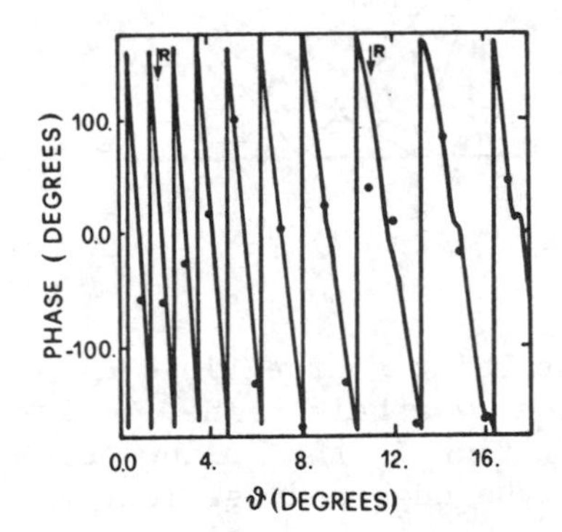

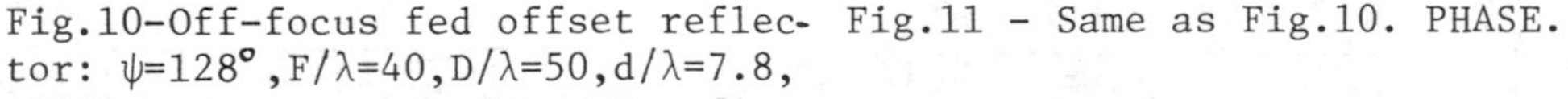

Fig.10-Off-focus fed offset reflector: ψ=128°,F/λ=40,D/λ=50,d/λ=7.8, -10dB taper on the edges. R:reflection boundary. AMPLITUDE .

Fig.11 - Same as Fig.10. PHASE.

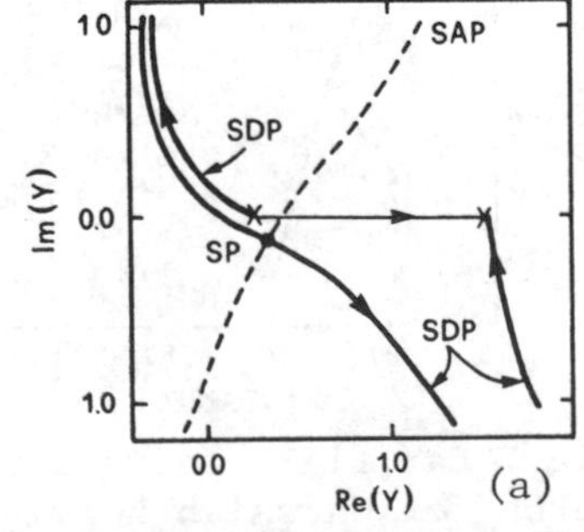

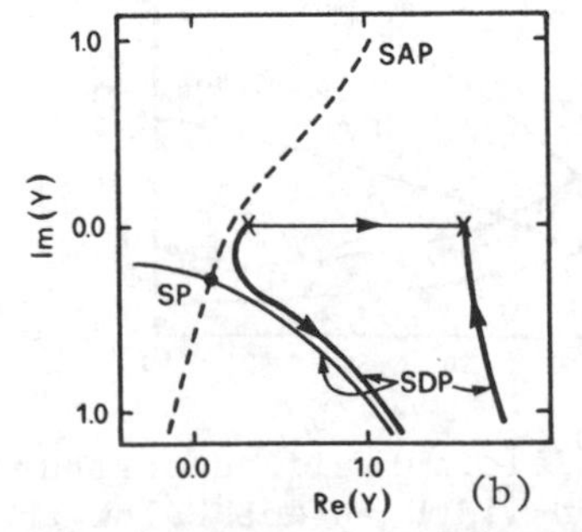

Fig.12 Same reflector as in Fig.10.Integration paths (a) for θ=10° and (b) for θ=12°. x---x : original contour ; ——►—— : deformed contour;SP:saddle point;SAP (SDP):steepest ascent (descent) path.

IMAGING AND PROPAGATION OF BEAMS IN METALLIC OR DIELECTRIC WAVEGUIDES

Edward F. Kuester and David C. Chang

Department of Electrical and Computer Engineering
University of Colorado
Boulder, Colorado 80309 USA

ABSTRACT

Multimode dielectric or metallic waveguides have the property that a narrow beam launched into the waveguide does not randomly disperse, but forms repetitive cohesive patterns at certain points along the axis of the guide. This imaging property is the result of constructive interference between modes, and is mathematically related to the hybrid ray-mode formulations of Felsen and his colleagues. This paper will review the origins and applications of the imaging properties of uniformly filled metallic or dielectric waveguides.

1. INTRODUCTION

Multimode waveguides occur frequently in optical communications, underwater acoustics and other areas. Often a source of small extent (compared to the transverse dimensions of the waveguide) provides the excitation for the guide, resulting in a large number of modes being produced with significant amplitude. Field computation in such a case may be done by evaluating the amplitude of each mode, and summing all the modes together. In many applications, hundreds or even thousands of modes may be involved, and a large degree of cancellation of terms in the mode sum may occur. This can result in the accumulation of roundoff error when computer calculation is done, especially for large propagation distances.

The work of Felsen and his colleagues (1) on hybrid ray-mode expansions suggests that a more efficient representation in a ray expansion may be available, or even a yet more efficient combination of ray and mode terms. So long as a large number of mode terms can be traded for a relatively small number of ray terms, this method provides significant improvement, but for extremely large values of the axial distance z (on the order of a^2/λ, where a is the waveguide width and λ the wavelength) the number of rays required becomes comparable to the number of modes, and no improvement is obtained. The Gaussian beam excitation is handled within this framework by the use of complex source points (2), but this incurs additional difficulties when the beam is paraxial with the waveguide axis, which behaves almost as a focal-type caustic for the rays at large propagation distances.

A third approach to the problem tackles the partial differential equation by direct numerical means--using finite element or fast Fourier transform techniques (3). Once again, behavior at large distances is difficult to treat due to uncertain error accumulation in the process.

This paper summarizes an extension to the hybrid techniques which is based on imaging properties of waveguides: at certain distances from the source plane, a replica (or a specific combination or transform of such replicas) of the source will be produced in the paraxial approximation. Knowing the field at one of these imaging planes, the transform of the corresponding mode sum into a ray series can be used to calculate the field at locations sufficiently close to these planes. Since these planes are located at reasonably close intervals along the axis of the guide, it becomes possible to compute the field very accurately at nearly any point of the guide without any of the problems cited above.

2. FOURIER AND FRESNEL IMAGES IN WAVEGUIDES

2.1 Equivalence Between Metallic and Dielectric Guides

We limit our considerations to hollow (uniformly filled) metallic waveguides and step-index dielectric waveguides of various cross-sections. There is in fact a very general but simple equivalence between these two types of waveguide under the conditions we deal with here. When a large number of modes is present on a dielectric waveguide of any cross-sectional shape, the propagation constants and transverse field distributions of these modes are, to a good approximation, given by the propagation constants and longitudinal electric fields of the TM modes of a certain related metallic guide. The cross-section of the metallic guide is obtained from that of the dielectric guide by extending the core boundary normally outwards at each point by the distance $k_o^{-1}(n_1^2 - n_2^2)^{-\frac{1}{2}}$, where $k_o = \omega(\mu_o\varepsilon_o)^{\frac{1}{2}}$ is the free-space wavenumber,

n_1 is the core refractive index, and n_2 is the index of the cladding. This "effective cross-section" (4) is a means of accounting for the very slight penetration of the fields of most of the dielectric waveguide modes into the cladding, and allows problems involving multimode dielectric waveguides to be reduced to corresponding ones for metallic waveguides.

2.2 Parallel-Plate Waveguides

Let us first consider the parallel-plate waveguide illustrated in Fig. 1. The walls at $x = 0$ and $x = a$ are perfectly conducting, and some known source produces a given excitation or input field at the plane $z = 0$. For simplicity, we consider only the TE polarization so that $\bar{E} = \bar{a}_y E_y$ and

$$\left(\frac{\partial^2}{\partial x^2} + \frac{\partial^2}{\partial z^2} + k^2\right) E_y = 0 \tag{1}$$

for $z > 0$. Here $k = \omega\sqrt{\mu\varepsilon}$ is the wavenumber of the medium filling the guide, and a time dependence of $\exp(i\omega t)$ has been assumed.

By well-known techniques, the field $E_y(x,z)$ for $z > 0$ in the waveguide can be expressed in terms of the field $E_y(x,0)$ at an input plane ($z = 0$) by means of a Green's function $G(x,x';z)$:

$$E_y(x,z) = \int_0^a E_y(x',0)G(x,x';z)dx' \tag{2}$$

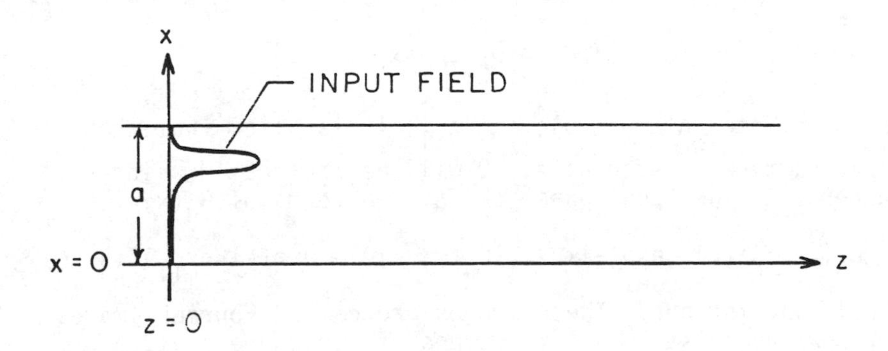

Figure 1: Parallel-plate waveguide

where G can be expressed as a modal expansion:

$$G(x,x';z) = \frac{2}{a} \sum_{m=1}^{\infty} \sin \frac{m\pi x}{a} \sin \frac{m\pi x'}{a} \exp(-i\beta_m z), \quad z \geq 0 \tag{3}$$

where $\beta_m = (k^2 - m^2\pi^2/a^2)^{\frac{1}{2}}$.

Now, for modes with small enough m -- the paraxial modes -- , we can approximate the propagation constant β_m by

$$\beta_m \simeq k - m^2\pi^2/2ka^2 . \tag{4}$$

The propagation factor of the mth mode is thus

$$\exp(-ikz + 2\pi i\, m^2 z/z_{11}) \tag{5}$$

where $z_{11} = 4ka^2/\pi$ is the so-called Fourier imaging distance of the guide. We can expect this paraxial approximation to be valid (a) if only modes for which eqn. 4 is accurate contribute significantly to eqn. 2, and (b) the value of z is not so large that significant error enters into eqn. 5. It can be shown that $kz \ll (ka)^4$ and $(ka)^2 \gg 1$ are necessary, though not sufficient, for (a) and (b) to be true (5). It will also be necessary that the characteristic transverse dimension w of the input field be such that $kw \gg 1$ as well. From eqns. 3,4 and 5 we see that the paraxial Green's function is given by

$$G_o(x,x';z) = \frac{1}{2a} \exp(-ikz) \sum_{m=-\infty}^{\infty} \exp(2\pi i m^2 z/z_{11}) \cdot \{\exp[-im\pi(x-x')/a] - \exp[-im\pi(x+x')/a]\}. \tag{6}$$

When z is an integral multiple of z_{11}, it is clear that whatever field pattern existed at z = 0 will be replicated, since all modes undergo an identical phase shift. In terms of $G_o(x,x',z)$,

$$G_o(x,x'nz_{11}) = \exp(-iknz_{11})G_o(x,x';0) = \exp(-iknz_{11})\delta(x-x')$$

where n is any integer. These images are called Fourier images and are known to exist in several types of structures (5). For electromagnetic waveguides, this phenomenon was first noticed by Rivlin and Shul'dyaev in 1968 (6). Nearly simultaneously, Weston (7,8) noticed the same thing happening in acoustic waveguides, though his analysis was somewhat more qualitative.

Now if $(ka)^2 >> 1$, we can expect that the Fourier imaging distance z_{11} is quite large, so that extrapolation to a distance of $z_{11}/2$, say, by ray methods will still encounter some numerical problems. Fortunately, another form of imaging occurs at the intermediate values of z,

$$z_{pq} \equiv \frac{q}{p} z_{11} \; ; \qquad p,q = \text{integers}, \quad p \neq 0 \tag{7}$$

At $z = z_{pq}$, there appear Fresnel images as we discuss below. Letting

$$E_m = \frac{2}{a} \int_0^a E_y(x,0) \sin \frac{m\pi x}{a} \, dx \tag{8}$$

we have from eqns. 2-5 that, in the paraxial limit,

$$E_y(x,z_{pq}) = e^{-ikz_{pq}} \sum_{m=1}^{\infty} E_m \, e^{2\pi i m^2 q/p} \sin \frac{m\pi x}{a} \; . \tag{9}$$

The factor $\exp(2\pi i m^2 q/p)$, while not the same for all m, is a periodic function in m of period p. Hence it can be expressed (9) as a Discrete Fourier Transform (DFT)

$$\begin{aligned} e^{2\pi i m^2 q/p} &= \sum_{n=0}^{p-1} c_n(p,q) e^{-2\pi i mn/p} \\ &= \sum_{n=0}^{p-1} c_n(p,q) \cos(2\pi mn/p) \end{aligned} \tag{10}$$

where the coefficients $c_n(p,q)$ are given by

$$c_n(p,q) = \frac{1}{p} \sum_{m=0}^{p-1} e^{2\pi i (m^2 q + mn)/p} = c_{-n}(p,q) \, . \tag{11}$$

Inserting eqn. 10 into eqn. 9 we get

$$E_y(x,z_{pq}) = \frac{1}{2} e^{-ikz_{pq}} \sum_{n=0}^{p-1} c_n(p,q) [\hat{E}_o(x+\frac{2na}{p}) + \hat{E}_o(x-\frac{2na}{p})] \tag{12}$$

where $\hat{E}_o(x)$ is the extension of the field $E_y(x,0)$ at the input plane to an odd function of x, periodic with period $2a$:

$$\hat{E}_o(x) = \sum_{m=1}^{\infty} E_m \sin \frac{m\pi x}{a} \tag{13}$$

Thus the Fresnel image of $E_y(x,0)$ at $z = z_{pq}$ is a set of shifted, possibly inverted replicas of the input field, each having the amplitude $c_n(p,q)$. These amplitudes, known as Gaussian sums, have properties which are summarized in (5). An example of Fresnel images for $p = 3$ is shown in Fig. 2.

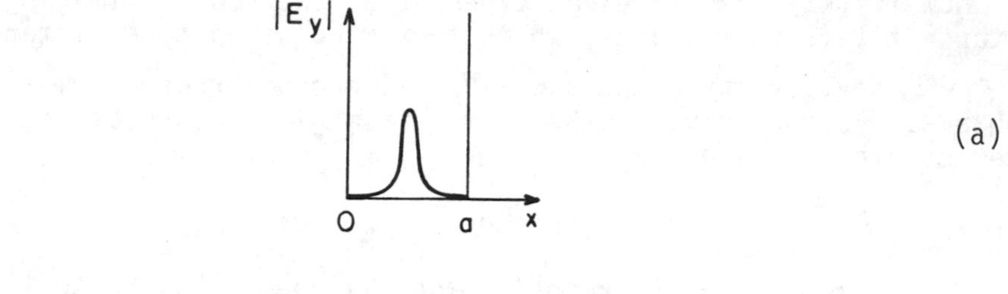

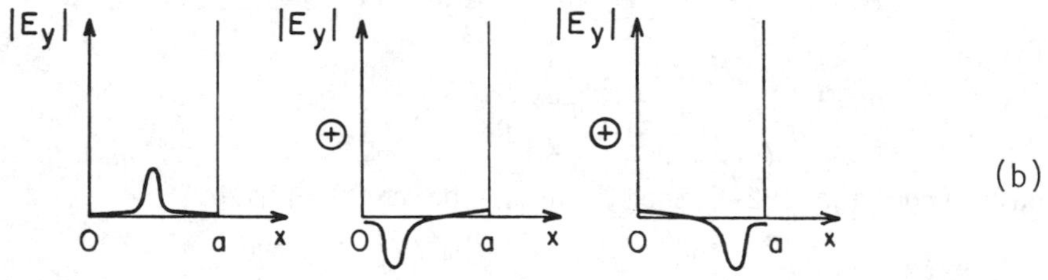

Figure 2: (a) The input field $E_y(x,0)$.
(b) The three components of the Fresnel image $E_y(x,z_{31})$.

2.3 Coupled Slab Waveguides

These same ideas apply also to the system (or normal) modes of two parallel dielectric slab waveguides (10). When the two slabs are identical, the system modes are either symmetric (equal fields at corresponding points of the guides) or antisymmetric (equal and opposite fields at corresponding points of the guides). In the paraxial case, each mode is very nearly a combination of one mode (the mth, say) from each guide in isolation, and according to known results from coupled mode theory, the propagation constants Γ_m for the system modes are

$$\Gamma_m \simeq k - \frac{m^2\pi^2}{2ka_e^2}\,[1 \pm 4e^{-Vd/a}/(V+2)] \tag{14}$$

where $a_e = a(1+2/V)$ is the effective width of each slab (whose width is a), $V = ka\,(1-n_2^2/n_1^2)^{\frac{1}{2}}$ is the normalized frequency, $k = k_o n_1$ is the wavenumber in the core as before, and d is the distance between the two slabs. The upper sign in eqn. 14 pertains to antisymmetric system modes, and the lower one to symmetric ones. From eqn. 14 we can see that if the input field to the two slabs is resolved into a symmetric and an antisymmetric part, then the symmetric part will undergo Fourier and Fresnel imaging with an imaging distance

$$z_{11}^{s} = \frac{z_{11}^{e}}{1 - 4e^{-Vd/a}/(V+2)} \tag{15}$$

and the antisymmetric part will do the same with an imaging distance of

$$z_{11}^{a} = \frac{z_{11}^{e}}{1 + 4e^{-Vd/a}/(V+2)} \tag{16}$$

with $z_{11}^{e} = 4ka_e^2/\pi$. Since these imaging distances are not the same, and since the phase shifts associated with the symmetric and antisymmetric parts are different, there results a complex set of interference patterns in which both images and transfer of energy between guides play a part. Complete power transfer between guides can occur only when the input field in one guide is symmetric about the center of the guide; this does not occur otherwise. In Fig. 3, we illustrate the complexity of the power transfer process for multimode slabs.

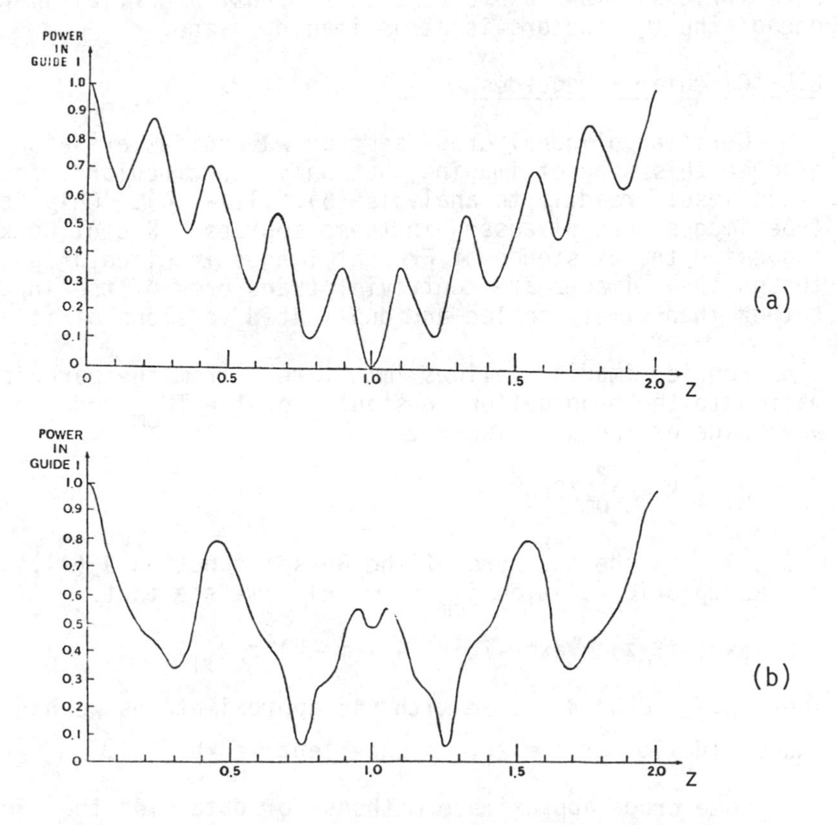

Figure 3: Power distribution in guide #1 with (a) a centered Gaussian beam at the input plane; (b) an off-centered beam at the input plane - see (10).

2.4 Rectangular Waveguides

Rectangular waveguides whose side lengths (or effective side lengths) are rational multiples of each other can also exhibit this type of imaging phenomenon (6),(11). More generally, the imaging process can take place separately with respect to either the horizontal or vertical direction of the waveguide. By separation of variables techniques, we can show, for example that the paraxial Green's function corresponding to the E_z field in a metallic rectangular waveguide of sides a and b (i.e., the complete TM field) is given by

$$G_{ab}(x,x';y,y';z) = G_{oa}(x,x';z)G_{ob}(y,y';z) \tag{17}$$

where G_{oa} is given by eqn. 6, and G_{ob} is given by eqn. 6 with x,x' replaced by y,y' and a replaced by b. Clearly the imaging distances z_{11} for G_{oa} and G_{ob} are generally different, but eqn. 17 nevertheless allows us to compute at least a partial imaging when one of the two factors is at an imaging plane.

2.5 Circular Waveguides

Certain polygonal cross-section waveguides are also known to produce this type of imaging, but only the circular cross-section lends itself readily to analysis (6), (12)-(14). Only Fourier-type images were discussed in these sources. Recent work (15) has suggested the existence of Fresnel images in circular guides, though these images are nontrivial transforms of the input field rather than simply scaled and duplicated versions of it.

Fourier imaging follows immediately from the paraxial approximation to the propagation constant; for the TM_{om} modes of a metallic waveguide of radius b, we have

$$\beta_m \simeq k - j_{om}^2/2kb^2 \tag{18}$$

where j_{om} is the mth zero of the Bessel function $J_o(x)$. Invoking the asymptotic relation $j_{om} \sim (m - \frac{1}{4})\pi$, we see that

$$\exp(-i\beta_m z) \simeq \exp(-ikz + 4\pi i(m - \tfrac{1}{4})^2 z/z_{11}) \tag{19}$$

where $z_{11} = 8kb^2/\pi$. Even with the approximations we have made, image fidelity at $z = z_{11}$ is excellent if $kb >> 1$.

Some crude approximate methods for obtaining the field at $z = z_{11}/2$ are discussed in (15); exact computations based on the mode sum indeed show a well-defined focused image, but one which is transformed from one near the center to one concentrated in a ring around the outer boundary as shown in Figure 4. More work is needed to provide a simple yet accurate means of computing these Fresnel images.

3. FIELD COMPUTATIONS AT NON-IMAGE POINTS

At points intermediate between image planes, we evaluate the field using the equivalent image representation of the field at a nearby image plane. If in the case of a parallel-plate waveguide, the input field is a Gaussian beam, the field will be the result of the propagation of the Fresnel image field at $z = z_{pq}$ (including its images in the waveguide walls) just as if it were propagating and broadening in an infinite homogeneous medium. If p were increased, we could come arbitrarily close to an image plane, at the expense of a large number p of image beams. For small p on the other hand, we may be quite far from an image plane, and the broadening of these beams in getting to z may require an inordinately large number of reflected images to be summed up. These considerations lead us to choose the next larger integer than a/w as our optimum value of p (5). Examples of beam propagation through a parallel-plate waveguide can be found in (5).

4. CONCLUSION

The technique described in section 3 will carry over easily to a rectangular guide and to the coupling between parallel dielectric slabs. The extension to circular waveguides is more difficult, and much work remains to be done on this problem. As it stands, the use of Fourier and Fresnel images can go far towards improving the efficiency of ray or image techniques for field calculations at extremely long propagational distances in multimode waveguides.

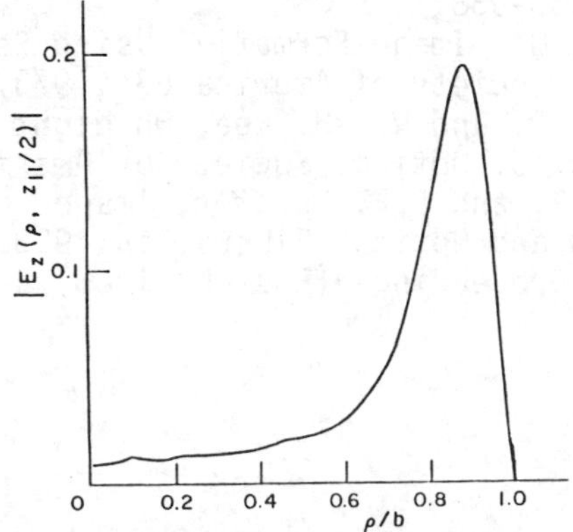

Figure 4: Fresnel image of a centered Gaussian beam at the input plane, observed at $z = z_{11}/2$ in a circular waveguide

REFERENCES

1. Felsen, L.B. and A.H. Kamel, Hybrid Ray-Mode Formulation of Parallel Plane Waveguide Green's Functions, IEEE Trans. Antennas and Propagation 29 (1981) 637-649.
2. Felsen, L.B. and S.-Y. Shin, Rays, Beams and Modes Pertaining to the Excitation of Dielectric Waveguides, IEEE Trans. Microwave Theory and Techniques 23 (1975) 150-161.
3. Yeh, C., Optical Waveguide Theory, IEEE Trans. Circuits and Systems 26 (1979) 1011-1019.
4. Kuester, E.F., Propagation Constants for Linearly-Polarized Modes of Arbitrarily-Shaped Optical Fibers or Dielectric Waveguides, Optics Letters 8 (1983) 192-194.
5. Chang, D.C. and E.F. Kuester, A Hybrid Method for Paraxial Beam Propagation in Multimode Optical Waveguides, IEEE Trans. Microwave Theory and Techniques 29 (1981) 923-933.
6. Rivlin, L.S. and V.S. Shul'dyaev, Multimode Waveguides for Coherent Light, Radiophysics and Quantum Electronics 11 (1968) 318-321.
7. Weston, D.E., A Moiré Fringe Analog of Sound Propagation in Shallow Water, J. Acoustical Society of America 32 (1960) 647-654.
8. Weston, D.E., Sound Focusing and Beaming in the Interference Field due to Several Shallow-Water Modes, J. Acoustical Society of America 44 (1968) 1706-1712.
9. Apostol, T.M., Introduction to Analytic Number Theory (New York, Springer, 1976).
10. Kuester, E.F., G.S. Dow and D.C. Chang, Coupling and Imaging of Gaussian Beams in Parallel Dielectric Slab Waveguides, Archiv Elektronik Übertragungstechnik 36 (1982) 427-435.
11. Voges, E. and R. Ulrich, Self-Imaging by Phase Coincidences in Rectangular Dielectric Waveguides, in 6th European Microwave Conference Proceedings (Rome, Italy, 1976) pp. 447-451.
12. Andrychuk, D., A Multi-Image Optical System, Applied Optics 3 (1964) 933-938.
13. Bryngdahl, O., Image Formation Using Self-Imaging Techniques, J. Optical Society of America 63 (1973) 416-419.
14. Bryngdahl, O. and W. -H. Lee, On Light Distribution in Optical Waveguides, J. Optical Society of America 68 (1978) 310-315.
15. Mahnad, A.R. and E.F. Kuester, Image Formation in Circular Waveguides and Optical Fibers, in 1983 International Microwave Symposium Proceedings (Boston, 1983) pp. 122-124.

GAUSSIAN BEAM TRANSFORMATION THROUGH LENSES AND BY CURVED GRATINGS

J. Jacob and H.-G. Unger

Institut für Hochfrequenztechnik,
Technische Universität Braunschweig,
Postfach 3329, D-3300 Braunschweig

1 INTRODUCTION

In planar and integrated optics film waveguides serve to confine light in their fundamental mode. Usually the film waves are launched from lasers, fibres, planar strip guides, or strip-derived structures, and then have a nearly Gaussian intensity distribution in the transverse direction parallel to the film waveguide. Film lenses and film prisms as well as gratings may serve to transform or deflect such Gaussian beams in the fundamental film mode and also to refocuse them into a fibre guide or strip waveguide.

To analyse the transformation through lenses and by curved gratings the two-dimensional Gaussian beam is approximated by the field of a line source in an imaginary location /1/. The complex rays radiating from the complex source point may then be traced through a film lens or prism or their reflection in film gratings be determined by analytical continuation of these objects into the complex domain /1/, and by applying Fermat's principle and the complex extension of the geometrical theory of diffraction.

2 FIBRE TO FILM WAVEGUIDE CONNECTION

As a typical example for the excitation of a wave beam in the fundamental mode of the film waveguide we consider the butt joint of a single-mode fibre to the film waveguide in Fig. 1.

In the single-mode fibre the cross-sectional field distribution is nearly Gaussian /2/, so that for a given polarisation, the electric field may be written as

$$E = E_o \exp\left(-\frac{x^2 + y^2}{w_o^2}\right)\exp(-j\beta_F z)$$

where E_o is the maximum amplitude, w_o the spot size and β_F the propagation constant of the fundamental mode in the fibre. A harmonic time dependence $\exp(j\omega t)$ is assumed.

We assume that the thickness, the indices of refraction and the position of the single-mode film waveguide are chosen so as to give a maximum coupling efficiency in the y-direction between the fibre and the fundamental film mode. In the x-direction a two-dimensional Gaussian beam in the fundamental film mode with spot size w_o at its waist at z=0 is excited. For single-mode fibres with a relative index difference between o.1% and o.5% operated at the wavelength λ=1.5μm in free space the spot size is of the order $3\mu m < w_o < 7.5\mu m$.

At some distance from the fibre-to-film joint the y-dependence of the field in and near the film is that of its fundamental mode and does not change in the film. The fundamental mode with its phase constant β propagates as a plane uniform wave in a homogeneous medium with the index of refraction

$$n = \beta/k$$

called the effective index of refraction of the mode of the film waveguide at the wavelength λ; $k=2\pi/\lambda$ is the wavenumber in free space. We consider only fields polarised in the y-direction

$$E_y = E_y(x,z)$$

which correspond to the TM_o film mode. In the two-dimensional model, only the field dependence on the x- and z- coordinates needs to be considered.

The two-dimensional Gaussian beam may be approximated by the field of a line source of electric current I flowing in y-direction, and located at the imaginary point /1/

$$\bar{x} = 0, \quad \bar{z} = -jb, \quad \text{with } b = \frac{w_o^2\beta}{2}. \tag{1}$$

At any real point (x,z) with the complex distance

$$\bar{\rho} = \sqrt{x^2 + (z + jb)^2}$$

from the source point this line source generates the electric field

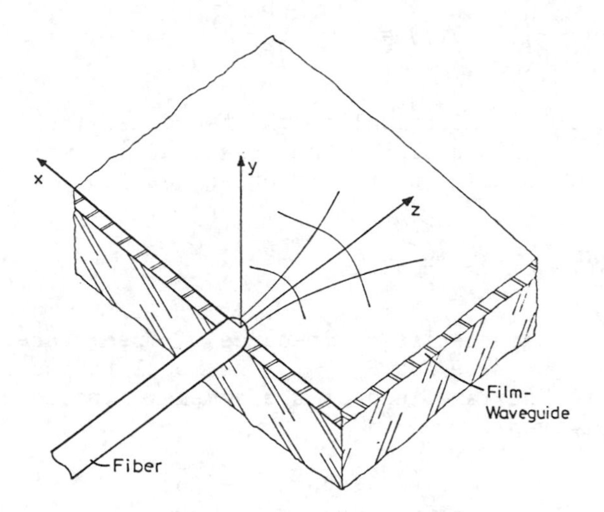

Fig. 1: Single mode fibre butt-joined to a single mode film waveguide.

$$E_y = - \frac{\beta^2 I}{4\omega n^2 \varepsilon_o} H_o^{(2)} (\beta\bar{\rho})$$

where $H_o^{(2)}$ is the zeroth order Hankelfunction of the second kind and ε_o is the free space permittivity. The bar on any of the length or coordinate symbols denotes their complex character.

Using the asymptotic approximation of the Hankelfunction

$$H_o^{(2)} (\beta\bar{\rho}) \simeq \sqrt{\frac{2j}{\Pi\beta\bar{\rho}}}\, e^{-j\beta\bar{\rho}} \quad \text{for } |\beta\bar{\rho}| >> \frac{1}{2}$$

and the definition of the near-to-the-axis region according to

$$x^2 << z^2 + b^2,$$

one obtains the near-to-the axis field

$$E_y = - \frac{\beta^2}{\omega n^2 \varepsilon_o} I \sqrt{\frac{1}{4\Pi\beta^2 w_o w}} \exp(\frac{w_o^2\beta^2}{2}) \exp\left[-j\beta z - j\frac{\beta x^2}{2R(z)} - \frac{x^2}{w^2(z)} + j\frac{1}{2}\arctan\frac{z}{b}\right] \quad (2)$$

with

$$w(z) = \sqrt{w_o^2 + \left(\frac{2z}{w_o\beta}\right)^2} \quad , \; R(z) = z + \frac{w_o^4\beta^2}{4z} \quad .$$

This field expression is identical to the two-dimensional paraxial Gaussian beam with amplitude E_o on the z-axis at its waist at z=0 when the current of the line source is chosen as /1/

$$I = - E_o n\eta_o \sqrt{\Pi} 2w_o \exp\left(-\frac{w_o^2\beta^2}{2}\right) ,$$

$\eta_o = \sqrt{\varepsilon_o/\mu_o}$ is the characteristic impedance of free space. In Eqn. 2 $w(z)$ is the local spot size and $R(z)$ is the local radius of curvature of the phase front. The far field of the complex line source for $|\beta\bar{\rho}| >> 1/2$ is now given by

$$E_y = \frac{E_o}{2} \sqrt{\beta} w_o (1+j) \exp\left(-\frac{w_o^2\beta^2}{2}\right) \frac{e^{-j\beta\bar{\rho}}}{\sqrt{\bar{\rho}}} = A \frac{e^{-j\beta\bar{\rho}}}{\sqrt{\bar{\rho}}} \qquad (3)$$

where A is a constant. Eqn. 3 describes a complex ray with cylindrical complex phase front propagating along a straight line from the imaginary source point to the real observation point.

3 GAUSSIAN BEAM REFRACTION AT A CURVED INTERFACE

If a curved boundary with a change in effective index occurs between the fibre joint and a real observation point the complex ray will intersect with this boundary at a certain complex point /1/: the complex point on the boundary lies on the analytic continuation of the given real boundary. One obtains the complex intersection point by applying Fermat's principle in order to find the refracted complex ray between the imaginary source point and the real observation point /1/. The complex source point method allows us now to use the geometrical theory of diffraction to determine how the beam with its complex phase curvature is transformed by the boundary. Eqn. 3 is in a suitable form to apply the geometrical theory of diffraction.

When, as shown in Fig. 2, a beam with a phase front of radius of curvature $\overline{\rho_i}$ is incident on a boundary with local radius of curvature $\overline{\rho_c}$ at an angle $\overline{v_i}$, then the refracted beam leaves the boundary with an angle $\overline{v_t}$ and a radius of curvature $\overline{\rho_t}$ of its phase front. The second order approximation of the geometrical theory of diffraction yields /3/

$$\bar{\rho}_t = \left[\bar{\rho}_i \bar{\rho}_c \cos \bar{v}_t\right] / \left[(\bar{m} - 1)\bar{\rho}_i + \bar{m}\rho_c \cos \overline{v_i}\right] \qquad (4)$$

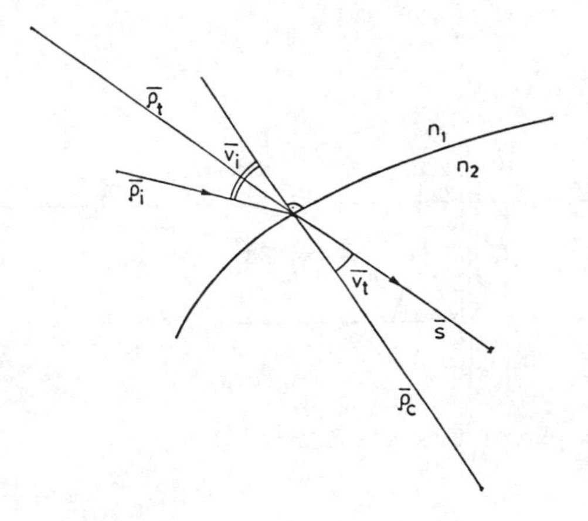

Fig. 2: Transformation of a complex ray by a curved boundary with effective index change.

with

$$\bar{m} = \frac{n_1 \cos \bar{v}_i}{n_2 \cos \bar{v}_t}$$

and, resulting from Fermat's principle

$$n_1 \sin \bar{v}_i = n_2 \sin \bar{v}_t \ .$$

In the complex extension of the theory the quantities $\overline{\rho_i}$, $\overline{\rho_t}$, $\overline{\rho_c}$, $\overline{v_i}$, $\overline{v_t}$ are in general also complex. With an incident field in the form of Eqn. 3, after it has travelled over a complex distance $\bar{s}$ from the boundary, the analytic expression for the field is

$$E(\bar{\rho}_i + \bar{s}) = TA\frac{1}{\sqrt{\bar{\rho}_i}} \sqrt{\frac{\bar{\rho}_t}{\bar{\rho}_t + \bar{s}}}\ e^{-jk\ (n_1\bar{\rho}_i + n_2\bar{s})}\ , \qquad (5)$$

where T is the transmission coefficient of a plane wave incident with the angle $\overline{v_i}$ on a plane boundary.

4 TRANSFORMATION OF THE GAUSSIAN BEAM BY A FILMLENS

As a typical example for the transformation of a Gaussian beam by a film lens we consider the arrangement in Fig. 3, in which the lens is designed to refocuse an incident beam under ideal circumstances. To this end it transforms the divergent phase front (1) of the

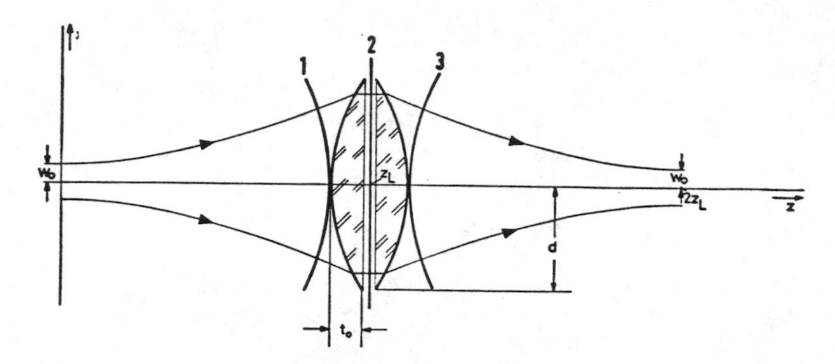

Fig. 3: Lens as phase transformer.

incident beam into a convergent phase front (3) with opposite curvature behind the lens. The phase front (2) in the middle of the lens is plane. The ideal lens should be infinitely thin to perform exactly this phase transformation. We consider real lenses with finite thickness $2t_o$, so designed that they transform the parabolic phase front of the paraxial Gaussian beam properly. For this a film region with raised effective index of refraction n_1 and parabolic boundaries can be used. If the lens boundaries are specified according to

$$z = z_L \pm (t_o - ax^2), \quad |x| < d \tag{6}$$

then the condition for the phase transformation together with the second order approximation Eqn. 2 for the paraxial beam yield

$$a = \frac{n}{2R(z_L)(n_1 - n)} \quad . \tag{7}$$

The local radius of curvature of the lens boundaries is

$$\rho_1 \; (x) = \pm \frac{1}{2a}(1 + (2ax)^2)^{3/2} \quad . \tag{8}$$

The upper signs in Eqns. 6 and 8 refer to the front side of the lens and the lower signs to its back side. The analytic continuation of Eqns. 6 and 8 is obtained by letting x become complex.

To assure that the power density of the Gaussian beam at the edges of the lens is less than 2% of its maximum on the z-axis the lens is chosen to be $d=1.42\; w(z_L)$ wide.

Fig. 4 shows a complex ray propagating from the imaginary source point to the real observation point A(x,z) and intersecting with the lens boundaries at the complex points $P_1\; (\bar{x}_1,\bar{z}_1)$ and $P_2\; (\bar{x}_2, \bar{z}_2)$. The complex phase shift along this ray path is

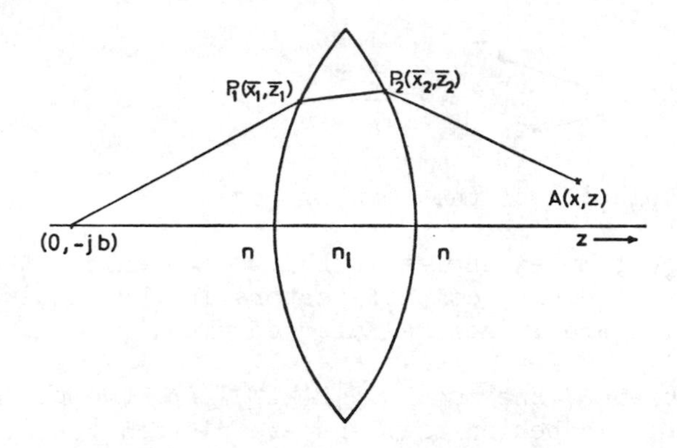

Fig. 4: Complex ray path trough the film lens.

$$\phi_A = nk\sqrt{\bar{x}_1^2 + (a\bar{x}_1^2 + z_L - t_o + jb)^2} + n_1 k\sqrt{(\bar{x}_1 - \bar{x}_2)^2 + (a(\bar{x}_2^2 + \bar{x}_2^2) - 2t_o)^2} + nk\sqrt{(\bar{x}_2 - x)^2 + (z_L + t_o - a\bar{x}_2^2 - z)^2} .$$

According to Fermat's principle the phase shift ϕ_A must be stationary relative to the points P_1 and P_2. This yields the following system of nonlinear equations

$$\begin{aligned} &\mathrm{Re}\,(\partial\phi_A/\partial\bar{x}_1) = 0 \\ &\mathrm{Im}\,(\partial\phi_A/\partial\bar{x}_1) = 0 \\ &\mathrm{Re}\,(\partial\phi_A/\partial\bar{x}_2) = 0 \\ &\mathrm{Im}\,(\partial\phi_A/\partial\bar{x}_2) = 0 \end{aligned} \qquad (9)$$

for the four unknown variables Re $(\bar{x}_1)$, Im $(\bar{x}_1)$, Re $(\bar{x}_2)$ and Im $(\bar{x}_2)$. This set of equations has to be solved numerically. The computer time needed for this numerical solution is however quite moderate. Any errors in $\bar{x}_1$ and $\bar{x}_2$ cause only second order errors in the stationary phase shift ϕ_A.

When the ray path is known the field at the point A(x,z) can be calculated explicitly by evaluating Eqns. 4 and 5 for both boundaries. We assume that the transition from the base film to the lens region is tapered smoothly enough such that no reflections occur. Thus the product of the transmission coefficients for the front side and the back side of the lens becomes $T_1 . T_2 = 1$.

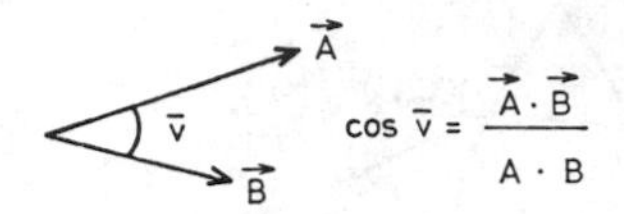

Fig. 5: Calculation of the complex cosine.

The cosines of the complex angles in Eqn. 4 are calculated according to Fig. 5 where $\vec{A}$, $\vec{B}$ denote complex vectors in the required direction and the moduli A, B are in general also complex.

In Figures 6 and 7 the modulus $|E(2z_L)|$ and the phase shift $\varphi(x)-\varphi(0)$ of the field behind the lens are plotted versus x/w_o for w_o=7.5µm and w_o =3µm. The field amplitudes are normalised by taking E_o =1, and the effective index of refraction of the base film is chosen to be n=1.4. The distance from the fibre-to-film joint to the film lens is z_L=10b. At $z=z_L$ the paraxial Gaussian beam has spread out to $w(z_L) \approx 10w_o$. The film lens with $n_1 - n = 0.1$ and $d/w(z_L) = 1.42$ is given a thickness according to Eqn. 7. At the operating wavelength λ = 1.5µm the ratio d/t_o amounts to 2.23 for w_o = 7.5µm but only to 0.88 for w_o = 3µm: the strong deviation from the ideal infinitely thin lens ($d/t_o \rightarrow \infty$) becomes more pronounced with decreasing w_o. The film lens should refocuse the beam at $z = 2z_L$ yielding a plane phase front there with $\varphi(x)-\varphi(0)=0$ and a modulus $|E(2z_L)|=|E(z=0)|$ equal to the incident field at z=0. The deviation from this behaviour is mainly due to the finite thickness of the film lens so that the spreading of the beam while it is propagating through the film lens is no more negligible.

As the beam is not yet focused at $z=2z_L$, we search for the coordinate $z=2z_L+\Delta z$ where it is. From the calculated field at two points x=0 and $x=x_p$, $z=2z_L$ one obtains

$$w(2z_L) = x_p/\sqrt{\ln(|E(0)|/|E(x_p)|} \tag{10}$$

and the phase curvature

$$R(2z_L) = \frac{\beta x_p^2}{2(\varphi(x_p) - \varphi(0))}$$

of the transformed beam at $z=2z_L$, assuming that the field distribution is still Gaussian. The focus shift is thus

$$\Delta z = \left(\frac{4R}{w^4\beta^2} + \frac{1}{R}\right)^{-1} . \tag{11}$$

Actually the beam behind the lens is not exactly Gaussian anymore and Eqn. 11 is only an approximation for the focal shift.

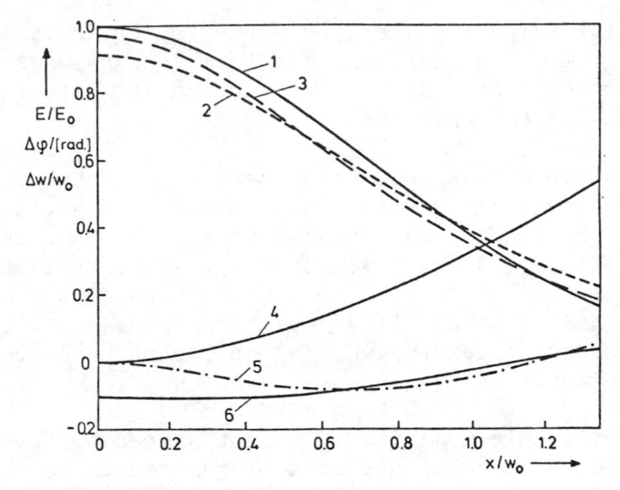

Fig. 6: w_o = 7.5 μm; λ = 1.5 μm; n = 1.4; n_1 = 1.5;
b_o = 164.9 μm; d/t_o = 2.23,
z_L = 1649 μm; Δz = 82.7 μm .

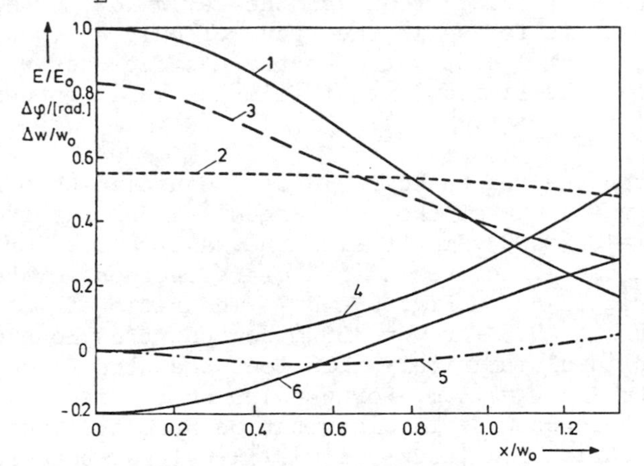

Fig. 7: w_o = 3 μm; λ = 1.5 μm, n = 1.4; n_1 = 1.5;
b_o = 26.4 μm; d/t_o = 0.88
z_L = 264 μm; Δz = 100μm .

Figs:6 and 7: Refocused field behind the film lens

1	$\|E(0)\|/E_o$	4	$\varphi(x) - \varphi(o)$ at $z = 2z_L$
2	$\|E(2z_L)\|/E_o$	5	$\varphi(x) - \varphi(o)$ at $z = 2z_L+\Delta z$
3	$\|E(2z_L+\Delta z)\|/E_o$	6	$[w(2z_L+\Delta z)-w_o]/w_o$

The dash-dotted lines in Figs. 6 and 7 give the phase fronts at $z=2z_L+\Delta z$. Their deviation $\varphi(x)-\varphi(0)$ from zero is of higher order. The phase fronts at $z=2z_L+\Delta z$ are hence as close to a plane phase front as they will ever come under the given circumstances. For this reason we are justified to call Δz the focal shift and $z=2z_L+\Delta z$ the actual location of the focal plane. The higher order deviation of the phase fronts in the focus at $z=2z_L+\Delta z$ from a plane phase front shows that the parabolic form of the lens boundaries does not guarantee an exact phase transformation.

The refocused field distribution at $z=2z_L+\Delta z$ is nearly as well confined to the beam axis as was the incident field distribution, but, as Fig. 6 and 7 show, it is not anymore exactly Gaussian: from the plots of $[w(2z_L+\Delta z) - w_o]/w_o$ versus x/w_o, where Eqn. 10 was evaluated at $z = 2z_L+\Delta z$ with $x_p=x$, one obtains an idea of the deviation from the Gaussian distribution. This deviation increases with decreasing w_o and becomes excessive when w_o approaches the order of only a few wavelengths.

If an outgoing single mode fibre is butt-joined to the film waveguide at the focus at $z=2z_L+\Delta z$, or a planar single mode strip waveguide is placed there, the refocused field distribution should beGaussian with its spot size w_o matched to the fundamental mode of the outgoing waveguide in order to give a maximum coupling efficiency between the focused beam and the waveguide. Since this is not exactly possible for w_o in the order of only a few wavelengths (i.e. $3\mu m < w_o < 7.5\mu m$ for $\lambda=1.5\mu m$) the coupling efficiency will be reduced. This shows the limitations of film lenses in integrated optics.

For comparison the refocused field of a Gaussian beam incident with a spot size $w_o= 100\mu m$ at the wavelength $\lambda=1.5\mu m$ has been calculated for $z_L=10b = 293.2mm$, taking the same values for the ratio $d/w(z_L)$ and the effective indices of refraction in the slab and in the lens region as in Fig. 6 and 7. The ratio of lens thickness to lens width is 29.5 for $w_o= 100\mu m$ so that this lens comes much closer to an ideal infinitely thin lens than the lenses in the devices for $w_o=3\mu m$ and $w =7.5\mu m$. For $w_o= 100\ \mu m$ the relative amplitude deviation of the refocused field distribution from the incident Gaussian field distribution $|E(2z_L)-E(0)|/|E(0)|$ has been found to be less than $5x10^{-4}$ for any x in the range $0 < x < 1.42w_o$. Further the phase difference $|\varphi(x)-\varphi(0)|$, giving the deviation of the phase front at $z=2z_L$ from the plane phase front of the incident field, remains less than $2.3x10^{-3}$ rad. This example shows that Gaussian beams with spot sizes sufficiently large compared with the wavelength are transformed nearly without distortions by parabolic film lenses.

5 GAUSSIAN BEAM DEFLECTION AND TRANSFORMATION BY CURVED GRATINGS

If a film grating is placed in the path of a Gaussian beam in the fundamental film mode, and if the grating lines have such a distance from each other that the individual reflections from each line interfere constructively with each other, the whole beam or part of it may be reflected or deflected. If the grating lines are curved appropriately the reflected or deflected beam may in addition be transformed in its curvature and also be refocused.

Fig. 8 shows such a curved grating upon which the Gaussian beam I is incident making an angle of incidence ϑ of its axis with the line z normal to the first grating line at the point of intersection. Also shown in Fig. 8 is a reflected Gaussian beam II which has nominally the same distribution as the incident beam but is the mirror image of the incident beam with respect to the above mentioned line z normal to the grating. The particular grating that will at least approximately transform the incident beam into this reflected beam (or into any other nearly Gaussian beam) is defined by the interference pattern of the two beams.

In terms of their individual coordinates x_1, z_1 and x_2, z_2 in Fig. 8 the fields of the two Gaussian beams have the following complex source point presentations

$$\text{beam I} \qquad E_I = \frac{A}{\sqrt{\bar{\rho}_1}} e^{-j\beta\bar{\rho}_1} \text{ with } \bar{\rho}_1 = \sqrt{x_1^2 + (z_1 + jb)^2} \tag{12}$$

$$\text{beam II} \qquad E_{II} = \frac{A}{\sqrt{\bar{\rho}_2}} e^{-j\beta\bar{\rho}_2} \text{ with } \bar{\rho}_2 = \sqrt{x_2^2 + (z_2 + jb)^2} \,. \tag{13}$$

For any real point of observation the complex distance $\bar{\rho}_1$ of beam I has the modulus

$$|\bar{\rho}_1| = \left[(x_1^2 + z_1^2 - b^2)^2 + (2z_1 b)^2\right]^{1/4} \tag{14}$$

and the phase

$$\varphi_{\rho 1} = \frac{1}{2} \arctan \frac{2z_1 b}{x_1^2 + z_1^2 - b^2} \,. \tag{15}$$

With the modulus $|\bar{\rho}_1|$ and the phase $\varphi_{\rho 1}$ of $\bar{\rho}_1$ the phase φ_1 of the field E_I at any real point of observation follows as

$$\varphi_1 = -\beta|\bar{\rho}_1| \cos \varphi_{\rho 1} - \frac{1}{2} \varphi_{\rho 1} \tag{16}$$

and likewise follows the phase φ_2 of E_{II} from $|\bar{\rho}_2|$ and $\varphi_{\rho 2}$.

The grating lines are defined by the interference pattern of E_I and E_{II} and thus are given by

$$\varphi_1 + \varphi_2 - \varphi_{ci} = 0 \tag{17}$$

where the Bragg-condition requires for the phase shift between neighbouring lines i-1 and i:

$$\varphi_{ci} = \varphi_{c(i-1)} - 2\pi \,. \qquad (18)$$

Eqn. (17) describes by way of the analytic expressions (14), (15) and (16) the grating lines in their real location.

The complex source point method requires the analytic continuation of the grating lines into the complex domain. It can be obtained by simply allowing the coordinates in the grating equation (17) to become complex. For the complex coordinates $\overline{x_1}$, $\overline{z_1}$ and $\overline{x_2}$, $\overline{z_2}$ the quantities $|\overline{\rho_1}|$ and $\varphi_{\rho 1}$ according to Eqn. (14) and (15) respectively as well as $|\overline{\rho_2}|$ and $\varphi_{\rho 2}$ represent no longer the moduli and phases of $\overline{\rho_1}$ and $\overline{\rho_2}$, nevertheless will the grating equation (17) be analytically extended with them into the complex domain, because for all real coordinates Eqn. (17) represents the desired interference pattern.

We now need to determine the complex point on the analytic continuation of a grating line, at which the complex ray reflects, when it is to reach a certain real point of observation. In order to find this complex point of reflection we transform from the beam coordinates x_1, z_1 and x_2, z_2 to the coordinates x, z of Fig. 8 by means of the coordinate transformations

$$\begin{pmatrix} x+a \\ z+c \end{pmatrix} = [M_1] \begin{pmatrix} x_1 \\ z_1 \end{pmatrix} \text{ and } \begin{pmatrix} x-a \\ z+c \end{pmatrix} = [M_2] \begin{pmatrix} x_2 \\ z_2 \end{pmatrix} \quad . \qquad (19)$$

The grating equation (17) may then be expressed in terms of $\overline{x}$ and $\overline{z}$ and written as

$$F(\overline{x},\overline{z}) = 0 \qquad (20)$$

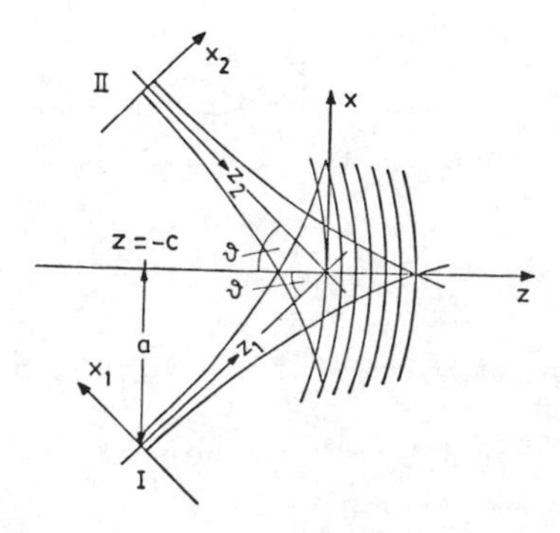

Fig. 8: Reflection of a Gaussian beam by a curved grating.

thus continuing it again analytically into the complex domain.

The complex phase shift ϕ from the complex source point of the incident beam to a complex point on the analytic continuation of a particular grating line, and from there to the real point of observation, may now also be expressed in terms of $\bar{x}$, $\bar{z}$ by using Eqn. (12) and the coordinate transformations (19). The actual complex point of reflection is then found by invoking Fermat's principle in the form

$$\frac{d\phi}{d\bar{x}} = \frac{\partial\phi}{\partial\bar{x}} - \frac{\partial F/\partial\bar{x}}{\partial F/\partial\bar{z}}\frac{\partial\phi}{\partial\bar{z}} = 0 \quad . \tag{21}$$

The complex equations (20) and (21) contain 4 conditions, which determine the real and imaginary components of $\bar{x}$ and $\bar{z}$ for the respective point of deflection. The local complex radius of curvature of the grating line at this complex point of deflection may now also be determined by evaluating

$$\bar{\rho}_c = \frac{\left[(\partial F/\partial\bar{x})^2 + (\partial F/\partial\bar{z})^2\right]^{3/2}}{\begin{vmatrix} \frac{\partial^2 F}{\partial\bar{x}^2} & \frac{\partial^2 F}{\partial\bar{x}\partial\bar{z}} & \frac{\partial F}{\partial\bar{x}} \\ \frac{\partial^2 F}{\partial\bar{x}\partial\bar{z}} & \frac{\partial^2 F}{\partial\bar{z}^2} & \frac{\partial F}{\partial\bar{z}} \\ \frac{\partial F}{\partial\bar{x}} & \frac{\partial F}{\partial\bar{z}} & 0 \end{vmatrix}} \tag{22}$$

Fig. 9 shows a grating line with its local radius of curvature $\bar{\rho}_c$ and the incident complex ray with its complex source point distance $\bar{\rho}_i$. At the point of reflection a vector normal to the grating line has the $(\bar{x}, \bar{z})$ components

$$(\partial F/\partial\bar{x} \; , \;\; \partial F/\partial\bar{z})$$

and makes an angle $\bar{v}_i$ with the incident complex ray that follows from Fig. 5.

With these quantities the geometrical theory of diffraction yields for the reflected beam the following complex source point distance from the point of reflection on the grating line

$$\bar{\rho}_r = \frac{\bar{\rho}_i\,\bar{\rho}_c \cos\bar{v}_i}{2\bar{\rho}_i + \bar{\rho}_c\cos\bar{v}_i} \quad . \tag{23}$$

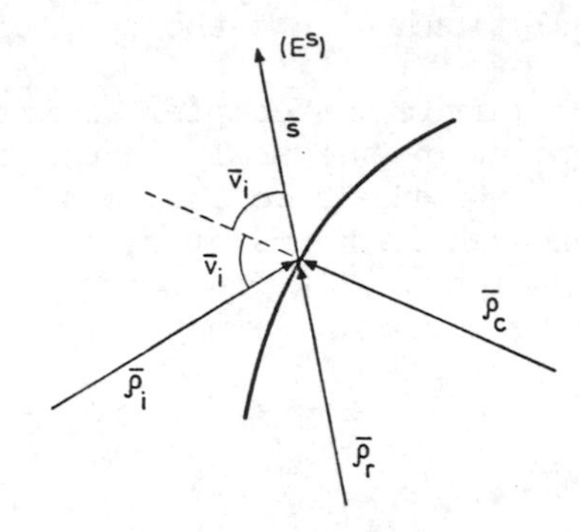

Fig. 9: Partial reflection of a complex ray by a curved grating line.

With it the reflected field from one particular grating line has the complex source point presentation

$$E^S = R_i \frac{A}{\sqrt{\bar{\rho}_i}} \sqrt{\frac{\bar{\rho}_r}{\bar{\rho}_r + \bar{s}}}\ e^{-j\beta(\bar{\rho}_i + \bar{s})} \tag{24}$$

where $\bar{s}$ is the complex distance from the point of reflection on the complex extension of the grating line to the real point of observation.

Still unknown in Eqn. (24) is the ratio R_i of the reflected field amplitude to the field amplitude that the incident beam would have at the point of reflection, were it not weakened by reflections at previous grating lines. If the beam were incident from an infinitely far removed source point on to straight grating lines, then it would correspond to a plane uniform wave incident on a plane grating. The incident wave would under these conditions decay into the grating according to

$$|E_i| = |E_o| \frac{\cosh \varkappa(L-z)}{\cosh \varkappa L} \tag{25}$$

while the reflected wave would grow from zero at the backside of the grating according to

$$|E_r| = |E_o| \frac{\sinh \varkappa(L-z)}{\cosh \varkappa L}\ . \tag{26}$$

This continuous decay of the incident beam into the grating, as well as the continuous growth of the reflected beam out of the grating come about by the mutual coupling of both with the coupling coefficient $\varkappa$, that accounts for the multiple reflections which incident and reflected beam components experience at the grating lines. Although the multiple reflections are the actual cause for the interaction between incident and reflected beams, it appears from Eqn. (26) as if the grating lines between z and z+Δz contribute with

$$\Delta|E_r| = |E_o| \frac{\sinh \varkappa(L-z) - \sinh \varkappa(L-z-\Delta z)}{\cosh \varkappa L} \quad (27)$$

to the reflected beam. We therefore let

$$R_i = \frac{\sinh \varkappa(L-z) - \sinh \varkappa(L-z-\Delta z)}{\cosh \varkappa L} \quad (28)$$

in Eqn. (24) and obtain with it a reasonable approximation for the contribution to the reflected beam not just from one grating line, but from all the grating lines between z and z+Δz. For the evaluation of Eqn. (26) together with Eqn. (28) the total depth L of the grating will usually be specified, with grating lines strong enough to convert most of the power of the incident beam into the reflected beam. We therefore adjust the coupling coefficient $\varkappa$ accordingly. In the numerical examples which follow we always choose $\varkappa$ so that

$$|E_r(o)/E_o| = \tanh \varkappa L = 0.99 \quad (29)$$

specifying with it that only 2% of the incident beam power will not be converted into the reflected beam.

To examine the practical situation where the incident beam is excited by the transversely fundamental mode of an index-guided diode laser and must be launched into a single-mode fibre, we have assumed for the spot sizes at the waist of the incident as well as the reflected beam

$$w_o = 4 \text{ µm}$$

and taken the vacuum wavelength to be λ = 1.3 µm. The effective index of the fundamental film waveguide mode was assumed to be N = 1.92 and refers to the TM_o mode of a film waveguide that consist of a As_2S_3 film which is 0.35 µm thick and for monolithic integration deposited on an InP-substrate with SiO_2 as a 2 µm thick buffer layer sandwiched between substrate and film. The grating was placed so that its first line intersects the axis of the incident beam at an angle of 45°, thus causing the reflected beam to have its axis at right angle with the axis of the incident beam. The distance from the waist of the incident beam to the intersection of its axis with the first grating line was chosen at 1.5 mm so that according to Eqn. (2) its spot size has widened to w = 81 µm. Such a wide beam will appear to be necessary because for the reflected beam to maintain the Gaussian distribution and be refocused to the small spot size w_o at its waist, the incident beam should not penetrate much deeper into the grating than it is wide. Otherwise the many reflections which superimpose in the reflected beam come from a too widely spaced range to still interfere with each other properly.

On the other hand the grating can have only a limited modulation index for the effective index of the fundamental mode and needs a large number of lines for the incident beam to decay sufficiently.

From the results of the numerical evaluation the relative field amplitude and the relative phase of the reflected beam, and of contributing components, in the plane where it should nominally be refocussed and have its waist, are plotted versus the relative distance x_2/w_o from the nominal axis of the reflected beam in Figs. 10 to 12. The solid line represents as a reference the amplitude distribution of the incident beam in its waist. The dashed lines are amplitude and phase distribution of the reflected beam. The dotted line is the relative amplitude distribution of that particular contribution to the reflected beam that has been calculated to originate from the group of grating lines which are located at $z=L/10$, the dash-dotted line is the corresponding contribution from $z=L$. Both contributions are plotted in the Figs. without the reflection factor R_i according to Eqn. (28) being applied to them.

Fig. 10 shows results for a grating only L = 25 μm deep, into which the incident beam penetrates only to a depth which is a small fraction of its width. The reflected beam amplitude differs only little from its perfect reference and is slightly shifted against it. Its phase distribution indicates a tilt of the phase front as if the reflected beam comes not from the location where the incident beam meets the first grating line, but from a location within the grating. The depth of penetration of the incident beam into the grating, which this tilt indicates and the shift of the reflected beam corresponds to a kind of Goos-Haenchen shift which is associated with the Bragg-reflection of wave beams /4/.

The contribution to the reflected amplitude distribution from $z=L/10$ merges with the reference distribution and a dotted curve does therefore not appear in Fig. 10. The dash-dotted curve for the contribution from $z=L$ is shifted against the reference distribution and some what narrower with the peak value correspondingly large.

These deviations from the reference distribution become more pronounced when the grating has less modulation and is made deeper so that the incident beam penetrates farther into it. Figs. 11 and 12 show these effects for L = 100 and 200 μm respectively. But even for L = 200 μm when the incident beam penetrates farther into the grating, than it is wide, is the reflected beam at its nominal waist still reasonably well focused; only its skew phase distribution indicates that the reflection comes effectively from deeper inside the grating.

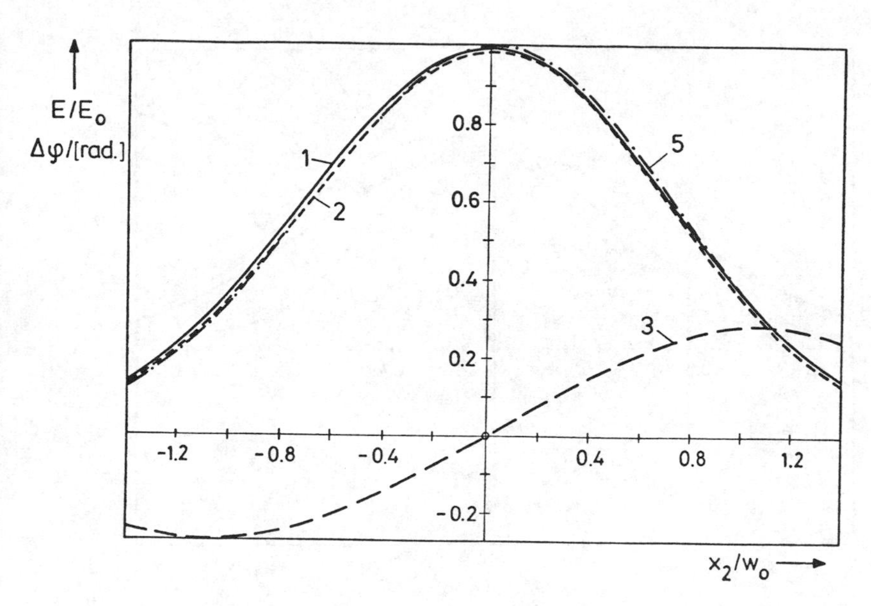

Fig. 1o: L = 25 μm, 52 grating lines .

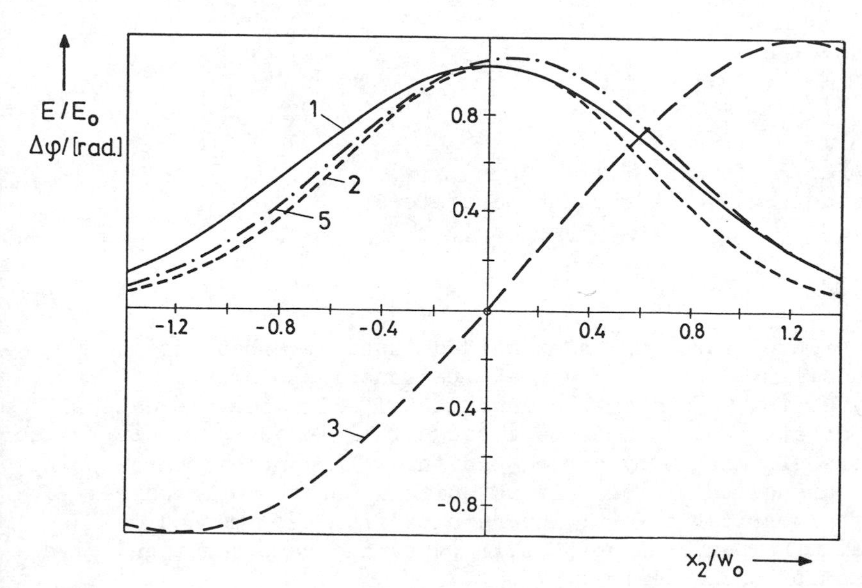

Fig. 11: L = 100 μm, 213 grating lines .

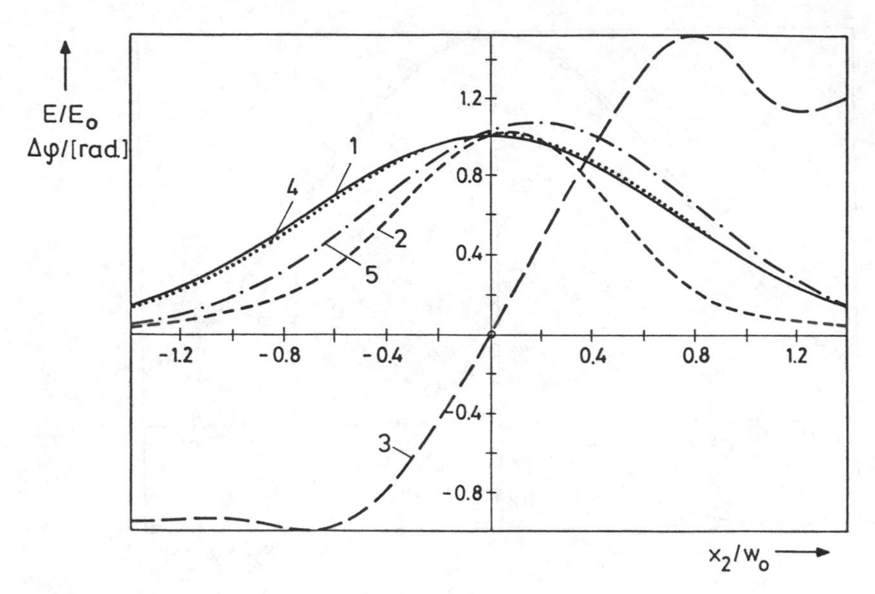

Fig. 12: L = 200 μm, 435 grating lines .

Figs. 1o, 11 and 12: Reflected field from the focussing grating:

1 Gaussian
2 $|E_r|$
3 $\varphi_r(x_2) - \varphi_r(0)$
4 $|\Delta E_r|/R_i$ from z = L/10
5 $|\Delta E_r|/R_i$ from z = L
(See the text for the other parameters)

6 CONCLUSION

When single-mode fibres or index guided lasers are butt-joined to single-mode film waveguides they excite Gaussian beams with relatively narrow waists in the fundamental film mode. To calculate the transformation of such beams through film lenses or their reflection and transformation by curved gratings the complex source point method appears quite useful. It is quite accurate but requires only modest numerical effort. The results from this method may thus explain the performance of film waveguide components and give guide lines for their design.

REFERENCES

/1/ G. Ghione, I. Montrosset and R. Orta. Complex Rays, this book.

/2/ A.W. Snyder. Understanding Monomode Optical Fibers, Proc. of the IEEE, vol. 69, no. 1, January 1981, p.6 .

/3/ L.B. Felsen and N. Marcuwitz. Radiation and Scattering of Waves, Prentice-Hall Inc., Englewood Cliffs, N.J., 1973 .

/4/ J. Jacob and H.-G. Unger. A Goos-Haenchen Effect for Bragg Reflection, to be published in AEÜ.

GAUSSIAN BEAMS IN ANISOTROPIC ELASTIC MEDIA AND APPLICATIONS

Andrzej Hanyga[1,2]

[1] Institute of Geophysics, 00973 Warszawa, Pasteura 3, Poland

[2] On leave of absence, now at NTNF/NORSAR, Norway

1 INTRODUCTION

We discuss here an alternative method of dealing with caustics in dynamic ray tracing (DRT). Instead of tracing WKBJ signals which propagate exactly along rays we shall trace signals carried by asymptotic solutions of special form concentrated near a ray. Such solutions, known as Gaussian beams (GBs), do not exhibit any singularity at the caustics. An asymptotic solution of a wavefield produced in an inhomogeneous medium by a point source now has the form of a linear superposition of GBs built on all the rays leaving the source. The analytic forms of the asymptotic solution thus constructed as well as of the DRT equations are valid in the whole domain of the solution without any need of modification in the vicinity of caustics. This is one of the advantages of the Gaussian beam expansions over the WKBJ method.

The wavefield at a point $\underset{\sim}{x}_1$ is the superposition of the contributions of a series of "generalized rays", specified by the original wave species of the signal leaving the source as well as by its subsequent conversions at the interfaces of the medium. As a result solution of a two-point boundary value problem for every relevant "generalized ray" is unavoidable. In the case of DRT of WKBJ signals in 3 dimensions in a complicated medium the difficulties inherent in numerical solution of the two-point boundary value problem may be quite prohibitive. In the case of the GB expansion method the wavefield at $\underset{\sim}{x}_1$ is a superposition of GB centered at rays which intersect a small but finite neighborhood of $\underset{\sim}{x}_1$.

Analytically, the GB expansion of a wavefield is an integral over a set of parameters specifying the GBs (or their central rays). In numerical practice this integral is approximated by a finite sum of GBs multiplied by the appropriate weight factors. In Sec. 5 we show that this approximation introduces an upper bound on frequencies for a fixed error bound and a fixed number of GBs. Since GBs are asymptotic solutions of the wave equation for high frequencies the approximation of a solution by a finite number of GBs is contained within a prescribed error bound at most for a finite interval of frequencies. Notwithstanding some practical successes of GB expansions ([2]) the problems of an optimal choice of a finite number of GBs and of estimating the error at least near a prescribed receiver location $\underset{\sim}{x}_1$ are still open.

In the spirit of [7] we develop the GB method for inhomogeneous anisotropic elastic media. Basic notations of that paper are preserved. In particular we need the ray-centered coordinates defined in Sec. 3 of that paper. In Sec. 2 we present the parabolic equation describing waves propagating predominantly in the direction of a ray. For isotropic elastic media such an equation was derived in [10], [3]. The derivation of the parabolic equation for an anisotropic medium is described in outline. For details the reader is referred to [8].

In Sec. 3 we proceed to construct the Hermite-Gaussian beam solutions of the parabolic equation. A Hermite-Gaussian beam is constructed by solving a DRT system which describes the evolution of the amplitude and shape of the beam along its central ray. DRT of GBs involves the equations of the kind previously obtained for WKBJ signals but the solutions which we seek now are complex-valued.

In Sec. 4 we expand the field of a point source in terms of GBs. The expansion provides the initial data for the DRT equations. Finally, in Sec. 5 we return to the questions raised in this section.

We shall often refer to formulae in [7]. Thus (In) denotes the formula (n) in [7]. Basic notions of Sec. 2 of [7] will be used without further explanation.

2 THE PARABOLIC EQUATION FOR AN ANISOTROPIC ELASTIC MEDIUM

By the <u>parabolic equation</u> we mean a system of partial differential equations which governs the elastic wavefields having the following three properties: (1) the wavefield is concentrated near a ray $\tilde{\underset{\sim}{x}}(s)$ of species σ; (2) the waves propagate predominantly in the direction of $\tilde{\underset{\sim}{x}}(s)$; (3) the wavefield satisfies eqs. (I1) in the asymptotic sense for high frequencies.

Let $\tilde{\underset{\sim}{x}}(s)$, $\tilde{\underset{\sim}{p}}(s)$ be a bicharacteristic satisfying eqs. (I11). We now proceed to describe in brief outline the derivation of the parabolic equation for an anisotropic elastic medium with prestress, for a wavefield concentrated in a nbhd of $\tilde{x}(s)$. For details the interested reader is referred to [8].

In the first step eqs. (I1) are transformed to a curvilinear ray-centered coordinate system (τ, y_1, y_2) associated with the ray $\tilde{x}(s)$. For definiteness we assume that condition (I26) is satisfied along an arc $s_1 < s < s_2$ of the ray under consideration and that the ray-centered coordinate system is defined by eqs. (I27).

In the next step the transverse coordinates y_a, a=1,2, are rescaled according to the formula $y_a = \omega^{-\frac{1}{2}} z_a$ and a tentative asymptotic expansion

$$\underset{\sim}{u}(\underset{\sim}{x}) = e^{i\omega\tau}[\underset{\sim}{u}_o(\tau,\underset{\sim}{z}) + \omega^{-\frac{1}{2}}\underset{\sim}{u}_1(\tau,\underset{\sim}{z}) + \omega^{-1}\underset{\sim}{u}_2(\tau,\underset{\sim}{z}) + \ldots] , \quad (1)$$

$\underset{\sim}{z} = (z_1, z_2)$ is substituted in the resulting equations. We now examine the coefficients of ω^2, $\omega^{3/2}$ and ω.

The coefficient of ω^2 yields the equation

$$[\tilde{p}_i(\tau)\tilde{p}_j(\tau)\underset{\sim}{B}^{ij}(\tilde{x}(\tau)) - \rho(\tilde{x}(\tau))\underset{\sim}{E}]\ \underset{\sim}{u}_o = 0 \quad (2)$$

whence we deduce that

$$\underset{\sim}{u}_o(\tau,\underset{\sim}{z}) \equiv \sum_{\nu=1}^{k_\sigma} f_\nu(\tau,\underset{\sim}{z})\underset{\sim}{r}_\sigma^{(\nu)}(\tilde{x}(\tau),\tilde{p}(\tau)) \quad (3)$$

Equating the coefficient of $\omega^{3/2}$ to zero we find a formula expressing the coefficients h_χ, $\chi \neq \sigma$, of the expansion

$$\underset{\sim}{u}_1(\tau,\underset{\sim}{z}) = \sum_{\sigma,\nu} h_\sigma^{(\nu)}(\tau,\underset{\sim}{z})\ \underset{\sim}{r}_\sigma^{(\nu)}(\tilde{x}(\tau),\tilde{p}(\tau)) + \sum_{\substack{\chi \\ \chi\neq\sigma}}{}' h_\chi(\tau,z)\underset{\sim}{r}_\chi(\tilde{x}(\tau),\tilde{p}(\tau)) \quad (4)$$

in terms of f_ν and $\partial f_\nu/\partial z_a$. Equating the coefficient of ω to zero and substituting the coefficients h_χ from the preceding formula we obtain a complicated equation. Using some identities this equation is shown to have the following form

$$i\frac{\partial f_\mu}{\partial \tau} + i\ V^a{}_b\ z_b \frac{\partial f_\mu}{\partial z_a} + \tfrac{1}{2}\ G^{ab}\frac{\partial^2 f_\mu}{\partial z_a\ \partial z_b} - \tfrac{1}{2}\ H_{ab}\ z_a z_b\ f_\mu +$$

$$+\frac{i}{2}\sum_{\nu=1}^{k_\sigma} E_\mu^\nu f_\nu = 0 \text{ , summation over a,b=1,2,} \tag{5}$$

where $i = \sqrt{-1}$,

$$V^a{}_b(\tau) = w^a{}_b - \tilde{p}_b\, G^{ac}\, H_c, \qquad a,b=1,2;\ \text{sum over } c=1,2, \tag{6}$$

$$H_c(\tau) = \frac{\partial c_\sigma}{\partial x^i}(\tilde{\underset{\sim}{x}}(\tau),\tilde{\underset{\sim}{p}}(\tau))\, P^i{}_c, \quad c=1,2;\ \text{sum over } i=1,2,3, \tag{7}$$

$$H_{ab}(\tau) = \frac{\partial^2 c_\sigma}{\partial p_i\, \partial p_j}(\tilde{\underset{\sim}{x}}(\tau),\tilde{\underset{\sim}{p}}(\tau))\, P^i{}_a(\tau) P^j{}_b(\tau), \ a,b=1,2; \tag{8}$$

$$\text{sum over } i,j\leqslant 3,$$

$$E_\mu^\nu(\tau) = \frac{1}{\tilde{\rho}} \widetilde{\langle \underset{\sim}{r}_\sigma{}^{(\mu)}} |\widetilde{\underset{\sim}{B}^{k\ell}}{}_{,k}\tilde{p}_\ell \widetilde{\underset{\sim}{r}_\sigma{}^{(\nu)}\rangle} - \frac{\widetilde{\partial c_\sigma}}{\partial p_k}\frac{\widetilde{\partial c_\sigma}}{\partial x^k}\delta_\mu^\nu +$$
$$+ 2\left[\widetilde{\langle \underset{\sim}{r}_\sigma{}^{(\mu)}} | \frac{\widetilde{\partial \underset{\sim}{r}_\sigma{}^{(\nu)}}}{\partial x^i}\rangle \frac{\widetilde{\partial c_\sigma}}{\partial p_i} - \widetilde{\langle \underset{\sim}{r}_\sigma{}^{(\mu)}} | \frac{\widetilde{\partial \underset{\sim}{r}_\sigma{}^{(\nu)}}}{\partial p_j}\rangle \frac{\widetilde{\partial c_\sigma}}{\partial x^j}\right] , \tag{9}$$

$$\mu,\nu\leqslant k_\sigma,\ \text{sum over } i,j,k,\ell\leqslant 3,$$

$G^{ab}(\tau)$ and $w^b{}_\ell(\tau)$ are given by eqs. (I29), and the tilda denotes the substitution of $\underset{\sim}{x}=\tilde{\underset{\sim}{x}}(\tau)$, $\underset{\sim}{p}=\tilde{\underset{\sim}{p}}(\tau)$.

If $G^{ab}(\tau)$ is positive definite (e.g., if $c_\sigma(\underset{\sim}{x},\cdot)$ is a convex function for every $\underset{\sim}{x}$), then eq. (5) is a parabolic equation of the Schrödinger type. Hence the phenomena of unidirectional wave propagation are reversible with respect to the transformation $\tau\to-\tau$.

3 GENERALIZED HERMITE-GAUSSIAN BEAMS

We now consider a family of exact solutions of (3), (5). With this in view we substitute an expression

$$f_\nu(\tau,\underset{\sim}{z};\underset{\sim}{w}) = g_\nu(\tau)\exp\{i[\tfrac{1}{2}S_{ab}(\tau)z_a z_b - z_a w_b\, T_{ab}(\tau) +$$
$$+\tfrac{1}{2}\, w_a w_b\, U_{ab}(\tau)]\}, \quad \nu=1,\dots,k_\sigma,\ \underset{\sim}{z}=(z_a),\ \underset{\sim}{w}=(w_a) \tag{10}$$

which involves two real parameters w_1,w_2, in eq. (5). The resulting equation has the form

$$K + \sum_{r,s=1}^{4} \Gamma_{rs} v_r v_s = 0 \text{ with } v_a = z_a,\ v_{a+2} = w_a \text{ for } a=1,2.$$

Setting $v_r=0$, $r=1,\ldots,4$, we get the equation $K=0$, or

$$\frac{dg_\mu}{d\tau} = -\tfrac{1}{2}(G^{ab}S_{ab}\delta^\nu_\mu + E^\nu_\mu)g_\nu \tag{11}$$

The equation $\sum_{r,s} \Gamma_{rs}v_r v_s=0$ implies that $\Gamma_{rs}=0$, hence

$$\frac{dS_{ab}}{d\tau} = - V^c{}_a S_{bc} - V^c{}_b S_{ac} - G^{cd} S_{ac} S_{bd} - H_{ab} \tag{12}$$

$$\frac{dT_{ab}}{d\tau} = -V^c{}_a T_{cb} - G^{cd} S_{ac} T_{db} \tag{13}$$

$$\frac{dU_{ab}}{d\tau} = -G^{cd} T_{ca} T_{db} \tag{14}$$

($a,b=1,2$; summation over $c,d=1,2$).

We note that eq. (12) is identical with (I31). Also eq. (11) is essentially the equation satisfied by the function $J_\sigma^{-\frac{1}{2}} g_\sigma^{(\mu)}$, provided J_σ and $g_\sigma^{(\mu)}$ satisfy eqs. (I30) and (I33), resp., as may be shown after some tricky calculi. We shall however be interested in complex-valued solutions of eq. (12) which satisfy the condition

$$\sum_{a,b=1}^{2} (\mathrm{Im} S_{ab}(\tau))\xi_a \xi_b > 0 \tag{15}$$

for all $\tau \in [s_1,s_2]$ and arbitrary real ξ_1,ξ_2 which do not vanish simultaneously.

In order to find a solution of eq. (12) satisfying condition (15) we consider the following system of linear differential equations

$$\frac{d\underset{\sim}{Q}}{d\tau} = \underset{\sim}{G}\,\underset{\sim}{R} + \underset{\sim}{V}\,\underset{\sim}{Q}, \quad \frac{d\underset{\sim}{R}}{d\tau} = - {}^t\underset{\sim}{V}\,\underset{\sim}{R} + \underset{\sim}{H}\,\underset{\sim}{Q}, \tag{16}$$

where $\underset{\sim}{Q}$, $\underset{\sim}{R}$ are unknown 2×2 matrices with complex-valued entries, $\underset{\sim}{G}=[G^{ab}]$, $\underset{\sim}{V}=[V^a{}_b]$, $\underset{\sim}{H}=[H_{ab}]$, with the first index from left numbering the rows of the matrix. By ${}^t\tilde{A}$ we denote the transpose of $\tilde{A}$.

A straightforward calculus shows that the matrix

$$\tilde{S} = \tilde{R}\,\tilde{Q}^{-1}, \quad \tilde{S} = [s_{ab}] \tag{17}$$

satisfies eq. (12), provided $\tilde{Q}$ is invertible.

We now quote without proof the following result of [6].

<u>Theorem</u>

Suppose that for some value τ_o of τ the matrix $\tilde{Q}(\tau_o)$ is invertible while the matrix $\tilde{S}(\tau_o) = \tilde{R}(\tau_o)\tilde{Q}(\tau_o)^{-1}$ is symmetric and satisfies condition (15).

Then the solution $\tilde{Q}(\tau)$ of eqs. (15) is invertible for each τ while $\tilde{S}(\tau)=\tilde{R}(\tau)\tilde{Q}(\tau)^{-1}$ is symmetric and has a positive definite imaginary part for each τ. □

We note that the Riccati equation (12) need not <u>in general</u> have solutions which are defined for all τ. The theorem above specifies the conditions under which $\tilde{Q}(\tau)$ is reversible and $\tilde{S}(\tau)$ satisfies (12),(15) as well as the symmetry condition for each τ.

We now note that

$$\frac{d \log \det \tilde{Q}}{d\tau} = \mathrm{tr}(\tilde{G}\,\tilde{S}) + \mathrm{tr}\,\tilde{V}\ , \tag{18}$$

tr denoting the trace of a matrix. Hence

$$\frac{dg_\mu}{d\tau} = -\tfrac{1}{2}\,\frac{d \log \det \tilde{Q}}{d\tau}\,g_\mu - \sum_\nu \tfrac{1}{2}(E_\mu^\nu - \mathrm{tr}\tilde{V}\,\delta_\mu^\nu)g_\nu \tag{19}$$

and g_μ is regular for all τ if the hypotheses of the theorem are satisfied. In particular for $k_\sigma=1$ we have

$$g(\tau) = g_1(\tau) = [\det \tilde{Q}(\tau)]^{-\frac{1}{2}} \exp\{-\tfrac{1}{2}\int_0^\tau d\sigma[E_1^1(\sigma) - \mathrm{tr}\,\tilde{V}(\sigma)]\} \tag{20}$$

One easily shows that $\tilde{Q}(\tau)^{-1}$ and ${}^t\tilde{T}(\tau)$ satisfy an equation of the form $d\tilde{X}/d\tau=\tilde{X}\,\tilde{A}(\tau)$ with the same coefficient matrix $\tilde{A}(\tau)$. Hence there is a constant matrix $\tilde{C}$ such that

$${}^t\tilde{T}(\tau) = \tilde{C}\,\tilde{Q}(\tau)^{-1} \tag{21}$$

On account of (14) we also have

$$\underset{\sim}{U}(\tau) = \underset{\sim}{U}(0) - \underset{\sim}{C} \int_0^\tau d\sigma\, \underset{\sim}{Q}(\sigma)^{-1}\, \underset{\sim}{G}(\sigma)\, {}^t\underset{\sim}{Q}(\sigma)^{-1}\, {}^t\underset{\sim}{C} \tag{22}$$

We now define the generalized Hermite-Grad polynomials $He_{m_1m_2}(\underset{\sim}{u};\underset{\sim}{A})$ in terms of the generating function

$$\exp\{i[\tfrac{1}{2}\, U_{ab}w_aw_b - T_{ab}z_aw_b]\} = \sum_{m_1,m_2} (w_1)^{m_1} (w_2)^{m_2}\, He_{m_1m_2}(-i\, {}^t\underset{\sim}{T}\underset{\sim}{z}; \frac{i}{2}\underset{\sim}{U}), \quad \underset{\sim}{z}=(z_1,z_2) \tag{23}$$

cf. [1].

We have thus constructed a family of exact solutions of the parabolic equation

$$f_\nu(\tau,\underset{\sim}{z}) \equiv g_\nu(\tau)\exp[\frac{i}{2} S_{ab}(\tau)z_az_b]\, He_{m_1m_2}(-i\, {}^t\underset{\sim}{T}(\tau); \frac{i}{2}\underset{\sim}{U}(\tau)) \tag{24}$$

with S_{ab}, T_{ab}, U_{ab}, g_ν given by eqs. (17), (21), (22) and (19) and with S_{ab} satisfying condition (15) for all τ. The corresponding asymptotic expression for the wavefield is given by eq. (3).

The solutions (24) are referred to as generalized Hermite-Gaussian beams. For $m_1=m_2=1$ the last factor of the rhs of (24) drops out and the corresponding solution is a Gaussian beam (GB).

We note that on account of the theorem quoted above these solutions are everywhere regular, including the caustic regions.

The phase factor $\exp\{i\omega[\tau+\frac{1}{2}S_{ab}(\tau)y_ay_b]\}$ looks very much like $\exp(i\omega S)$ with the eikonal S replaced by its Taylor expansion to second order in y_a. Indeed, τ is the value of S at $\underset{\sim}{x}=\tilde{\underset{\sim}{x}}(\tau)$, cf. [7]. For the first order term in the expansion we have $[\partial S(\tilde{\underset{\sim}{x}}(\tau))/\partial x^i]P^i{}_a(\tau)y_a = \tilde{p}_iP^i{}_ay_a=0$. The functions $S_{ab}(\tau)$ satisfy the same equations as the functions $[\partial^2 S(\tilde{\underset{\sim}{x}}(\tau))/\partial x^i\partial x^j]P^i{}_aP^j{}_b$ although they are now complex-valued. Furthermore, it is shown in [6] that the matrices $Q_{a\alpha}(\tau)$, $R_{a\alpha}(\tau)$ satisfy the same equations as the matrices $\tilde{Q}_{a\alpha}=(\partial\bar{x}_i/\partial u_\alpha)\partial y_a/\partial x^i$ and $\tilde{R}_{a\alpha}= P^i{}_a\partial\bar{p}_i/\partial u_\alpha$, evaluated at $u_\alpha=0=y_a$, except that they have complex-valued entries. We recall that (s, u_1, u_2) are the ray coordinates for a family of rays in which the ray $\tilde{\underset{\sim}{x}}(\tau)$ is embedded: $\tilde{\underset{\sim}{x}}(s) = \bar{\underset{\sim}{x}}(s,0,0)$, $\tilde{\underset{\sim}{p}}(s)=\bar{\underset{\sim}{p}}(s,0,0)$, $s=\tau$ on $\tilde{\underset{\sim}{x}}(s)$. We also note that the matrices $\tilde{Q}$, $\tilde{R}$ are related to the second-order derivatives of the real eikonal $\tilde{S}$ associated with the family $\tilde{x}(s,u_1,u_2), \tilde{p}(s,u_1,u_2) = \nabla\tilde{S}(\tilde{x}(s,u_1,u_2))$ by eq. (17). We may resume these observations in the form of the following conclusions.

(*) The DRT system (19), (12) is a complex extension of the DRT system (I31), (I33).

(**) Signals are traced along a real bicharacteristic $\tilde{x}(s)$, $\tilde{p}(s)$ but all the matrices involving the derivatives of the bicharacteristic with respect to the transverse ray coordinates u_1, u_2 are replaced by their complex-valued counterparts.

Heuristically (**) means that the real bicharacteristic $\tilde{x}, \tilde{p}$ is embedded in a complex field with infinitesimal extension in the imaginary directions. The corresponding eikonal is complex except on $\tilde{x}(s)$. This situation is reminiscent of the notion of Lagrangian germs introduced in [12] as well as of another idea put forward in [11] in the context of GBs. It is hardly possible to extend the field of rays to a finite neighborhood of the real space in the complex space unless the functions $B^{rs}_{kl}(x)$ are real analytic and admit a complex analytic extension.

The GBs constructed above are defined only in a neighborhood of the ray $\tilde{x}(s)$, $s_1<s<s_2$, in which the ray-centered coordinate system (τ, y_1, y_2) is defined. They may be extended to the whole space by multiplying them by a smooth cutoff function ϕ which equals 1 for $|y_a|<d$, and zero for $|y_a|>d+\varepsilon$, $\varepsilon>0$, with $d+\varepsilon$ sufficiently small. The new solution satisfies eqs. (I1) asymptotically with an additional error $O(e^{-\omega d^2})$.

If the condition$(\partial c_\sigma/\partial p_3)(\tilde{x}(s),\tilde{p}(s)) \neq 0$ fails to be satisfied for $s>s_2$, say, then a new $\tilde{G}B$ solution is constructed for $\tau>s_2$ in terms of the appropriate ray-centered coordinates in such a way that the boundary conditions at $\tau=s_2$ are satisfied in the asymptotic sense. A "global" GB is then built over the whole ray $\tilde{x}(s)$ by patching up the GBs constructed above.

4 APPLICATIONS OF DYNAMIC RAY TRACING OF GAUSSIAN BEAMS IN SEISMOLOGY

Suppose that the field produced by a seismic source is known in its neighborhood. We may then calculate the wavefield produced by this source in an inhomogeneous anisotropic medium by the following procedure.

(1) The wavefield near the source is decomposed into a superposition of GBs built over every ray $\tilde{x}(s,u_1,u_2)$ leaving the source (the parameters u_1,u_2 specify the rays). This provides a weight

factor and/or the initial conditions (at $\tau=s=0$) for each GB.

(2) The GBs are traced along the respective rays by solving the ordinary differential equations (11), (12) along with (I11).

For a point source $\underset{\sim}{x}_o$ in a homogeneous anisotropic medium characterized by the elasticities $B_{kl}^{rs}(\underset{\sim}{x}_o)$ we find that the asymptotic solution is

$$\underset{\sim}{u}_o(\underset{\sim}{x};\omega) \sim \frac{1}{4\pi i\rho(\underset{\sim}{x}_o)} \sum_{\sigma} \sum_{\nu=1}^{k_\sigma} \sum_{r} \exp\left[i\omega \frac{|\underset{\sim}{\eta}^{(r)}(\underset{\sim}{x})\cdot\underset{\sim}{x}|}{c_\sigma(\underset{\sim}{x}_o,\underset{\sim}{\eta}^{(r)}(\underset{\sim}{x}))}\right] \times$$

$$\exp\left[-\frac{i\pi}{4}\,\mathrm{sgn}\,\underset{\sim}{B}_r\right] \frac{c_\sigma(\underset{\sim}{x}_o,\underset{\sim}{\eta}^{(r)}(\underset{\sim}{x}))^{-1}}{|\underset{\sim}{\eta}^{(r)}(\underset{\sim}{x})\cdot\underset{\sim}{x}|} \,|\det \underset{\sim}{B}_r|^{-\frac{1}{2}} \times \tag{25}$$

$$\langle \underset{\sim}{s}|\underset{\sim}{r}_\sigma^{(\nu)}(\underset{\sim}{x}_o,\underset{\sim}{\eta}^{(r)}(\underset{\sim}{x}))\rangle \underset{\sim}{r}_\sigma^{(\nu)}(\underset{\sim}{x}_o,\underset{\sim}{\eta}^{(r)}(\underset{\sim}{x})) \quad ,$$

where the unit vectors $\underset{\sim}{\eta}^{(r)}(\underset{\sim}{x})$ are the solutions of the equation

$$\underset{\sim}{x} = \underset{\sim}{x}_o+\lambda \frac{\partial c_\sigma(\underset{\sim}{x}_o,\underset{\sim}{p})}{\partial \underset{\sim}{p}} \qquad \text{for some } \lambda>0 , \tag{26}$$

$\underset{\sim}{B}_r$ is the 2×2 matrix

$$\sum_{i,j}^{3} \frac{\partial\eta_i}{\partial v_\alpha} \frac{\partial^2 c_\sigma}{\partial p_i\,\partial p_j} \frac{\partial\eta_j}{\partial v_\beta} \quad , \qquad \alpha,\beta=1,2$$

evaluated at $(\underset{\sim}{x}_o,\underset{\sim}{\eta}^{(r)}(\underset{\sim}{x}))$, and $v_1=\zeta=\cos\theta$, $v_2=\phi$, with θ,ϕ denoting the two angular spherical coordinates (the latitude and the longitude) specifying the directions of the unit slowness vectors $\underset{\sim}{\eta}$. The load at $\underset{\sim}{x}_o$ is assumed to be a simple concentrated force $\underset{\sim}{f}=\rho(\underset{\sim}{x}_o)\delta(\underset{\sim}{x}-\underset{\sim}{x}_o)e^{i\omega t}\underset{\sim}{s}$.

It is shown [6], [8] that eq. (26) has a finite number of solutions $\underset{\sim}{p}=\underset{\sim}{\eta}^{(r)}(\underset{\sim}{x})$ of unit length, $r=1,\ldots m_\sigma<\infty$, provided

$$\mathrm{rank}\left[\frac{\partial^2 c_\sigma}{\partial p_i\,\partial p_j}\right] = 2 \tag{27}$$

(note that $(\partial^2 c_\sigma/\partial p_i \partial p_j)p_i=0$ on account of Euler's identities). Condition (27) is generally satisfied by anisotropic elastic media. The index r specifies various wavefront arrivals at $\underset{\sim}{x}$ of waves of the same species σ.

We now set up a linear superposition of GBs built on all the rays leaving the source $\underset{\sim}{x}_o$. The medium is again assumed homogeneous. The rays are specified by the direction of the slowness vector $\underset{\sim}{n}(\zeta,\phi)$ on the ray. The slowness is then given by the expression $\underset{\sim}{p}=c_\sigma(\underset{\sim}{x}_o,\underset{\sim}{n})^{-1}\underset{\sim}{n}$ and the corresponding ray is given in the parametric form $\underset{\sim}{x}=\underset{\sim}{x}_o+\tau(\partial c_\sigma/\partial \underset{\sim}{p})(\underset{\sim}{x}_o,\underset{\sim}{n})$. For definiteness we assume that the amplitude of each GB satisfies the initial condition $g_\sigma^{(\nu)}(0,\zeta,\phi)=1$. The linear superposition now takes the form

$$\underset{\sim}{u}(\underset{\sim}{x}) \sim \sum_\sigma \sum_{\nu=1}^{k_\sigma} \int_{-1}^{1} d\zeta \int_0^{2\pi} d\phi\, a_\sigma^{(\nu)}(\zeta,\phi) g_\sigma^{(\nu)}(\tau,\zeta,\phi) \times \exp\{i\omega[\tau+\tfrac{1}{2}S_{ab}^{(\sigma)}(\tau)y_a y_b]\}\, \underset{\sim}{r}_\sigma^{(\nu)}(\underset{\sim}{x}_o,\underset{\sim}{n}(\zeta,\phi)) \tag{28}$$

where $\tau=c_\sigma(\underset{\sim}{x}_o,\underset{\sim}{n})^{-1}\underset{\sim}{n}\cdot(\underset{\sim}{x}-\underset{\sim}{x}_o) \equiv \tau(\underset{\sim}{x},\zeta,\phi)$ and $y_a=x_a-(x_o)_a-[x_3-(x_o)_3] \times (\partial c_\sigma/\partial p_a)(\underset{\sim}{x}_o,\underset{\sim}{n})/(\partial c_\sigma/\partial p_3)(\underset{\sim}{x}_o,\underset{\sim}{n}) \equiv y_a(\underset{\sim}{x},\zeta,\phi)$ are the ray coordinates of $\underset{\sim}{x}$ with respect to the ray (ζ,ϕ) (for simplicity we restrict our attention to those rays for which $(\partial c_\sigma/\partial p_3)(\underset{\sim}{x}_o,\underset{\sim}{n})\neq 0$). The weight factors $a_\sigma^{(\nu)}(\zeta,\phi)$ are determined by evaluating (28) by the saddle point method and comparing the result with (25).

According to the saddle point method the main contributions come from those saddle points at which the imaginary part of the phase attains its lowest value zero. Such points satisfy the equations $y_a=0$, $a=1,2$, hence eq. (26), and it turns that any point satisfying eq. (26) is a saddle point. Comparison with (25) yields the formula

$$a_\sigma^{(\nu)}(\underset{\sim}{n}) = \frac{\omega}{8\pi^2 i\rho(\underset{\sim}{x}_o)} \frac{1}{c_\sigma(\underset{\sim}{x}_o,\underset{\sim}{n})^3} \langle \underset{\sim}{s} | \underset{\sim}{r}_\sigma^{(\nu)}(\underset{\sim}{x}_o,\underset{\sim}{n})\rangle \tag{29}$$

Since the main contributions to the integrals (28) come from the points satisfying $y_a=0$, no restrictions on the initial data for eq. (12) can be obtained in this way. In particular the width controlling functions $\mathrm{Im}S_{ab}(\tau)$ remain at our disposal.

The asymptotic solution for a point source in an inhomogeneous medium is given by an expression of the form (28) in which $g_\sigma^{(\nu)}$, $S_{ab}^{(\sigma)}$ have been obtained by integrating eqs. (11) and (12) with the initial data $g_\sigma^{(\nu)}(0,\zeta,\phi)=1$ and some initial data for $S_{ab}^{(\sigma)}(\tau)$, and the weight factors are given by (29).

5 CONCLUDING REMARKS

In numerical calculi the integrals in (28) must be approximated by finite sums of the form

$$\sum_{k=1}^{N} \Delta_k a_\sigma^{(\nu)}(\zeta_k,\phi_k) g_\sigma^{(\nu)}(\tau_k,\zeta_k,\phi_k) \times \exp\{i\omega[\tau_k+\tfrac{1}{2}S_{ab}(\tau_k)y_a^k y_b^k]\}\, \underset{\sim}{r}_\sigma^{(\nu)}(\underset{\sim}{x}_o,\underset{\sim}{n}(\zeta_k,\phi_k)) \tag{30}$$

with Δ_k=the area of the k-th element L_k of the partition of unit sphere into N elements, $(\zeta_k,\phi_k) \in L_k$, $\tau_k(\underset{\sim}{x},\zeta_k,\phi_k)$, $y_a^k = y_a(\underset{\sim}{x},\zeta_k,\phi_k)$. If it is enough to calculate the wavefield at a single receiver $\underset{\sim}{x}_1$, then it is sufficient to pick out from (30) only those GBs whose central rays intersect a small neighborhood of $\underset{\sim}{x}_1$ whose linear dimensions are of order $O(\omega^{-\frac{1}{2}})$.

It is easy to see that for $\omega\to\infty$ expression (30) tends to zero everywhere except on the central rays of the N GBs. Hence it is asymptotic to the zero solution of (I1). On the other hand the error introduced by replacing (28) with (30) grows with the oscillation of the integrand. The latter grows with $\omega\to\infty$ (since the beams get narrower) and with $|\underset{\sim}{x}-\underset{\sim}{x}_o|\to\infty$ (because the beams diverge away from the source. For fixed upper bounds on the error, on the number N of beams as well as on $|\underset{\sim}{x}-\underset{\sim}{x}_o|$ we thus obtain a restriction on the frequency from above. Since GBs are high-frequency asymptotic solutions of (I1), there is a lower bound on frequencies too. Thus expressions of the form (30) may approximate the true solution at most for a <u>finite</u> interval of frequencies.

The growth of the error with $|\underset{\sim}{x}-\underset{\sim}{x}_o|\to\infty$ would imply that the approximation is not uniform with respect to $|\underset{\sim}{x}-\underset{\sim}{x}_o|$. This error growth may be partially offset by the spreading of the GBs. This phenomenon can be visualized by an explicit solution of eq. (12) in a homogeneous medium. In this case $\underset{\sim}{H}=0$, $\underset{\sim}{V}=0$ and the solution $\underset{\sim}{S}(\tau,\underset{\sim}{n})$ of (12) with the initial condition $\underset{\sim}{S}(0,\underset{\sim}{n})=i\underset{\sim}{K}(\underset{\sim}{n})$, $\underset{\sim}{K}(\underset{\sim}{n})>0$, is given by the formula $\underset{\sim}{S}(\tau,\underset{\sim}{n})=i\underset{\sim}{K}(\underset{\sim}{n})[1+i\tau\underset{\sim}{K}(\underset{\sim}{n})\underset{\sim}{G}(\underset{\sim}{n})]\{1+\tau^2[\underset{\sim}{K}(\underset{\sim}{n})\underset{\sim}{G}(\underset{\sim}{n})]^2\}^{-2}$. For $\tau\to\infty$ the imaginary part of $\underset{\sim}{S}(\tau,\underset{\sim}{n})$ is $O(\tau^{-4})$ and the beam width is $O(\tau^2)$.

REFERENCES

1. Arnaud, J.A. Beam and Fiber Optics (New York, Academic Press, 1976).
2. Červený, V. Synthetic Body Wave Seismograms for Laterally Varying Structures by the Gaussian Beam Method. Geophys. J.R. astr. Soc. 73 (1983) 389-426.
3. Červený, V., M.M. Popov and I. Pšenčík. Computation of Wave Fields in Inhomogeneous Media - Gaussian Beam Approach. Geophys. J.R. astr. Soc. 70 (1982) 109-128.

4. Červený, V. and I. Pšenčík. Gaussian Beams and Paraxial Ray Approximation in Three-Dimensional Elastic Inhomogeneous Media, to appear.
5. Fedoryuk, M.V. The Saddle Point Method (Moscow, Nauka, in Russian).
6. Hanyga, A. Linear and Non-linear DRT Systems and Their Complex-Valued Solutions. Acta Geophysica Polonica, to appear in 1983.
7. Hanyga, A. Numerical Computation of Elastic Wavefields in Anisotropic Media in the Presence of Caustics (this volume).
8. Hanyga, A. Point Source in an Anisotropic Elastic Medium. Pageoph, to appear in 1984.
9. Hanyga, A. Dynamic Ray Tracing on Lagrangian Manifolds. Geophys. J.R. astr. Soc., to appear in 1984.
10. Kirpichnikova, N.Ya. Construction of Solutions Concentrated Close to Rays for the Equations of Elasticity Theory in an Inhomogeneous Isotropic Space, in: Mathematical Problems of the Theory of Diffraction and Wave Propagation, vol. 1, V.M. Babich, ed.; AMS translation, 1974.
11. Klimeš, L. The Relation between Gaussian Beams and Maslov Asymptotic Theory. Studia Geophys. et Geodet., to appear.
12. Maslov, V.B. Complex WKB Method for Non-linear Equations (Moscow, Nauka 1977, in Russian).

PART III

TRANSIENT PROPAGATION AND DIFFRACTION

THE SINGULARITY EXPANSION REPRESENTATION OF SURFACE CURRENT ON A PERFECTLY CONDUCTING SCATTERER

L. Wilson Pearson

Department of Electrical Engineering
University of Mississippi
University, MS 38677

1. INTRODUCTION

The term "singularity expansion" was coined by Baum [1] to apply to the characterization of the scattering phenomenon in terms of the complex natural resonances of the scattering object. The use of the concept of complex natural resonance appears in the literature from time to time previous to Baum's 1971 work (e.g. Schelkunoff [2]), but it is Baum's work that motivated a substantial effort during the 1970's to put the concept on a sounder mathematical footing and to apply it in a variety of applications and to a variety of geometries.

The singularity expansion characterizes the scattering phenomenon in terms of fundamental quantities intrinsic to a given scattering object--quantities in terms of which one may expand the surface current density induced on the object by a given incident field. Thence in turn one can express the scattered fields in terms of this surface current density. The quantities entering into the expansion are the complex natural resonances (or "poles") and corresponding natural current modes. These quantities depend on the global features of the scattering object, so that in the framework of alternative representations, they are well suited to the description of the scattering phenomenon at frequencies ranging over the first few resonant frequencies of the object. Equivalently, in the transient regime, they are well suited to computation of the induced current density at moderate and late times, but are ill-suited to early-time computation at least when time variations in the excitation waveform occur on time scales that are small compared with transit time across the object.

Mathematicians have given some attention to a rigorous theory of the singularity expansion for the scalar (acoustic) scattering case (e.g. see [3] and the bibliography therein), but the larger part of what has proceeded has appeared in the engineering science literature. As a consequence rigorous support is lacking in defining a number of features of the expansion and is supplanted by ad hoc arguments and at times only by numerical observation. In the present exposition we take the pragmatic point of view of the engineering scientist, calling on results which are felt to be trustworthy though at times non-rigorous. Where sound mathematical results are available, they are pointed out, however.

2. THE FORMAL SINGULARITY EXPANSION AND RAMIFICATIONS

2.1 The Singularity Expansion in the Frequency and Time Domains

We are concerned with the surface current density induced on a scattering object by an incident electromagnetic field as indicated generically in Figure 1. For present purposes we view the current on the scatter as being dicated by the electric field integral equation, though other formulations defining it are equally applicable. Defining the Laplace transform of the field quantities as

$$E(\mathbf{r},s) = \int^{\infty} e(\mathbf{r},t)\, e^{-st}\, dt, \tag{1}$$

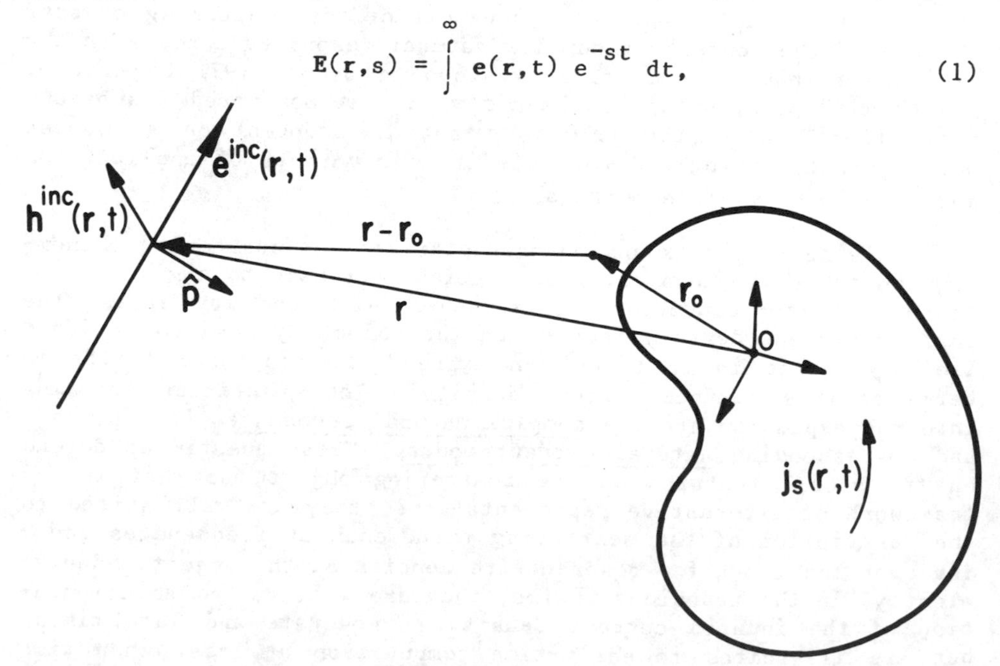

Figure 1. A scattering object under time-dependent excitation. The point $\mathbf{r}_0$ is the "space-time origin" through which the wavefront passes at time t = 0.

we may write this integral equation as

$$\langle \underline{\mathbf{Z}}(\mathbf{r},\mathbf{r}',s);\mathbf{J}(\mathbf{r}',s)\rangle = \tan \mathbf{E}^{inc}(\mathbf{r},s), \quad \mathbf{r} \in \mathbf{S}, \tag{2}$$

where $\underline{\mathbf{Z}}$ is the two-dimensional dyadic kernel to the integral equation, $\mathbf{J}$ is the surface current density residing on the object, $\mathbf{E}^{inc}$ is the incident electric field, and "tan" means "tangential part of." The symmetric product notation is used to denote surface integration of the dot product of the elements over the object. Viz.

$$\langle \mathbf{F}(\mathbf{r});\mathbf{G}(\mathbf{r})\rangle = \int_S \mathbf{F}(\mathbf{r})\cdot\mathbf{G}(\mathbf{r})\, dS \quad , \tag{3}$$

with the integration being over the common variable when multiple argument functions are involved. The incident field in (2) is presumed to be a propagating plane wave for simplicity of discussion. Specifically, it is assumed to be representable as

$$\mathbf{E}^{inc}(\mathbf{r},s) = F(s)\, \mathbf{E}_0\, e^{-s\hat{\mathbf{p}}\cdot(\mathbf{r}-\mathbf{r}_0)/c} \quad , \tag{4}$$

where $F(s)$ is the transform of the time-history of the incident wave and is presumed to be a rational function, $\hat{\mathbf{p}}$ is a unit vector in the direction of propagation of the wavefront, $\mathbf{E}_0$ is a constant vector such that $\hat{\mathbf{p}}\cdot\mathbf{E}_0=0$ and denotes the polarization and magnitude of the electric field, and $\mathbf{r}_0$ is a "space-time origin" in terms of which this incident field is expressed.

The singularity expansion for the solution to (2) is written formally as

$$\mathbf{J}(\mathbf{r},s) = F(s) \sum_\alpha H_\alpha(\mathbf{r},s)\, \mathbf{J}_\alpha(\mathbf{r}) \left[\frac{1}{s - s_\alpha} + P_\alpha(s)\right] + \mathbf{J}_e(\mathbf{r},s). \tag{5}$$

The poles s_α are the complex natural resonances intrinsic to the given scattering object, the $\mathbf{J}_\alpha$ are the "natural modes" associated with these resonances, and the H_α are the so-called coupling coefficients. The P_α are convergence polynomials required by the Mittag-Leffler theorem in forming an expansion such as (5), and $\mathbf{J}_e$ is an entire function of s, which is also required in the Mittag-Leffler expansion.

The expansion (5) is based on a Mittag-Leffler expansion theorem which has as its hypothesis only that $\mathbf{J}(\mathbf{r},s)$ be a meromorphic function of s. An early result due to Marin and Latham [4] establishes the meromorphicity of the surface current density function on a finite-extent, perfectly-conducting scatterer residing in a lossless medium. The degree of generality required under the weak hypothesis of meromorphicity alone led to a great deal of confusion about the polynomials and entire function in (5) through the early years of development of the singularity expansion. In a subsequent section we discuss what is known about $\mathbf{J}(\mathbf{r},s)$ to restrict the form of the expansion further and thereby obviate the need for $\mathbf{J}_e$.

The time-domain counterpart to (5) is obtained through the Laplace transform inversion procedure and is written as

$$\mathbf{j}(\mathbf{r},t) = f(t) * \left[\Sigma\ \eta_\alpha(\mathbf{r},t)\ \mathbf{J}_\alpha(\mathbf{r})\ e^{s_\alpha t} + \mathbf{j}_e(\mathbf{r},t)\right] \tag{6}$$

where the $\eta_\alpha(\mathbf{r},t)$ are time-domain coupling coefficients, the precise definition of which which we leave unspecified for the moment. The term $\mathbf{j}_e$ is the inverse transform of $\mathbf{J}_e$ in (5). Leaving the influence of $F(s)$ as a convolution with its transform obviates the need to include a summation over the poles of $F(s)$.

The presence of the poles s_α in (5) is manifested through resonances in the frequency domain expansion when s is replaced by the angular frequency variable $j\omega$, while the poles indicate a temporal current which contains a superposition of damped exponentials in (6). (The real parts of the s_α must be negative as a result of energy considerations, and the poles must occur in complex conjugate pairs in order that $\mathbf{j}(\mathbf{r},t)$ be purely real.)

2.2 Relationship to Eigenfunction Expansion

The singularity expansion can be identified through the eigenfunction expansion of the solution to the integral equation (3). We define eigenvalue/eigenfunction pairs for the integral equation through

$$\langle \underline{\mathbf{Z}}(\mathbf{r},\mathbf{r}',s);\mathbf{J}_n(\mathbf{r}',s)\rangle = \lambda_n(s)\mathbf{J}_n(\mathbf{r},s), \quad \mathbf{r} \in \mathbf{S}. \tag{7}$$

The eigenfunction expansion for the solution to (3) is thus written as

$$\mathbf{J}(\mathbf{r},s) = \Sigma \frac{1}{\lambda_n(s)} \frac{\langle \mathbf{J}_n(\mathbf{r},s);\mathbf{E}^{inc}(\mathbf{r},s)\rangle}{\langle \mathbf{J}_n(\mathbf{r},s);\mathbf{J}_n(\mathbf{r},s)\rangle} \mathbf{J}_n(\mathbf{r},s) \ . \tag{8}$$

The resonance phenomenon is embodied in (8) by virtue of the zeros of the eigenvalues $\lambda_n(s)$ with respect to the complex frequency variable s. Viz. the poles of $\mathbf{J}$ are s_{np} such that

$$\lambda_n(s_{np}) = 0. \tag{9}$$

The pole set for a given object expressed with the single index α previously is ordered into subsets so that each pole subset is associated with the nth eigenvalue of the integral equation for the object. This association is indicated for the example of a spherical scatterer in Figure 2. Only the second quadrant of the complex s-plane is shown since no poles lie in the right half plane and since the poles arise in conjugate symmetric pairs. The eigenvalue subsets of poles are those lying along common arcs indicated by solid lines in the Figure.

Whether the formal eigenfunction expansion (8) is rigorously applicable and under what circumstances is a question which must be considered carefully. Dolph, et. al. [5], and Ramm [6] raise this question since the electric field integral equation (2) involves a non-Hermitian-symmetric kernel (rather it is spatially-symmetric) and the applicability of (8) hinges on the operator equation's being self-adjoint. When the operator is non-self-adjoint, the domain of the operator is not necessarily spanned by the eigenvectors, and one must include so-called "root vectors" in order to have a complete basis for the solution [5], [6].

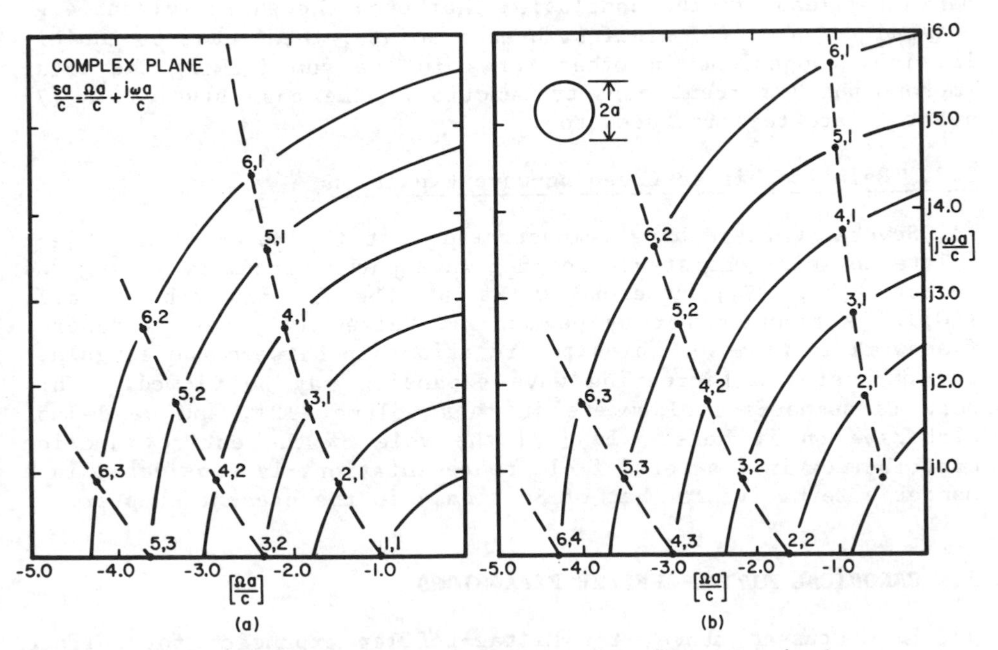

Figure 2. Pole set for a perfectly conducting spherical scatterer (a) poles associated with eigenmodes excited by an incident field TE to r, (b) by a field TM to r. The solid lines connect poles associated with a given eigenvalue.

A case study by Wilton, et. al. [7] sheds light on this issue. Root vectors must be included in the basis for the domain space when two or more eigenvalues are degenerate. Eigenvalue degeneracy can indeed occur for isolated points in the s-plane, and these points are branch points for the eigenvalues. It is observed in [7] that in network theory branch points appear in eigenvalues for the impedance operator of the network except for symmetric networks. This suggests that in the distributed counterpart, the electric field integral equation, object symmetry bears on whether branch points arise. Earlier observations regarding the eigenfunction expansion (8) were drawn from the cases of a spherical scatterer and of a wire-loop scatterer--both symmetric structures. The cases of acoustic scattering from a prolate spheroid and electromagnetic scattering from an elliptic cylinder are considered in [7] and are observed to manifest branch points in the eigenvalues which recede to infinity when the eccentricities of the respective structures are limited to zero, recovering symmetric structures.

The conclusions drawn from the case study, though they are non-rigorous ones, are that the eigenfunction expansion (8) applies for the electric field integral equation except at isolated branch points in the s-plane, while the meromorphicity result of Marin [4] leads to the conclusion that even though individual eigenvalue terms of (8) manifest branch points, these must be annihilated by companions in other terms in the sum forming the total (meromorphic) current density function. The case studies in [7] substantiate this feature, too.

2.3. Relationship to Creeping-wave Expansion

Several workers have demonstrated that the poles of an object relate to self-consistent creeping wave paths circumnavigating the object ([8], [9], numerous works by Uberall and others (c.f. [10]). A recent paper by Heyman and Felsen [11] sets a general framework in terms of which the relationship between the singularity expansion and creeping wave expansion may be viewed. This work is summarized elsewhere in this volume [12], and we do not elaborate on it here. In [12] the role of the entire function contribution in scattered field representations is considered in a manner alternative to that of Section 4 in the present chapter.

3. CANONICAL MITTAG-LEFFLER EXPANSIONS

As discussed above, the Mittag-Leffler expansion for surface current given in (5) is based on a hypothesis that **J** be a meromorphic function of the frequency variable s. If a knowledge of the large-s asymptotic behavior of a function is known, then a more-tightly-defined Mittag-Leffler expansion may be constructed.

Work by Wilton [13] and by Pearson [14] has defined a better-restricted expansion, which we refer to here as the "canonical Mittag-Leffler expansion" for an induced current.

3.1 Asymptotic Behavior of the Resolvent Kernel to the Integral Equation

It is useful to direct our attention for this discussion to the "resolvent kernel" for the electric field integral equation (2). This resolvent is denoted $\underline{\mathbf{Z}}^{-1}$ and is defined by

$$<\underline{\mathbf{Z}}(\mathbf{r},\mathbf{r}_0,s);\underline{\mathbf{Z}}^{-1}(\mathbf{r}_0,\mathbf{r}',s)> = \underline{\mathbf{I}}\ \delta(\mathbf{r}-\mathbf{r}') \quad , \tag{15}$$

where $\underline{\mathbf{I}}$ is a two dimensional identity dyadic lying tangential to the surface of the scatterer. It follows that the solution to (2) is expressible as

$$\mathbf{J}(\mathbf{r},s) = <\underline{\mathbf{Z}}^{-1}(\mathbf{r},\mathbf{r}',s);\mathbf{E}^{inc}(\mathbf{r}',s)> \quad . \tag{16}$$

Our turning attention to the Mittag-Leffler expansion for the resolvent kernel is predicated upon the fact that its large-s asymptotic behavior is coordinate independent, while $\mathbf{E}^{inc}$ and in turn $\mathbf{J}$ manifest asymptotic behaviors which depend on the choice of coordinate origin and phase-time origin $\mathbf{r}_0$ in (4).

No rigorous determination of the required asymptotic behavior of $\underline{\mathbf{Z}}^{-1}$ has been made, to date. An estimate of this asymptotic form has been deduced by Wilton [13] for the case of a convex perfectly conducting object residing in a lossless medium and is expressible as

$$\underline{\mathbf{Z}}^{-1} \sim \underline{\mathbf{P}}_{\substack{R\\L}}(s)\ e^{\mp s|\mathbf{r}-\mathbf{r}'|/c} \quad , \text{Re } s \to \pm\infty, \tag{17}$$

where $\underline{\mathbf{P}}$ are dyadics which are algebraic in s, and where c is the velocity of light in the medium. This result follows from approximating the integral operator in (2) with its matrix counterpart using a method of moments scheme. The approximation to the inverse operator thus follows from constructing the inverse to the matrix using Cramer's rule and is amenable to asymptotic analysis as the discretization scheme is made finer and as $|s| \to \infty$. This result is not rigorous since the limit of the matrix operator as the discretization scheme is rendered finer does not go over to the original integral operator. However, the matrix formulation can be made arbitrarily accurate for a bandlimited excitation and the approximate results are satisfying in engineering applications.

The orders of the dyadic polynomials $\underline{\mathbf{P}}$ in (17) dictate the order of the convergence polynomials in the Mittag-Leffler expansion of $\underline{Z}^{-1}$ [14] and have not been considered in Wilton's development in [13]. If they are assumed to be of zeroth order, the following expansion results [14]

$$\underline{\mathbf{Z}}^{-1}(\mathbf{r},\mathbf{r}',s) = \sum_{\alpha} \beta_{\alpha} \mathbf{J}_{\alpha}(\mathbf{r})\mathbf{J}_{\alpha}(\mathbf{r}') \left[\frac{1}{s - s_{\alpha}} + \frac{1}{s_{\alpha}} \right] \quad (18)$$

where

$$\beta_{\alpha} = \langle \mathbf{J}_{\alpha}(\mathbf{r});\underline{Z}'(\mathbf{r},\mathbf{r}',s_{\alpha});\mathbf{J}_{\alpha}(\mathbf{r}')\rangle^{-1} \quad (19)$$

is a normalization constant arising from the residue evaluation. The prime on $\underline{\mathbf{Z}}$ indicates differentiation with respect to s. The zeroth order convegence polynomials present here result from the assumed zeroth order for the dyadic polynomials in (17). This form correctly models terminal properties of objects excited as antennas and has proven to be consistent with other physically correct behavior, thus providing some confidence in the zeroth order choice. The absence of an entire function in (18) is a consequence of the exponential decay of (17) in both the right and the left halves of the s-plane.

3.2 Canonical Current Expansions

The current expansion in terms of the resolvent kernel expansion (18) follows from (16) and can be couched in a multitude of forms depending on the way in which one carries out adjustment of the asymptotic properties of the series part of the current expansion. Two forms prove to be particularly useful, and their construction is outlined here. They are termed the "frequency independent coupling coefficient" expansion and the "frequency dependent coupling coefficient" expansion in the present discussion.

The frequency independent expansion is developed by substituting the Mittag-Leffler expansion for the resolvent (18) into the symmetric product (16), rearranging the exponentials as shown below:

$$\mathbf{J}(\mathbf{r},s) = F(s)\, e^{-s\hat{\mathbf{p}}\cdot(\mathbf{r}-\mathbf{r}_0)/c}$$
$$\times \left\langle \sum_{\alpha} \beta_{\alpha} \mathbf{J}_{\alpha}(\mathbf{r})\, \mathbf{J}_{\alpha}(\mathbf{r}') \left[\frac{1}{s - s_{\alpha}} + \frac{1}{s_{\alpha}}\right] ; \mathbf{E}_0\, e^{-s\hat{\mathbf{p}}\cdot(\mathbf{r}'-\mathbf{r})/c} \right\rangle . \quad (20)$$

The arrangement of the exponentials ensures that the integrand of the symmetric product is

$$\sim \quad [\ ]\ e^{-s[\hat{\mathbf{p}}\cdot(\mathbf{r}-\mathbf{r}')\pm|\mathbf{r}'-\mathbf{r}|]/c} \quad , \qquad \text{Re } s \to \pm\infty .$$

Clearly,

$$\hat{\mathbf{p}}\cdot(\mathbf{r}-\mathbf{r}')\pm|\mathbf{r}-\mathbf{r}'| \quad \begin{cases} \geq 0, & \text{Re } s \to \infty \\ \leq 0, & \text{Re } s \to -\infty \end{cases} , \tag{21}$$

so that (20) admits to expansion in the same form as (18) and, after a change in order of summation and symmetric product integration yields

$$\mathbf{J}(\mathbf{r},s) = F(s)\, e^{-s\hat{\mathbf{p}}\cdot(\mathbf{r}-\mathbf{r}_0)/c} \sum_\alpha H^i_\alpha \mathbf{J}_\alpha(\mathbf{r}) \left[\frac{1}{s - s_\alpha} + \frac{1}{s_\alpha} \right] , \tag{22}$$

where the coupling coefficient is given by

$$H^i_\alpha = \beta_\alpha \langle \mathbf{J}_\alpha(\mathbf{r}'); \mathbf{E}_0 e^{-s_\alpha \hat{\mathbf{p}}\cdot(\mathbf{r}'-\mathbf{r})/c} \rangle \tag{23}$$

and is distiguished by the presence of s particularized to pole values as a result of the residue evaluation in the expansion.

We term the second useful form the "frequency dependent coupling coefficient" (though it is in fact one of many frequency dependent forms conceivable) and construct the expansion by forming from (18) and (16)

$$\mathbf{J}(\mathbf{r},s) = F(s) \langle \sum_\alpha \beta_\alpha \mathbf{J}_\alpha(\mathbf{r})\, \mathbf{J}_\alpha(\mathbf{r}') \left[\frac{1}{s - s_\alpha} + \frac{1}{s_\alpha} \right] ; \mathbf{E}_0\, e^{-s\hat{\mathbf{p}}\cdot(\mathbf{r}'-\mathbf{r}_0)/c} \rangle . \tag{24}$$

Distributing the second term in the symmetric product through the series and changing the order of summation and integration yields

$$\mathbf{J}(\mathbf{r},s) = F(s) \sum_\alpha H_\alpha(s)\, \mathbf{J}_\alpha(\mathbf{r}) \left[\frac{1}{s - s_\alpha} + \frac{1}{s_\alpha} \right] , \tag{25}$$

where the coupling coefficient is given by

$$H_\alpha = \beta_\alpha \langle \mathbf{J}_\alpha(\mathbf{r}');\mathbf{E}_0 e^{-s\hat{\mathbf{p}}\cdot(\mathbf{r}'-\mathbf{r}_0)/c} \rangle \tag{26}$$

in which explicit dependence on frequency appears since no residue expansion is executed in this development.

These two forms have counterpart time-domain singularity expansions, and the physical interpretation of the two forms is best reserved for discussion in the time domain. The expansion in terms of a frequency independent coupling coefficient form manifests the obvious merit of computational simplicity since the coefficient does not have to be recomputed for each new frequency. It has also found utility in the synthesizing of equivalent circuits for receiving structures [15], [16]. The time-domain form counterpart to (26) is observed to yield an expansion which is much less sensitive to numerical error, however.

3.3 Expansions in the Time Domain

The singularity expansion in the time domain has been a source of confusion and, at times, controversy. The source of this confusion is that one has substantial lattitude in the way in which the coupling coefficient is constructed--i.e. there is no unique coupling coefficient form. Recently, a paper by Pearson, Wilton and Mittra [17], which was intended to make the source of this non-uniqueness clear, appeared.

In the previous section are put forth two choices for frequency domain coupling coefficients which have proven useful. By factoring exponentials representing different phase delays, in the same manner that the exponential is set in the front in (20), one may generate new coupling coefficients almost arbitrarily.

The lattitude in coupling coefficient construction is most easily interpreted in the time-domain expansion, the subject of the present section. While it is possible to proceed operationally from (22) and (23) and from (25) and (26) to arrive at counterpart time domain forms, we take a different course in order to expose clearly the source of the lattitude of which we speak. The remainder of this section follows in essence the discussion in [17].

The transient surface current response of the scattering object is given by the Laplace inversion of (16):

$$\mathbf{j}(\mathbf{r},t) = \int_{C_b} \langle \underline{\mathbf{Z}}^{-1}(\mathbf{r},\mathbf{r}',s);\mathbf{E}^{inc}(\mathbf{r}',s)\rangle \, e^{st} \, ds \quad , \tag{27}$$

where C_b denotes the Bromwich contour. Changing the order of the spatial and frequency integrations and expressing the spatial integration explicitly, we obtain

$$\mathbf{j}(\mathbf{r},t) = \int_{\text{body}} \int_{C_b} \underline{Z}^{-1}(\mathbf{r},\mathbf{r}',s)\cdot\mathbf{E}_0 e^{-s\hat{\mathbf{p}}\cdot(\mathbf{r}-\mathbf{r}')/c} F(s)\, e^{st}\, ds\, dS' \quad . \quad (28)$$

Using (17), we observe that in (28)

$$\text{Integrand} \sim \underline{\mathbf{P}}_{\substack{R\\L}}(s)\cdot\mathbf{E}_0\ e^{s[t-T_\pm(\mathbf{r},\mathbf{r}')]} \quad , \quad \text{Re } s \to \pm\infty, \quad (29)$$

where

$$T_\pm(\mathbf{r},\mathbf{r}') = [\hat{\mathbf{p}}\cdot(\mathbf{r}'-\mathbf{r}_0) \pm |\mathbf{r}-\mathbf{r}'|]/c \quad . \quad (30)$$

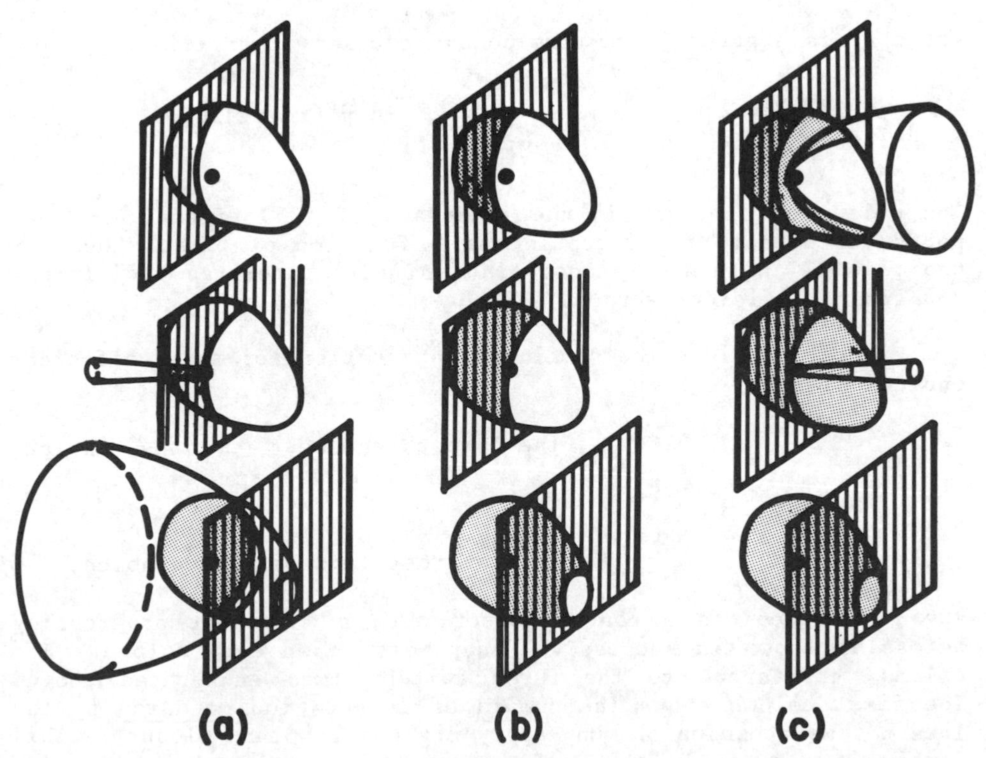

Figure 3. Evolution of the domain of integration for the generalized time-domain coupling coefficient as the wavefront traverses the object. (a) $T_s=T_+$, the latest permissible time, (b) $T_s= 0$, yielding integration behind the wavefront, and (c) $T_s= T_-$, the earliest permissible time. In each column "snapshots" are shown from top to bottom as the wavefront advances.

For $t<T_+$, the integrand decays exponentially in the right half of the s-plane, while for $t>T_-$, it manifests exponential decay in the left half plane. We may evaluate (28) using Jordan's theorem, obtaining zero by closing C_b with an infinite semicircle in the right half plane or obtaining a sum of residues at the poles of $\underline{Z}^{-1}$ by closing with an infinite semicircle in the left half plane. If we define a "switch time" T_s at which we go over from right half plane closure to left half plane closure, with

$$T_- < T_s < T_+, \tag{31}$$

then the Laplace inversion yields

$$\mathbf{j}(\mathbf{r},t) = f(t) * \sum_\alpha \eta_\alpha(\mathbf{r},t)\, \mathbf{J}_\alpha(\mathbf{r})\, e^{s_\alpha t} \,, \tag{32}$$

where η_α is a generalized time domain coupling coefficient

$$\eta_\alpha(\mathbf{r},t) = \int_{\text{Body}} u[t-T_s(\mathbf{r},\mathbf{r}')]\, \mathbf{J}_\alpha(\mathbf{r}')\cdot\mathbf{E}_0 e^{-s_\alpha \hat{\mathbf{p}}\cdot(\mathbf{r}'-\mathbf{r}_0)/c}\, dS' \quad . \tag{33}$$

The unit step function in the integrand of (33) effectively supports the domain of integration as a function of time. Thus any choice of T_s within the constraints of (31) generates a different coupling coefficient through (33).

Figure 3 depicts the domain of integration for three different choices of T_s:

$T_s = T_+$, the latest (and most cautious) choice;

$T_s = 0$; and

$T_s = T_-$, the earliest (and boldest) choice.

The Figure depicts a convex object with a wavefront progressing across it shown in successive "snapshots" down each column. The columns correspond to the three switch times enumerated above. The first column shows the domain of integration <u>mandated</u> by the latest time at which one may apply right half plane closure. This domain is seen to be the region on the surface of the object included within a paraboloid of revolution behind the wavefront whose axis passes through the observation point and is normal to the wavefront. The vertex of the paraboloid lies half way between the observation point and the wavefront, and the <u>latus rectum</u> of the paraboloid expands as the wavefront advances so that the paraboloid opens up and eventually embraces the entire object.

The rightmost column depicts the domain of integration allowed by the earliest time for left half plane closure. This domain is defined by a paraboloid complementary to the one described above in that it advances in front of the wavefront, and integration of the coupling coefficient (33) is permitted behind this paraboloid. The middle column in the Figure indicates that the choice $T_s = 0$ corresponds to integration of (33) over the region where the wavefront has passed, an intuitively appealing choice.

The choice $T_s = 0$ leads to the explicit form for the coupling coefficient below.

$$\eta_\alpha(t) = \int_{I(t)} \mathbf{J}_\alpha(\mathbf{r}')\cdot\mathbf{E}_0 e^{-s_\alpha \hat{\mathbf{p}}\cdot(\mathbf{r}'-\mathbf{r}_0)/c} dS', \tag{34}$$

where I(t) defines the "illuminated region" of the object--i.e. that portion of the object over which the wavefront has passed. This form has traditionally been termed the "Class 2" coupling coefficient and represents the operational inverse transform of the frequency-dependent coupling coefficient in (26).

The operational inversion of the frequency independent form (23) multiplied by the exponential factor in front of the series in (22) yields

$$\eta_\alpha(\mathbf{r}) = u[t-\hat{\mathbf{p}}\cdot(\mathbf{r}-\mathbf{r}_0)/c] \int_{Body} \mathbf{J}_\alpha(\mathbf{r}')\cdot\mathbf{E}_0 e^{-s_\alpha \hat{\mathbf{p}}\cdot(\mathbf{r}'-\mathbf{r}_0)/c} dS'. \tag{35}$$

In this form, the expansion is turned on just as the wavefront passes the observation point $\mathbf{r}$. This corresponds to applying right half plane closure of the Bromwich contour up until the time that the wavefront arrives at the observation point (column one in Figure 4), reverting to left half plane closure (column three in Figure 4) just as the wavefront passes the observation point. This form is convenient since the domain of integration always spans the entire body, thereby obviating the need to recompute the coupling coefficient for each time step as the wavefront traverses the body. A recent paper by Michalski [18] provides numerical support for the efficacy of this form. One must be cautious in its use, however, because of the sensitivity to numerical error incumbent with the boldness embodied in integrating ahead of the wavefront as it traverses the object [19].

3.4 Ramifications of Early-Time Computation

The preceeding section deals with details of coupling coefficient construction which have bearing only in the early time while the wavefront is crossing the scatterer. That only the early time is influenced is evident from comparison of (34) and (35) since in both forms the domain of integration comprises the entire object as soon as the wavefront passes. Consequently, the section deals with differences in the time regime where the singularity expansion, being a global description of the scattering phenomenon, is ill-suited to computation in the first place. The construction depends on the non-rigorous asymptotic result of (17). The completeness of the singularity expansion is mathematically established only asymptotically for late time (see Melrose [20]).

While the form (20) provides a singularity expansion that is free of an exponential entire function, one may adjust the exponential term that is factored in such a way as to force the presence of an exponential entire function in the expansion of the symmetric product. In an operational inversion of the ensuing form (22) the exponential factor sets a turn-on time at $t=\mathbf{r}_0/c$--a time shown to arise naturally in arriving at (35). Any artificial adjustment to this natural turn-on time must be corrected with the addition of a term of bounded support in the time domain, the equivalent of which is an exponential entire function in the frequency domain.

The local character of a progressive wave ray expansion is appropriate to the early time computation of scattering response, and the hybrid formulation between the singularity expansion and an early-time ray description of [11] provides a systematic framework in terms of which one may blend the ray description and the singularity expansion. The time-varying coupling coefficient (33) is progressive wave representation, too, since the "illumninated region" of the object is excited by an advancing wavefront. Computations based on (32) and (33) are thus hybrid schemes in the sense that they exploit a progressive wave scheme in the early time.

One application area, that of equivalent circuit synthesis by way of the singularity expansion [15], [16], precludes any change of the mode of expansion as in the hybrid scheme or the use of time-varying coupling coefficients. It is desirable, of course, that the equivalent circuit models approximate the behavior of the receiving structure as accurately as possible over the entire duration of the time response. These models depend on the time-independent and frequency-independent forms of (34) and (23) so that the forgoing discussion does have practical bearing in this application.

4. EXPANSION OF SCATTERED FIELDS

The construction of far fields from the singularity expansion for surface current has been addressed only in special cases, in particular in [21]. Recent interest in the application of the singularity expansion representation for classification of radar targets from backscattered data puts an increased importance on this subject, however. A recent presentation by Morgan, et. al. [22] addresses the scattered field expansion in the context of this application.

The representations in the preceding section demonstrate that the singularity expansion for surface current can be written as a residue series alone--i.e. no entire function apart from the convergence polynomials are required in the Mittag-Leffler expansion. (Obviously, too, no "entire function contribution" need appear in a properly constructed time domain representation.) In contrast, the scattered field representations in the frequency and time domains cannot be constructed so as to be free of an entire function, or the inverse thereof in the time domain representation, except in the special case of forward scatter.

The following development proceeds along the same line as that in the preceding section, drawing directly on the estimate of the asymptotic form for the resolvent kernel (17). These results are therefore analytical, though non-rigorous. The arguments of [22] are analytical and proceed from the concepts of fundamental causal Green's functions. As a result, they may prove to be a rigorous alternative to the arguments in the present work.

To observe the analytical properties of the scattered fields, it suffices to consider the magnetic vector potential due to the surface currents residing on the scatterer--i.e.

$$\mathbf{A}(\mathbf{r},s) = \langle G(\mathbf{r},\mathbf{r}'',s),\mathbf{J}(\mathbf{r}'',s)\rangle \quad , \qquad (36)$$

where

$$G(\mathbf{r},\mathbf{r}',s) = \frac{e^{-s|\mathbf{r}-\mathbf{r}'|/c}}{4\pi|\mathbf{r}-\mathbf{r}'|} \quad . \qquad (37)$$

Expressing the current in terms of the resolvent kernel to the integral equation and using the usual far field asymptotic forms $|\mathbf{r}-\mathbf{r}'|\simeq r$ in the denominator of G and $|\mathbf{r}-\mathbf{r}'|\simeq r-\hat{\mathbf{a}}_r\cdot\mathbf{r}'$ in the exponent in G yields

$$\mathbf{A}(\mathbf{r},s) \simeq \frac{e^{-sr/c}}{4\pi r} \langle e^{s\hat{\mathbf{a}}_r\cdot\mathbf{r}''/c}, \underline{\mathbf{Z}}^{-1}(\mathbf{r}'',\mathbf{r}',s);\mathbf{E}^{inc}(\mathbf{r}',s)\rangle \quad , \qquad (38)$$

or with $\mathbf{E}^{inc}$ expressed explicitly and the exponents rearranged

$$\mathbf{A}(\mathbf{r},s) \simeq F(s)\frac{e^{-s(r-\hat{\mathbf{p}}\cdot\mathbf{r}_0)/c}}{4\pi r}$$

$$\times \ \langle e^{-s(\hat{\mathbf{p}}-\hat{\mathbf{a}}_r)\cdot\mathbf{r}''/c}, \underline{\mathbf{Z}}^{-1}(\mathbf{r}'',\mathbf{r}',s);\mathbf{E}_0 e^{-s\hat{\mathbf{p}}\cdot(\mathbf{r}'-\mathbf{r}'')/c}\rangle \ . \qquad (39)$$

The use of the far-field approximations in the foregoing development depends upon the coordinate origin's being located in or near the scatterer.

The asymptotic behavior of the integrand in the symmetric product in (39) is observed to be

$$\text{Integrand} \sim e^{-s[(\hat{\mathbf{p}}-\hat{\mathbf{a}}_r)\cdot\mathbf{r}'' \pm |\mathbf{r}''-\mathbf{r}'| + \hat{\mathbf{p}}\cdot(\mathbf{r}'-\mathbf{r}'')]/c}, \ \text{Re } s \to \pm\infty. \qquad (40)$$

The inequalities of (21) govern the second and third terms in the exponent of (40), and, their influence on the exponent is that of asymptotic decay. The dot product in the first term, however, can yield either a positive or negative value, causing growth of the integrand in one half plane and leading to the conclusion that (39) must contain an entire function in its Mittag-Leffler expansion. One exceptional case occurs--that of the observation point of the scattered field residing in the forward-scatter direction. Then $\hat{\mathbf{a}}_r = \hat{\mathbf{p}}$ and the critical term in the exponent vanishes so that (39) admits to expansion with no entire function apart from the convergence polynomials.

The physical interpretation of the conclusion drawn above is best explored in the time domain expansion of the scattered field. To this end we construct the Laplace inversion of (36)

$$\mathbf{A}(\mathbf{r},t) = \int_{C_B} \langle G(\mathbf{r},\mathbf{r}',s),\mathbf{J}(\mathbf{r}'',s)\rangle \, e^{st} \, ds \ . \qquad (41)$$

With the aid of (39) we write

$$\mathbf{A}(\mathbf{r},t) \approx \frac{1}{4\pi r} f(t-T_D) *$$

$$\int_{C_B} e^{st} \langle e^{-s(\hat{\mathbf{p}}-\hat{\mathbf{a}}_r)\cdot\mathbf{r}''/c}, \underline{\mathbf{Z}}^{-1}(\mathbf{r}'',\mathbf{r}',s); \mathbf{E}_0 e^{-s\hat{\mathbf{p}}\cdot(\mathbf{r}'-\mathbf{r}'')/c} \rangle \, ds \,, \quad (42)$$

where $T_D = (r-\hat{\mathbf{p}}\cdot\mathbf{r}_0)/c$. In (42)

$$\text{Integrand} \backsim e^{s[t-T^f_\pm]}$$

with

$$T^f_\pm(\mathbf{r}',\mathbf{r}'') = [(\hat{\mathbf{p}}-\hat{\mathbf{a}}_r)\cdot\mathbf{r}'' \pm |\mathbf{r}''-\mathbf{r}'| + \hat{\mathbf{p}}\cdot(\mathbf{r}'-\mathbf{r}'')]/c. \quad (43)$$

The delay term T_D in (42) expresses the time delay from time zero (when the wavefront passes through $\mathbf{r}_0$) until the wavefront reaches the coordinate origin and the delay for propagation over the distance r to the observation point. The delay expressed by T^f, which is dependent on the integration variables, accounts for the initial time of excitation of a given spatial increment of surface current and adjusts the delay to the far-field observation point relative to the delay over the path length r.

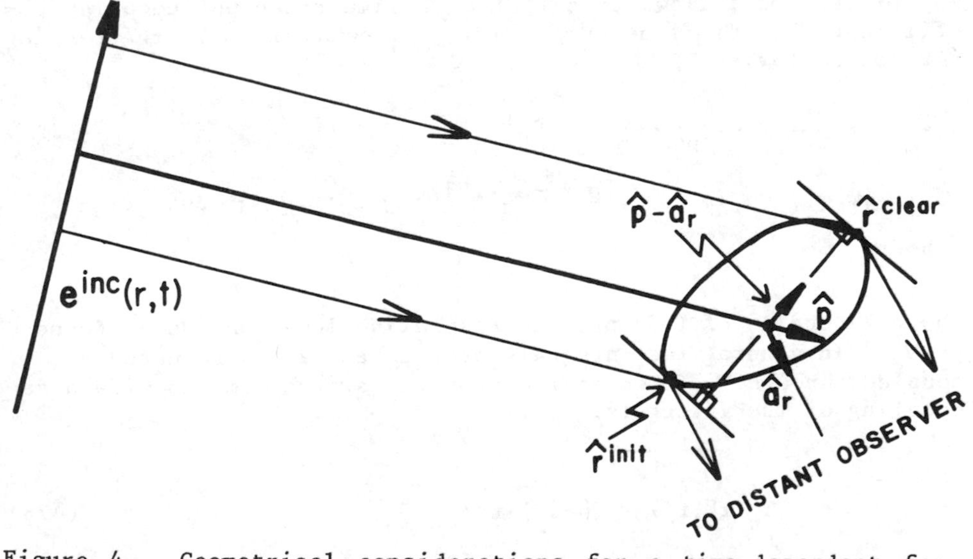

Figure 4. Geometrical considerations for a time-dependent far-field computation. Before the arrival of scattered energy from the point $\mathbf{r}^{init}$ the response in the observation direction is zero; and after the arrival of the first energy scattered from the point $\mathbf{r}^{clear}$ the response is representable in terms of the free oscillations of the object. A more complicated representation is required in the interval between.

For

$$t \leq \min_{\mathbf{r}''} (\hat{\mathbf{p}}-\hat{\mathbf{a}}_r)\cdot\mathbf{r}''/c \tag{44}$$

the integrand in (42) decays exponentially in the right half plane, and closure of the Bromwich contour with an infinite-radius semicircle there yields zero response--i.e. (44) expresses causality in the transient scattering phenomenon. A graphical construction is given in Figure 4, where the critical value of $\mathbf{r}''$ in (44) is labeled $\mathbf{r}''^{\text{init}}$.

For

$$t \geq \max_{\mathbf{r}''} (\hat{\mathbf{p}}-\hat{\mathbf{a}}_r)\cdot\mathbf{r}''/c \tag{45}$$

the integrand in (42) decays exponentially in the left half plane, and closure of the Bromwich contour there results in a residue series alone. The critical point for $\mathbf{r}''$ yielding this time is labeled $\mathbf{r}''^{\text{clear}}$ in Figure 4.

For times between the limits indicated in (44) and (45), one may proceed as follows to calculate a time dependent coupling coefficient for the scattered field by expanding and rearranging (42) as follows

$$\mathbf{A}(\mathbf{r},t) = \frac{1}{4\pi r} f(t-T_D) * \Sigma\, \beta_\alpha\, e^{s_\alpha t}$$

$$\times \int_{\text{body}} u[t-T_s^f(\mathbf{r}'',\mathbf{r}')]\, e^{-s_\alpha[\hat{\mathbf{p}}\cdot\mathbf{r}'-\hat{\mathbf{a}}_r\cdot\mathbf{r}'']/c}\, \mathbf{J}_\alpha(\mathbf{r}')\cdot\mathbf{E}_0\, \mathbf{J}_\alpha(\mathbf{r}'')\, dS'dS''. \tag{46}$$

where T_-^f and T_+^f of (43) provide respective lower and upper bounds on T_s^f. In general the integrals over $\mathbf{r}'$ and $\mathbf{r}''$ coordinates are coupled through T_s^f. Two choices for the switch time provide a decoupling of the integrals. Viz.

$$T_s^f(\mathbf{r}'',\mathbf{r}') = (\hat{\mathbf{p}}-\hat{\mathbf{a}}_r)\cdot\mathbf{r}''/c \quad , \tag{47a}$$

and

$$T_s^f(\mathbf{r}'',\mathbf{r}') = (\hat{\mathbf{p}}-\hat{\mathbf{a}}_r)\cdot\mathbf{r}'/c \quad , \tag{47b}$$

the first of which parallels the choice leading to (34) and the second the choice leading to (35). Then we may rewrite (46) in either of two ways with the integrals as a products:

$$\mathbf{A}(\mathbf{r},t) = \frac{1}{4\pi r} f(t-T_D) * \Sigma\, \beta_\alpha\, e^{s_\alpha t}$$

$$\times \int_{\text{body}} u[t-(\hat{\mathbf{p}}-\hat{\mathbf{a}}_r)\cdot\mathbf{r}''/c]\, e^{s_\alpha \hat{\mathbf{a}}_r\cdot\mathbf{r}''/c}\, \mathbf{J}_\alpha(\mathbf{r}'')\, dS''$$

$$\times \int_{\text{body}} e^{-s_\alpha \hat{\mathbf{p}}\cdot\mathbf{r}'/c}\, \mathbf{J}_\alpha(\mathbf{r}')\cdot\mathbf{E}_0\, dS' \quad . \tag{48a}$$

and

$$\mathbf{A}(\mathbf{r},t) = \frac{1}{4\pi r} f(t-T_D) * \Sigma\, \beta_\alpha\, e^{s_\alpha t}$$

$$\times \int_{\text{body}} e^{s_\alpha \hat{\mathbf{a}}_r\cdot\mathbf{r}''/c}\, \mathbf{J}_\alpha(\mathbf{r}'')\, dS''$$

$$\times \int_{\text{body}} u[t-(\hat{\mathbf{p}}-\hat{\mathbf{a}}_r)\cdot\mathbf{r}'/c]\, e^{-s_\alpha \hat{\mathbf{p}}\cdot\mathbf{r}'/c} \mathbf{J}_\alpha(\mathbf{r}')\cdot\mathbf{E}_0\, dS'. \tag{48b}$$

The symmetry present in the pairs $\mathbf{r}',\hat{\mathbf{p}}$ and $\mathbf{r}''$, $\hat{\mathbf{a}}_r$ appearing among the integrals in (48a) and (48b) is a consequence of reciprocity.

The second integral term in (48a) is simply the frequency/time-independent coupling coefficient of (23) with the phase referred to the geometric origin. (The delay between the space-time origin and the geometric origin is carried in T_D.) The second integral in (48b) is similar in form to the time-dependent coupling coefficient, but the support of the integration involves not only the advancing wavefront but also the observation direction.

The first-mentioned integrals in (48a) and (48b) may be viewed as far-field mode funtions, and in (48a) the support of the integrand varies with time during the early time accounting for a progressive-wave behavior of sources on the object as they become observable in the far field. In (48b) this role is fulfilled by the observation-direction dependence, which was mentioned above and which acts as an adjustment to the usual time-dependent coupling coefficient form.

Clearly, for sufficiently large t in (48a) and (48b) both domains of integration span the entire scatterer.

The conclusion drawn from the preceeding discussion is that a transient scattered field cannot be expressed purely as an exponential series until the scatterer's natural modes are established at the extreme boundary as expressed by (45). Equivalently, the Mittag-Leffler expansion of far-field quantities in the frequency domain, e.g. (39), must contain an entire function. Since for most scatterers of practical interest the transient scattered energy is contained in the time interval between the limits expressed in (44) and (45) and since the exponential model applies only after the end of that time interval, these observations cast still another doubt on the efficacy of systems identification for scattering target classification purposes.

5. CONCLUDING DISCUSSION

In this development we have attempted to review the singularity expansion formalism for perfectly conducting objects residing in a lossless medium and to call out the extent to which the representation is understood mathematically, at least on aspects which bear in engineering applications. Section 4 contains essentially new results which demonstrate that a representation of scattered fields solely as a singularity expansion is not possible: an entire function must be included in the frequency-domain expansion, and either time varying coefficients or the inverse transform of an entire function must be included in the time-domain expansion. That the resonant description alone is incomplete is likely to have adverse implications on the use of the singularity expansion as systems identification model in target classification applications.

A strict singularity expansion--one without time-varying coupling coefficients-- characterizes scattering response in terms of global features of the scattering object. Consequently, the representation is poorly suited to early time computations where local features provide the dominant influence on scattering phenomena. Indeed, the above-mentioned need to augment the expansion in order to represent scattered field response is a manifestation of this ill-suitedness.

In contrast, a progressive wave expansion provides a natural characterization where local features bear most heavily, but is cumbersome in later times when global resonances are established. The hybrid formalism of [11] and [12] provides a systematic scheme in terms of which to blend ray-optic models with the singularity expansion.

The singularity expansion adaptations that employ time-varying coefficients during the early time incorporate the progressive wave character of the illuminating field and the radiation field, thereby obviating the early-time difficulties of "strict" singularity expansions. These progressive wave features appear in the time-varying coupling coefficient definition (34) and in (48) by way of the Heaviside functions, which gate the domain of integration. The spatial terms in the arguments of these Heaviside functions supress current/field contributions until such times as causality in the surface equivalent scattering problem is honored. These spatial terms are intepretable in terms of distances traversed by the incident wavefront.

These adaptations of the singularity expansion to include time varying terms based on causality considerations therefore resemble the hybrid scheme in the sense that they embrace both progressive wave and resonance features. The effect of the explicit enforcement of causality in these forms is to suppress the early-time emphasis of high frequency resonances in the strict singularity expansion--high frequency resonances whose sole role is to provide the resolution necessary to resolve the wavefront turn on. These modified singularity expansions do not account for high-frequency effects in the systematic manner of the Heyman and Felsen hybrid approach, but where the spectrum of the excitation embraces only the first few resonances of the scatterer, they seem to prove efficient for computation. The relative merit of these two schemes remains a subject for evaluation.

6. ACKNOWLEDGEMENTS

The author's experience with the singularity expansion has been built through the conduct of work sponsored by the Air Force Office of Scientific Research, the Air Force Weapons Laboratory, the Physical Sciences Laboratory of New Mexico State University, and the Office of Naval Research. The new results reported here are a result of work sponsored by the Office of Naval Research under contract N00014-81-K-0256. The development of these new results was prompted by a discussion with Mr. Jon R. Auton with Effects Technology, Inc.

The author wishes to thank Profs. Donald R. Wilton of the University of Houston and K. A. Michalski of the University of Mississippi for their corrections and comments to this manuscript and for numerous discussions relative to the singularity expansion representation.

REFERENCES

1. C. E. Baum, On the Singularity Expansion Method for the Solution of Electromagnetic Interaction Problems, Interaction Note 88, Air Force Weapons Laboratory, Kirtland AFB, NM, Dec. 1971.
2. S. A. Schelkunoff, Theory of Antennas of Arbitrary Size and Shape, Proc. IRE 29 (1941) 493-521.
3. Electromagnetics 1 (1981) (No. 4, special issue on the Singularity Expansion Method).
4. L. Marin and R. W. Latham, Representation of Transient Scattered Fields in Terms of Free Oscillations of Bodies, Proc. IEEE 60 (1972) 640-641.
5. C. L. Dolph, V. Komkov, and R. A. Scott, A Critique of the Singularity Expansion and Eigenmode Expansion Methods, in Acoustic, Electromagnetic, and Elastic Wave Scattering--Focus on the T-Matrix Approach, V. K. Varadan and V. V. Varadan, Eds., (Oxford: Pergamon, 1980).
6. A. G. Ramm, Theoretical and Practical Aspects of the Singularity and Eigenmode Expansion Methods, IEEE Trans. Ant. and Propag. AP-28 (1980) 897-901.
7. D. R. Wilton, K. A. Michalski and L. W. Pearson, On the Existence of Branch-Points in the Eigenvalues of the Electric Field Integral Equation (EFIE) Operator in the Complex Frequency Plane, IEEE Trans. Ant. and Propag. AP-31 (1983) 86-91.
8. A. Q. Howard, A Geometric Theory of Natural Oscillation Frequencies in Exterior Scattering Problems, Interaction Note 378, Air Force Weapons Laboratory, Kirtland AFB, NM, Oct. 1979.
9. E. M. Kennaugh, The K-Pulse Concept, IEEE Trans. Ant. and Propag. AP-29 (1981) 327-331.
10. H. Uberall and G. C. Gaunaurd, The Physical Content of the Singularity Expansion Method, Appl. Phys. Lett. 39 (1981)362-364.
11. E. Heyman and L. B. Felsen, Creeping Waves and Resonances in Transient Scattering by Smooth Convex Objects, IEEE Trans. Ant. and Propag. AP-31 (1983) 426-437.
12. E. Heyman and L. B. Felsen, Wavefront Interpretations of SEM Resonances, Turn-on Times, and Entire Functions, these Proceedings.
13. D. R. Wilton, Large Frequency Asymptotic Properties of Resolvent Kernels, Electromagnetics 1 (1981) 403-412.
14. L. W. Pearson, Evidence that Bears on the Left Half Plane Asymptotic Behavior of the SEM Expansion for Surface Currents, Electromagnetics 1 (1981) 395-402.
15. K. A. Michalski, Synthesis of SEM-Derived Equivalent Circuits for Energy-Collecting Structures, Ph. D. Thesis, University of Kentucky, Lexington, KY, 1981.

16. K. A. Michalski and L. W. Pearson, Equivalent Circuit Synthesis for a Loop Antenna Based on the Singularity Expansion, to be published in IEEE Trans. Ant. and Propag. AP-32 (1984).
17. L. W. Pearson, D. R. Wilton, and R. Mittra, Some Implications of the Laplace Transform Inversion on SEM Coupling Coefficients in the Time Domain, Electromagnetics 2 (1982) 181-200.
18. K. A. Michalski, On the Class 1 Coupling Coefficient Performance in the SEM Expansion for the Current Density on a Scattering Object, Electromagnetics 2 (1982) 201-210.
19. C. E. Baum and L. W. Pearson, On the Convergence and Numerical Sensitivity of the SEM Pole-Series in Early-Time Scattering Response, Electromagnetics 1 (1981) 209-228.
20. R. B. Melrose, Singularities and Energy Decay in Acoustical Scattering, Duke Mathematical Journal 46 (1979) 43-59.
21. F. M. Tesche, The Far-Field Response of a Step-Excited Linear Antenna Using SEM, IEEE Trans. Ant. and Propag. AP-23 (1975) 834-838.
22. M. A. Morgan, M. L. Van Blaricum, and J. R. Auton, S-Plane Representations for Transient Electromagnetic Scattering, 1983 Spring National Radio Science Meeting, Houston, TX, May, 1983.

WAVEFRONT INTERPRETATION OF SEM RESONANCES, TURN-ON TIMES, AND ENTIRE FUNCTIONS

E. Heyman
Department of Electrical Engineering, Tel Aviv University, Israel

L.B. Felsen
Department of Electrical Engineering and Computer Science/
Microwave Research Institute
Polytechnic Institute of New York, Farmingdale, NY 11735 USA

ABSTRACT

By the Singularity Expansion Method (SEM), transient scattering is analyzed in terms of the damped oscillations corresponding to the complex resonant frequencies of the scatterer. Since the resonances describe global wave fields that encompass the scattering object as a whole, the SEM series representation encounters convergence difficulties at early observation times when portions of the object are as yet unexcited. Deficiencies in this representation must then be repaired by inclusion of an entire function in the complex frequency domain. The choice of the entire function is related intimately to the excitation coefficients, called coupling coefficients, of individual resonances and also to the "turn-on" and "switch-on" times of the SEM series. By using a traveling wave formulation in terms of progressing incident, reflected and diffracted wavefronts, these constructs in the SEM can be given a cogent physical interpretation. The wavefront analysis clearly identifies those portions of the entire function that are essential (intrinsic) and those that are removable. By combining wavefronts and resonances self-consistently, one may construct a hybrid field that avoids the difficulties at early times in the SEM formulation. These concepts are illustrated for two-dimensional scattering by a circular cylinder and a flat strip.

I. INTRODUCTION

Progressing waves (wavefronts) and oscillatory waves (resonances) may be employed as alternative building blocks to synthesize the transient fields scattered by an object.[1] In a progressing wave formulation,[2] one tracks a causal wavefront from the source to the scatterer, and thereafter observes contributions corresponding to multiple wavefront passes around the object and (or) multiple diffractions from scattering centers located on the object. This locally sensitive description becomes cumbersome at late observation times when many wavefronts have had time to reach the observer. Moreover, analytic tractability, by asymptotic considerations, usually is limited to moderate times after passage of a wavefront, with consequent difficulties when early arrivals are monitored at late times.

The oscillatory representation, formalized by the Singularity Expansion Method (SEM)[3,4], expresses global phenomena due to the scatterer as a whole. These global responses, the resonances, may be shown to account for the cumulative effect of multiple wavefront arrivals.[5] Therefore, this formulation is most convenient for late observation times whereas at early times, convergence problems may arise due to the many resonances needed to synthesize the abruptly changing local field near a wavefront or the causal zero field before the first wavefront has arrived. In the formal structure of the SEM, based on the solution of an integral equation formulation for the surface field, the SEM resonance series has had to be augmented by an entire function in the complex frequency plane to cope with these difficulties. The entire function, in turn, influences the excitation coefficients of the resonances, thereby introducing an arbitrariness which has, however, been clarified in recent studies.[4] Concerning the phenomenology, the cumulative effect of the many high-frequency resonances required at early times may actually be shown to produce a wavefront field.

A recently developed hybrid representation[5] avoids the difficulties associated with either of the above descriptions by incorporating the well-behaved portions of both - wavefronts at early times and resonances at later times - within a single self-consistent framework that is applicable and convenient for all times. The hybrid approach - illustrated in ref. 5 for two-dimensional smooth convex objects - formalizes the intimate relation between wavefronts and resonances as cumulative phenomena of the one in terms of the other, and thereby furnishes a cogent physical picture of the scattering process. The bilateral equivalence is established directly from an analysis based on the concepts of the Geometrical Theory of Diffraction (GTD) and avoids the use of integral equations entirely. The connection between wavefronts and resonances in terms of the physically motivated constructs of GTD furnishes an interpretation that has remained hidden within the integral equation format.

In the present investigation, the hybrid point of view is exploited to clarify the difficulties of the SEM representation at early times by relating the entire function to the wavefront-like nature of the response. Such an interpretation, while inherent in the previous study[5], was not related there specifically to the formal structure of the SEM. Moreover, source and observation points were there restricted to lie on the obstacle surface in order to demonstrate relevant concepts in their simplest form. The latter restriction is now removed, thereby making it possible to address the SEM-GTD connection for off-surface source and observer, a trivial extension within GTD but one with major implications for SEM.[4] The circular cylinder again serves as a prototype but the conclusions can be generalized to smooth convex shapes[5] although such extensions are not included here. The conclusions also hold for scatterers with edge discontinuities, as is illustrated by scattering from a flat strip. Emphasis here is on the basic ideas that establish the relation of the SEM resonances, their excitation coefficients, and the entire function with the wavefront structure of the scattered field, especially the ability, or not, to eliminate that function by appropriate choice of the excitation coefficients. This leads to a distinction between removable and non-removable (intrinsic) entire functions. Analytical details here are kept to a minimum but can be found in the cited references.

II. SEM FORMULATION

By the SEM, the scattered field in the frequency domain is described as

$$\hat{G}(\underset{\sim}{\rho},\omega) = \sum_{\alpha} \eta_{\alpha}^{(1)}(\underset{\sim}{\rho}) \frac{1}{\omega - \omega_{\alpha}} + \hat{f}(\underset{\sim}{\rho},\omega) \tag{1}$$

where ω_{α} is one in a set of complex resonance frequencies located in the lower half of the complex ω-plane, that produces in the time domain (with $\exp(-i\omega t)$ implied) a damped oscillation. The scattered field in the time domain results from Fourier inversion via

$$G(\underset{\sim}{\rho},t) = \frac{1}{2\pi}\int_{-\infty}^{\infty} d\omega\, e^{-i\omega t}\, \hat{G}(\rho,\omega), \tag{2}$$

with causality ensured by having the integration contour pass above all singularities in the complex ω-plane. The coefficients $\eta_{\alpha}^{(1)}$ are the residues at the complex frequency poles and are referred to in the SEM literature as "class 1 coupling coefficients"[4,6], a somewhat inappropriate designation since no intermode coupling is implied. They are actually the excitation coefficients of the space-time resonances corresponding to the natural resonance frequencies ω_{α}. These coefficients (of class 1) are identified by the fact that they are frequency (or time) independent. In the integral equation procedure of the SEM, they arise from surface current integrations over the <u>entire</u> obstacle surface. In what follows, we shall construct the scattered field explicitly by asymptotic diffraction theory without recourse to the integral equation. It is thereby implied that the interpretation of our results in terms of SEM coupling coefficients involves only "class 1" since integrations over portions of the scatterer, as required for "class 2"[4], have no counterpart in our treatment.

The pole series in (1) must generally be augmented by an entire function $\hat{f}(\rho,\omega)$ in the frequency domain[4,6] which, in the time domain, corresponds to a finite duration signal at early times. The role of the entire function and whether or when to include it, has caused confusion and controversy, but has recently been clarified[4,6]. It is closely linked to the turn-on time of the resonance series (the series of damped oscillations in the time domain). If the resonance series is turned on too early, convergence difficulties arise for the required synthesis of a null field before the first causal arrival. If the series is turned on too late, one should compensate for the early time field by an entire function in the frequency domain. In fact, the pole series in (1) may be modified via the formal identity[6]

$$\sum_\alpha \eta_\alpha^{(1)} \frac{1}{\omega - \omega_\alpha} = e^{i\omega T} \sum_\alpha \eta_\alpha^{(1)} \frac{e^{-i\omega_\alpha T}}{\omega-\omega_\alpha} + \sum_\alpha \eta_\alpha^{(1)} \frac{1-e^{i(\omega-\omega_\alpha)T}}{\omega-\omega_\alpha} \tag{3}$$

to express turn-on at $t = T$. Since the second sum on the right-hand side of (3) has no pole singularities, it is an entire function that contributes the field between $t = 0$ and $t = T$, and has been generated solely by a shift in the turn-on time of the pole series. There may, however, be another type of entire function which is intrinsic to the scattering process. The roles of these constituents in the SEM are examined below within the context of diffraction by a circular cylinder and by a strip, wherein the wavefronts play a crucial role that remains submerged in the conventional SEM approach.

III. SCATTERING BY A CIRCULAR CYLINDER

1. Analysis

We consider an impulsive line source at $\rho' = (\rho',\phi')$, $\phi' = 0$, and an observer at $\rho_0 = (\rho_0,\phi_0)$ located arbitrarily with respect to a circular cylinder of radius a; Neumann type (hard) boundary conditions $\partial G/\partial\rho = 0$ are assumed at $\rho = a$. To construct a wavefront representation for the scattered field in the time domain, one utilizes a Green's function $\hat{G}^\infty(\rho,\rho';\omega)$ defined in an infinitely extended angular domain $\rho = (\rho,\tilde{\phi})$, wherein the angular coordinate may assume all values $-\infty < \phi < \infty$. By separation of variables, $\hat{G}^\infty$ can be expressed as an angular wavenumber spectral integral[7]

$$\hat{G}^\infty(\rho,\rho';\omega) = \frac{1}{2\pi}\int d\mu\, e^{i\mu|\phi|}\, g_\rho(\rho,\rho';\mu;\omega) \tag{4}$$

involving the radial Green's function g_ρ. It can also be expanded in terms of angularly propagating radial eigenfunctions[7]

$$\hat{G}^{\infty}(\rho,\rho';\omega) = \sum_{p=1}^{\infty} G_p^{\infty}(\underset{\sim}{\rho},\underset{\sim}{\rho}';\omega) \tag{5}$$

The forms of the radial Green's function or the radial eigenfunctions may be found in ref. 7. For the Neumann boundary condition on the cylinder considered here, g_ρ has in the denominator the function $H_\mu^{(1)'}(ka)$, $k = \omega/v$ (v = free space wave speed), whose zeros $\mu_p(\omega)$ generate simple pole singularities in the complex μ-plane. The radial eigenfunctions, which correspond to the residues at these poles, therefore contain an angular propagation function $\exp[i\mu_p|\phi|]$. At high frequencies, μ_p has the asymptotic behavior

$$\mu_p(\omega) \sim ka + \alpha_p' \, e^{i\pi/3}(ka/2)^{1/3} \tag{6}$$

with α_p' denoting the p-th zero of $A_i'(-\alpha)$. The radial eigenfunctions themselves may be approximated asymptotically to yield the creeping rays of GTD, with appropriate diffraction, launching or attachment coefficients defined for various source and observer locations with respect to the scatterer.[8] While these interpretations are kept in mind, we shall not need the explicit asymptotic forms in what follows.

The field observed in the angularly periodic physical domain is synthesized by summing over the contributions from all images in the infinitely extended domain. It is suggestive to decompose G first into

$$G = G^+ + G^- \tag{7}$$

which functions account, respectively, for waves propagating in the positive and negative ϕ directions. They are given individually by

$$\hat{G}^{\pm}(\underset{\sim}{\rho}_0,\underset{\sim}{\rho}';\omega) = \sum_{j=0}^{\infty} \hat{G}^{\infty}(\underset{\sim}{\rho}_j^{\pm},\rho';\omega) \tag{8}$$

where the coordinates

$$\underset{\sim}{\rho}_j^{\pm} = (\rho_0,\phi_j^{\pm}), \phi_j^+ = \phi_0 + 2\pi j, \ \phi_j^- = \phi_0 - 2\pi(j+1) \tag{9}$$

locate the images of $\underset{\sim}{\rho}_0$ along the positive and negative directions, respectively, in the infinite angular space. The field due to a typical term on the right-hand side of (8), corresponding to the j-th image in (9), represents a wave contribution that reaches the observer at $\underset{\sim}{\rho}_0$ after $(j+1)$ passes around the cylinder. In the time domain, these passes are resolved as individual wavefront arrivals.

The series in (8) is readily summed into closed form since the index j affects only the angular phase term (see (9) and (4)). When the observer is located in the lit region, with $0 < \phi < \pi$ (see Fig. 1), all images $\underset{\sim}{\rho}_j^-$ are in the shadow region with respect to $\underset{\sim}{\rho}'$. When

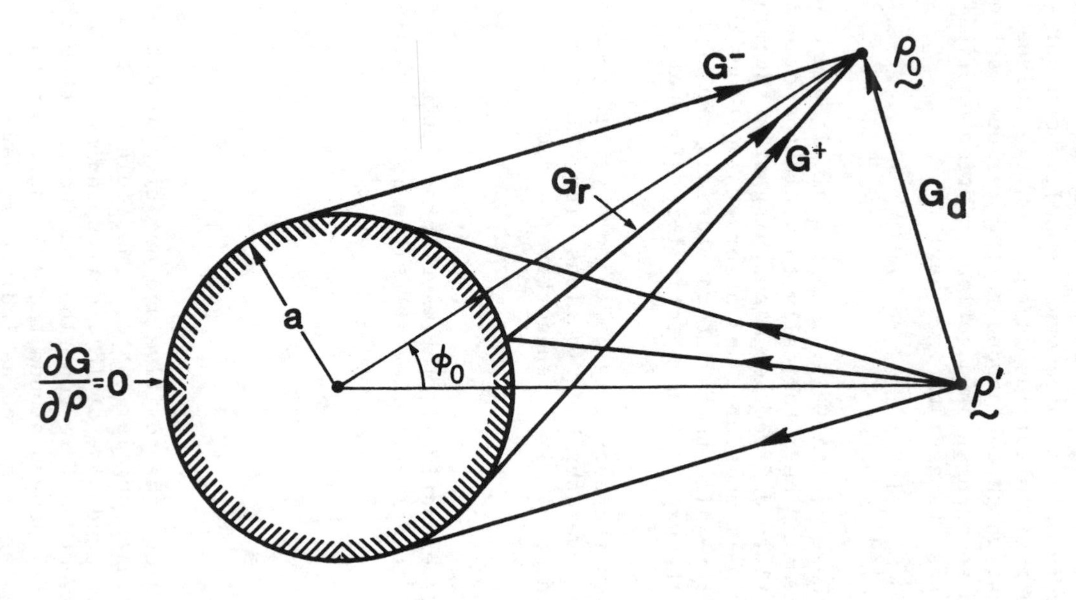

Fig. 1 Ray contributions for an observer in the lit region. G_d - direct ray, G_r - reflected ray, $G^{\pm}$ - diffracted fields due to positively and negatively revolving creeping waves.

(5) is used for the field due to each of these images with angular phase $\exp[i\mu_p|\phi_j^-|]$, one obtains:

$$\hat{G}^-(\rho_0,\rho';\omega) = \sum_p \hat{G}_p^\infty(\rho_0^-,\rho';\omega)/(1-e^{i2\pi\mu_p}) \qquad (10)$$

Note that the term in the numerator on the right-hand side of (10) is the field at ρ_0 corresponding to the first pass of the angularly propagating (creeping) wave, i.e., the first arrival of the negatively revolving wavefront of G^-.

The collective summation generates a resonance denominator, wherein $\exp(i\mu_p 2\pi)$ is the phase accumulation in a complete revolution. This denominator defines the SEM poles via the resonance condition[5]

$$\mu_p(\omega_{m,p}) = m, \quad m = \pm(2p-1), \pm 2p, \ldots. \qquad (11)$$

which states that for a resonance at frequency $\omega_{m,p}$, the phase accumulation in a complete revolution must be an integer multiple of 2π. For given p, the resonances lie on the p-th layer in the complex ω-plane (Fig. 2). Such a layer generates the singularity expansion of the angularly progressing p-th creeping wave. For given m, the resonances lie on arcs in the complex ω-plane, and define by this grouping the conventional angular harmonics that are oscillatory in the angular domain.

For collective treatment of contributions to $\hat{G}^+$, one notes that with $0 < \phi_0 < \pi$, $\rho_0^+ = \rho_0$ is in the illuminated region. To avoid convergence difficulties (see reference 7), the $j = 0$ term in (8) is therefore kept as a spectral integral whose high frequency asymptotic evaluation yields the direct and geometrically reflected ray contributions.[7] The collective summation begins with the $j = 1$ term, the first arrival of the positively revolving creeping wave at ρ_0. Thus,

$$\hat{G}^+(\rho_0,\rho';\omega) = \hat{G}^\infty(\rho_0,\rho';\omega) + \sum_p \hat{G}_p^\infty(\rho_0^+,\rho';\omega)/(1-e^{i2\pi\mu_p}) \qquad (12)$$

The first term on the right hand side is expressed as the spectral integral in (4). The numerator in the sum is again the wavefront field $\hat{G}_p^\infty$ corresponding to the first arrival of the positively revolving creeping wave at ρ_0 (see Fig. 1).

The transient field $G^+(t)$ corresponding to (12) is:

$$G^+(t) = G_d(t) + G_r(t) + \sum_p \{\sum_m G_{m,p}^+ + G_{p.c.}^+\} H(vt - D_1^+) \qquad (13)$$

where $H(\alpha) = 1$, $\alpha > 0$, but $H(\alpha) = 0$, $\alpha < 0$. The first two terms represent the direct and reflected ray contributions. Using the Transient Spectral Method (see ref. (9) and the references therein) for direct (asymptotic) inversion of the spectral integrals into the time domain, one may express these ray fields in closed form.

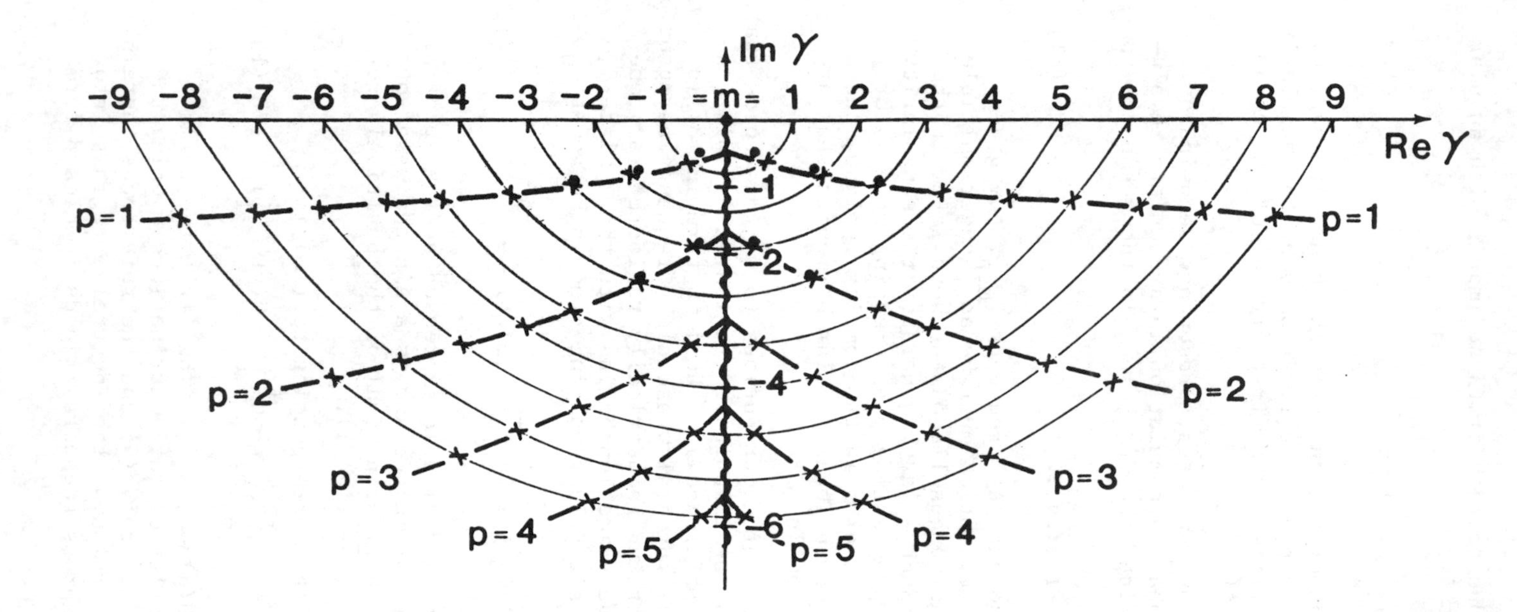

Fig. 2 Pole singularities (SEM resonances) in the lower half of the complex $\gamma = ka = \omega a/v$ plane, for a perfectly conducting circular cylinder.[5]

The third term in (13) is the contribution from the collective term in (12). This term becomes activated at $vt = D_1^+$, where v is the free space wave speed and D_1^+ is the ray path of the first diffracted wavefront ($j = 1$) that arrives at ρ_0 from the positive ϕ direction (Fig. 1). This turn-on time is determined from the asymptotic behavior of $G^+ \sim \exp[ikD_1^+]$ in the upper half of the complex ω plane, which permits closure of the path in (2) at infinity there, and therefore yields $G^+ \equiv 0$, for $vt < D_1^+$. After turn-on, closing the integration path in (2) at infinity in the lower half plane yields the field expressed in terms of the pole contributions $G^+_{m,p}$ and the branch cut integral contribution $G^+_{p,c}$ due to the branch point at $\omega = 0$.[5] In fact, the asymptotic behavior $\hat{G}^+ \sim \exp[ik(D_1^+ - 2\pi a)]$ in the lower half of the ω plane[5] validates (by closure of the integration contour) the formal application of the resonance expansion in (12) at a time interval ($2\pi a/v$) corresponding to a full revolution before the actual turn-on time. This series, however, is poorly convergent during this time interval since the higher order (high-frequency) resonances are dominant and many terms are needed to generate the null field before turn-on.

The turn-on of the resonance series can be delayed by summing the j series collectively from $j = J$ to infinity, keeping the first $j = 0,\ldots J-1$ terms intact as creeping wavefronts. The collective summations yields:

$$\sum_p \sum_{j=0}^{J-1} \hat{G}_p^\infty(\rho_j^+,\rho';\omega) = \sum_p \hat{G}_p^\infty(\rho_0,\rho';\omega)\,\frac{1-e^{i2\pi\mu_p J}}{1-e^{i2\pi\mu_p}} \tag{14}$$

which expression contains no pole singularities. It relates to, and interprets, the entire function in (3) as the contribution from the first J arrivals of the wavefronts.

2. Discussion

The following conclusions can be drawn from the analysis in Section III.1:

(a) For observation points in the lit region, the pole series is augmented by a spectral integral $\hat{G}^\infty(\rho_0,\rho';\omega)$ in (12) that yields the direct and reflected waves. This "quasi" entire function (since the only singularity is a branch point at $\omega = 0$[5]) contributes in the time domain before the turn-on of the resonances. It is intrinsic to the scattering process in that it contains early time local geometric information via the specularly reflected ray. In the shadow region there are no such wave constituents.

(b) The turn-on time of the resonance series is the arrival time of the first wavefront which is included in the cumulative treatment. It is the arrival time of the $j = 0$ wavefront for the collective expression in (10), of the $j = 1$ wavefront for the expression in (12), and of the $j = J$ wavefront for the expression in (14). We may write this condition as

$$t_{on} = D/v \tag{15}$$

where t_{on} is the turn-on time, and D is the ray path of the first wavefront included in the cumulative treatment. Each wave constituent generally has its own turn-on time. For the cylinder, there are two constituents G^+ and G^-, but more constituents may arise for more complicated configurations (four, for example, for the strip; see Fig. 3).

(c) The SEM switch-time t_s, usually defined as the time when the resonance series can be formally used[4], may be taken at a time corresponding up to a full revolution before the actual turn-on time of the resonances; i.e.,

$$(t_{on} - L/v) < t_s \leq t_{on} \tag{16}$$

where L is the circumference of the cylinder. In fact, since G is decomposed into wave constituents, each has its own t_{on} and t_s. The best choice of t_s is $t_s = t_{on}$. If G cannot be separated into wave constituents (e.g., for numerically specified residues), then it is best to choose $t_s = \min\{t_{on}\}$. Some of the wave constituents will now be non-causal and may therefore cause convergence problems at early times.

(d) For the special case when the observer is located on the cylinder, Michalski[10] has made the conjecture that the turn-on time for the resonance series of the induced current is the arrival time of the incident wavefront at the observer. However, we should note that, in the shadow region, this wavefront is not the creeping wavefront, which correctly describes the turn-on time. As mentioned already, the best description may be obtained by separating the field response G into its various wave constituents, each with its own t_{on} determined by the path traversed by its first wavefront arrival.

(e) If the resonance series is turned on too late, the entire function that compensates for the omitted early time field may be interpreted in terms of wavefront contributions as in (14). This entire function is different from the intrinsic entire function mentioned in (a).

IV. SCATTERING BY A STRIP

To apply the concepts introduced in Section III to an obstacle with edge discontinuities, we examine next diffraction by a strip of width a. We restrict ourselves here to the high-frequency domain, which is relevant for the early time phenomena under discussion, and which may be accommodated by GTD. The scattering process involves four wave constituents, to be identified by the index $\ell = 1,2,3,4$. In the first two which are sketched in Fig. 3,

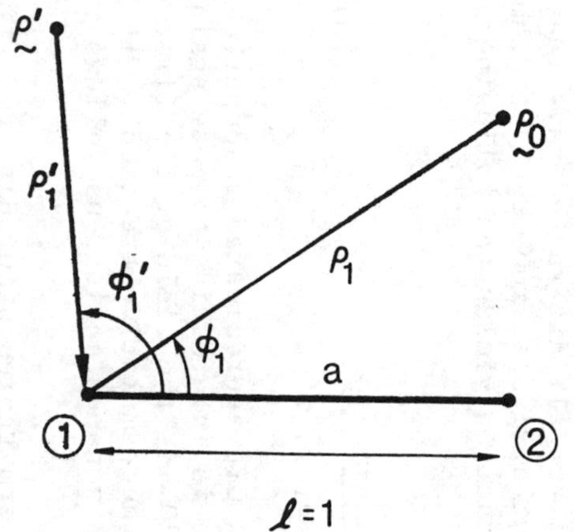

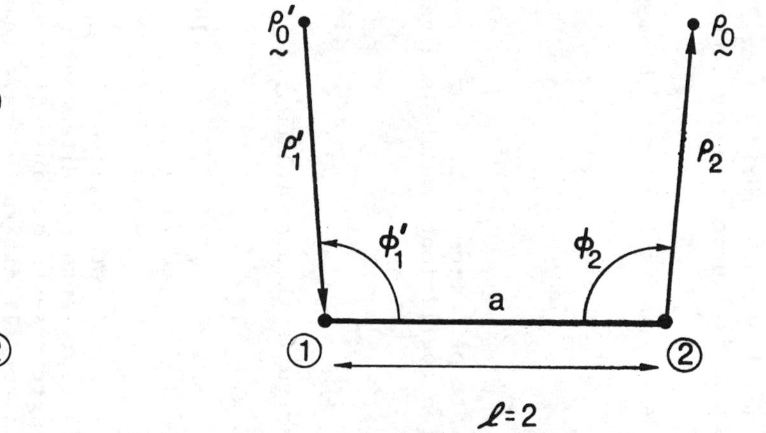

Fig. 3. Wave constituents $\ell = 1$ and $\ell = 2$ for diffraction by a flat strip. The edges are located at (1) and (2).

the incident ray hits edge no. 1 first. The ray diffracted toward the observer at $\underset{\sim}{\rho}_0$ comes from edge no. 1 or no. 2 for $\ell = 1$ or 2, respectively. For the other two wave constituents, the incident ray first hits edge no.2.

The GTD fields for these wave constituents involve as elements the free space Green's function

$$\hat{G}_0(\rho) = \frac{e^{ik\rho + i\pi/4}}{\sqrt{8\pi k \rho}} \tag{17}$$

which establishes the field at a distance ρ from a (true or virtual) line source, and the edge diffraction coefficient, which is given for the hard boundary condition by

$$D_h = -\left[\sec\left(\frac{\phi - \phi'}{2}\right) + \sec\left(\frac{\phi + \phi'}{2}\right)\right] \tag{18}$$

The GTD field for the $\ell = 1$ wave species is then as follows:

$$\begin{aligned}\hat{G}_1 = \hat{G}_o(\rho_1')\hat{G}_o(\rho_1)\{&D(\phi_1,\phi_1') \\ &+ D(0,\phi_1')D(0,0)D(\phi_1,0)[2\hat{G}_o(a)]^2 \sum_{j=0}^{\infty} [2\hat{G}_o(a)D(0,0]^{2j}\}\end{aligned} \tag{19}$$

with the coordinates $(\rho_1,\phi_1),(\rho_1',\phi_1')(\rho_2,\phi_2)$, etc. defined in Fig. 3. The first term inside the braces represents the direct diffraction at edge 1. The infinite series represents multiple diffractions between the edges and accounts for all diffracted rays leaving edge 1 after successive wavefront arrivals from edge 2. Summation of the series of multiple diffraction yields the following contribution to (19):

$$Q = \hat{G}_o(\rho_1')\hat{G}_o(\rho_1)\,\frac{D(0,\phi_1')\;D(0,0)D(\phi_1,0)}{1 - [\hat{G}_o(a)D(0,0)]^2} \tag{20}$$

The SEM poles generated by this GTD approximated resonant denominator[11] are shown in Fig. 4. It is interesting to note again that the term in the numerator of (20) is the field of the first wavefront which is incorporated in the cumulative treatment. It determines the turn-on time of the resonance series, as follows from (b) of Sec. III.2.

One may verify that the first term representing direct diffraction in (19) cannot be incorporated into the cumulative treatment. Thus, it plays the role of the intrinsic "quasi entire function" in the sense of (a) and (e) of Sec.III.2. This term has no pole singularities and conveys only local information about the first edge, in contrast to the global information conveyed by the SEM resonances. The reflected field, where it exists, also belongs to the quasi entire function category.

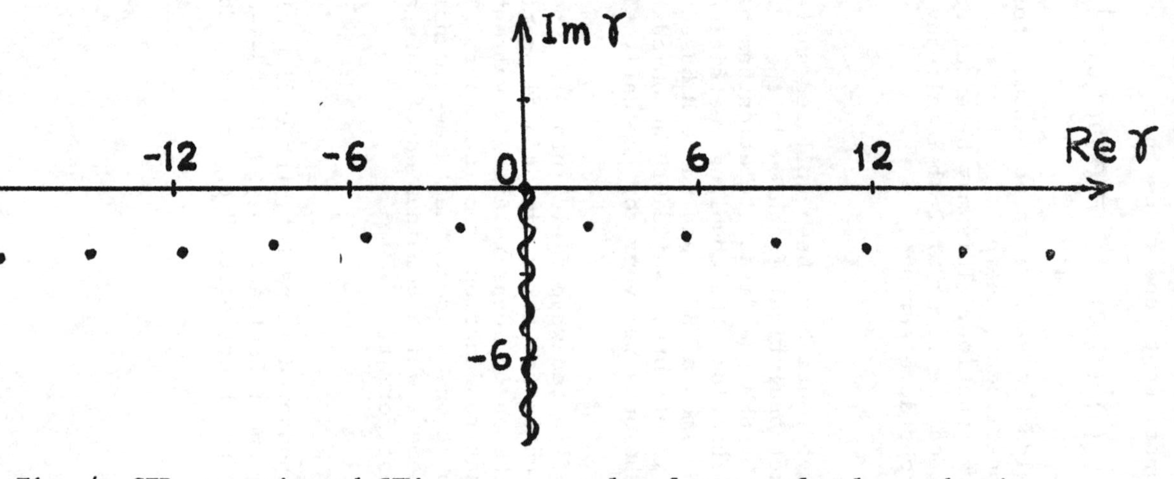

Fig. 4 GTD approximated SEM resonance poles for a perfectly conducting strip. $\gamma = ka$. Magnetic field polarized parallel to the edge.

Applying the same considerations to all ray contributions for the diffracted $\ell = 2$ wave constituent, one finds that

$$\hat{G}_2 = \hat{G}_o(\rho_1')\hat{G}_o(\rho_2)D(0,\phi_1')G_o(a)D(\phi_2,0) \sum_{j=0}^{\infty} [\hat{G}_o(a)D(0,0)]^{2j} \quad (21)$$

This series, when summed, yields the same resonant denominator as in (20), but for this wave constituent, there is no free term that appears as an intrinsic entire function. This may be expected from the ray picture in Fig. 3, since the first arrival is already global in that it has traversed the obstacle surface.

V. SUMMARY

In this presentation, wavefronts have been utilized systematically to analyze and interpret analytical features in the early time treatment of scattering by SEM. A direct relation has been established between the entire function, the coupling coefficients, and the turn-on and switch-on times of SEM, and the GTD based concepts of diffracted rays and wavefronts. While GTD accommodates only high frequencies, those are the ones relevant for the early time field.

The resolution of the diffracted wave field into GTD type constituents becomes more complicated for three-dimensional scatterers. Yet, it is anticipated that the basic ideas relating wavefronts to the intrinsic and removable entire functions remain intact. These aspects require further exploration, as does the three-dimensional application of the hybrid scheme[5], which combines wavefronts and resonances in a self-consistent format.

ACKNOWLEDGEMENT

This research has been sponsored in part by the Office of Naval Research, Electronics Branch, under Contract No. N00014-83-K-0214, and in part by the Joint Services Electronics Program under Contract No. F49620-82-C-0084.

REFERENCES

1. Felsen, L.B., "Progressing Fields, Oscillatory Fields, and Hybrid Combinations", this issue.

2. Friedlander, P.G., Sound Pulses, Cambridge University Press, 1958.

3. Baum, C.E., "The Singularity Expansion Method", Transient Electromagnetic Fields, L.B. Felsen ed., Springer-Verlag, N.Y. 1976.

4. Pearson, L.W., "The Singularity Expansion Method Representation of Surface Currents on a Perfectly Conducting Scatterer", this issue.

5. Heyman, E. and L.B. Felsen, "Creeping Waves and Resonances in Transient Scattering by Smooth Convex Objects", IEEE Trans. Antennas and Propagation, Vol. AP-31, pp. 426-437, 1983.

6. Pearson, L.W., D.R. Wilton and R. Mittra, "Some Implications of the Laplace Transform Inversion on SEM Coupling Coefficients in the Time Domain", Electromagn., Vol. 2, pp. 189-200, 1982.

7. Felsen, L.B. and N. Marcuvitz, "Radiation and Scattering of Waves", Prentice-Hall, 1973, Chap. 6.

8. Pathak, R.H. and R.G. Kouyoumjian, "An Analysis of Radiation from Apertures on Curved Surfaces by the Geometrical Theory of Diffraction", Proc. IEEE, Vol. 67, pp. 1438-1447, 1974.

9. Heyman, E.and L.B. Felsen, "Non-Dispersive Approximations for Transient Ray Fields in an Inhomogeneous Medium", this issue.

10. Michalski, K.A., "On the Class 1 Coupling Coefficient Performances in the SEM Expansion for the Current Density on a Scattering Object", Electromagn., Vol. 2, pp. 201-210, 1982.

11 Shirai, H., E. Heyman and L.B. Felsen, "Wavefronts and Resonances for Transient Scattering by a Perfectly Conducting Flat Strip", in preparation.

NON-DISPERSIVE APPROXIMATIONS FOR TRANSIENT RAY FIELDS IN AN INHOMOGENEOUS MEDIUM

E. Heyman
Department of Electrical Engineering, Tel Aviv University, Israel
L.B. Felsen
Department of Electrical Engineering and Computer Science/
Microwave Research Institute
Polytechnic Institute of New York, Farmingdale, NY, USA 11735

ABSTRACT

Time-harmonic wave propagation in an inhomogeneously layered medium may be expressed in terms of ray integrals comprised of local spectral plane waves associated with particular rays. At high frequencies, the spectral wavenumber integrands are non-dispersive, and the resulting ray integrals can be inverted into the time domain to yield the transient response in closed form. Two principal methods, by Cagniard-DeHoop and by Chapman, have been developed to deal with the inversion. The former has limited scope and the latter, while more broadly applicable, restricts the spectral wavenumbers to be real. The two methods, which generate results of dissimilar appearance, can be embedded within a spectral theory that accommodates each and also more general problems by allowing real and complex spectral wavenumber contributions. The theory is described, and is illustrated for caustic forming ray species, for which the Cagniard-DeHoop method is inapplicable and the Chapman method less convenient.

I. INTRODUCTION

The determination of the transient response due to a time-dependent source in a complicated environment poses a formidable task. By the conventional approach, the problem is first reduced to the time-harmonic regime and then solved there by whatever technique may be most appropriate. The choice of technique depends on the frequency range. At high frequencies, where the local wavelength is small compared with continuous changes in the environment, the field can be expressed as a superposition of local plane waves that follow geometrical ray trajectories from the source to the observer. Asymptotic ray theory, commonly referred to as the Geometrical Theory of Diffraction (GTD), provides a systematic scheme for tracking ray fields along direct, reflected, refracted and diffracted paths.[1,2] When carried out to higher orders of approximation, GTD yields for each ray field a formal asymptotic expansion in inverse powers of the large angular frequency ω. However, the determination of the coefficients in the higher order terms is usually so difficult as to render the method ineffective beyond its leading term. Moreover, in transition regions where simple GTD fails, the need for uniform approximations complicates even the behavior of the leading term.[3-5]

When the GTD fields are transformed into the time domain, one obtains a formal series in powers of $(t - t_o)$, where t_o is the arrival time of the wavefront along the ray path and t is the observation time thereafter.[6] This series generally diverges for long enough observation times and is therefore not useful for field calculations in that regime. To improve the representation, it is necessary to retain in the high-frequency fields not only the very narrow bundle of local plane waves that interfere constructively along the ray path but a broader range of plane wave spectral components. The resulting spectral integral, generally referred to as the "ray integral" or the "generalized ray", yields GTD by asymptotic (stationary phase) evaluation but accounts for a broader range of phenomena, including transition effects, when it is kept intact. For coordinate separable configurations, exact spectral integrals can be constructed by the method of separation of variables,[7] but techniques have recently been explored for constructing spectral integrals for rays even in complicated non-separable environments.[8-10]

It might be surmised that the time inversion of the ray integral poses difficulties that far transcend the direct closed form inversion of GTD. However, this is not always the case. For a class of non-dispersive wave fields, the transient response can actually be obtained exactly, and for all times, in closed form whereas the time-harmonic ray integral cannot be reduced except by asymptotic evaluation. While the non-dispersive property applies exactly only to a few prototype propagation and scattering problems, the spectral wave constituents in a more general environment may be approximable at high frequencies by local plane wave (WKB type)forms. The ray integral so obtained may then be explored for exact inversion. The resulting transient field which, in a sense, represents the closed form summation of the series derived from GTD, is convergent for <u>all</u> time although the validity of the formal expression is restricted to the high-frequency components of the signal. For each ray field, this generally implies observation times not too far behind the arrival of its wavefront, but considerably beyond the very short time accommodated by the "wavefront approximation" corresponding to GTD. However, by extracting the high-frequency non-dispersive portion from the ray integral for closed form inversion, the low-frequency dispersive remainder is then often amenable to more efficient treatment (even numerically) than the original ray integral.

In this presentation, we deal with propagation of source-excited wave fields in an inhomogeneous layered medium. Since the emphasis is on basic concepts built around wave spectra, no attempt is made to derive the non-dispersively approximated ray integrals for specific configurations. Further details may be found in a more comprehensive forthcoming manuscript. Here, in Section II, we merely describe the generic form and physical content of the ray integral. To deal with its inversion into the time domain, two principal approaches have been developed. Their advantages and limitations are discussed in Secs. III.1 and III.2 within the perspective of the plane wave spectra that contribute to the transient field. We then propose in Sec. III.3 a unified approach that accommodates both of the afore-mentioned methods within the same spectral format. This provides physical insight that is not as readily gleaned otherwise. Concluding remarks are made in Section IV.

II. SPECTRAL REPRESENTATION OF RAY FIELD

The class of problems treated here involves an impulsive line source in an inhomogeneous two-dimensional configuration which is assumed to be separable in an orthogonal (ℓ_1,ℓ_2) coordinate system. The ray integral of interest may represent reflected, refracted or multiple reflected-refracted ray species as in Fig. 1a. In the time harmonic regime, the field of each of these ray species may be expressed as a spectral integral of the form[7]

$$\hat{G}(\underset{\sim}{\rho},\underset{\sim}{\rho}';\omega) = \int_{-\infty}^{\infty} \phi_\nu(\ell_1,\omega)\bar{\phi}_\nu(\ell_1',\omega)g_\nu(\ell_2,\ell_2';\omega)\,d\nu \tag{1}$$

wherein $\underset{\sim}{\rho} = (\ell_1,\ell_2)$ and $\underset{\sim}{\rho}' = (\ell_1',\ell_2')$ are coordinates of the source and observation points, respectively, ϕ_ν is the one-dimensional eigenfunction in the ℓ_1 domain, $\bar{\phi}_\nu$ is its adjoint, and ν is the spectral parameter. The function g_ν corresponds to the characteristic Green's function in the ℓ_2 domain after a ray expansion has been performed.[11] In the high frequency regime, the WKB approximations for ϕ_ν and g_ν in (1) reduce the integrand so as to yield the "ray integral"

$$\hat{G}_{\pm}(\omega \gtrless 0) \sim (-ik)^m k \int_{C_{\pm}} A_{\pm}(\xi) e^{ikL_{\pm}(\xi)} d\xi \qquad (2)$$

wherein the positive and negative frequency ranges have been separated and assigned to quantities with subscripts + and -, respectively. In (2), $k = \omega/v$ is the wavenumber at the source location, v is the wave velocity there, and the frequency ω has been removed from the spectral parameter ν by the scaling transformation

$$\xi = \nu/k = \sin\theta, \qquad (3)$$

θ being the departure angle at the source, with respect to the ℓ_2 direction, of the local plane wave described by the integrand. From (1) and (3), it follows that the integration contours $C_{\pm}$ in (2) extend from $\mp\infty$ to $\pm\infty$, respectively, and they are chosen so that $\mathrm{Im}\,L_{\pm}(\xi) \gtrless 0$ on them, respectively. The amplitude functions $A_{\pm}(\underset{\sim}{\rho},\underset{\sim}{\rho}';\xi)$ and the phase functions $L_{\pm}(\underset{\sim}{\rho},\underset{\sim}{\rho}';\xi)$ must satisfy the symmetry relations

$$A_{+}(\xi) = A_{-}^{*}(\xi^{*}), \quad L_{+}(\xi) = L_{-}^{*}(\xi^{*}) \qquad (4)$$

in order to generate a real transient response, by Fourier inversion over ω, of the sum of $\hat{G}_{+}$ and $\hat{G}_{-}$ in (2). The asterisk denotes the complex conjugate.

In the high-frequency domain, where medium dispersion is negligible, the phase functions satisfy $L_{\pm} = L$ and, in the visible spectrum range that supports propagating local plane waves, L is real and represents the length between source point and observation point levels along the phase paths traversed by the local plane waves (Fig. 1b). At the endpoints $\xi = \pm\xi_c$ of the visible spectrum, $L(\xi)$ has branch point singularities. Within the visible spectrum, L may have one or more stationary points ξ_i determined by

$$L'(\xi_i) = 0 \qquad i = 1,2... \qquad (5)$$

These stationary points identify spectral parameters of geometrical ray paths between source and receiver, with the prime denoting the derivative with respect to the argument, and their number is indicative of the behavior of the ray family (ordinary, caustic forming, etc.).

The amplitude functions $A_{\pm}$ in (2) contain spectral plane wave reflection coefficients $R(\xi)$, which accumulate in multiplicative fashion for each encounter of a local spectral plane wave with a boundary. If the local plane wave for a given value of ξ is refracted back (has a turning point) before reaching a boundary, the reflection coefficient is replaced by $\exp(-i\pi/2)$ for each turning point.[11] When reflecting boundaries and the medium are lossless, the functions $A_{\pm}$ can be (a) real or (b) imaginary in the visible

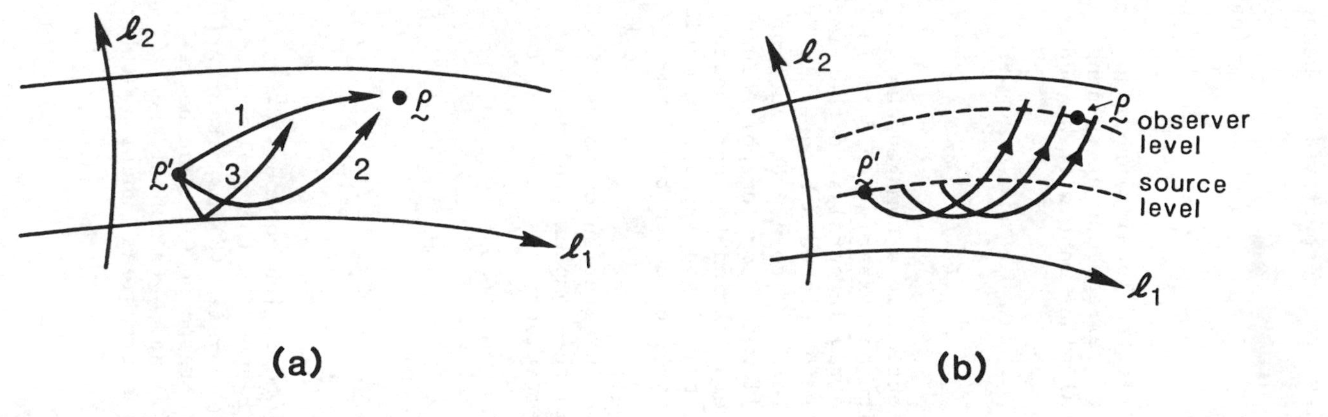

Fig. 1. Source at $\underset{\sim}{\rho}'$ and observer at $\underset{\sim}{\rho}$ in a medium with curved boundaries (if any) and variable refractive index, but such as to render the wave equation separable in the orthogonal coordinate system (ℓ_1, ℓ_2).

(a) Ray paths. Ray 1 is a direct ray, ray 2 is a turning ray, and ray 3 is a reflected ray.

(b) Local plane wave trajectories corresponding to $L(\xi)$. Stationary points ξ_i of $L(\xi)$ select trajectories that pass through both $\underset{\sim}{\rho}$ and $\underset{\sim}{\rho}'$.

spectrum. From (4), these possibilities are, respectively, characterized by

$$A_+(\xi) = A_-(\xi) = A_+^*(\xi^*) \tag{6a}$$

$$A_+(\xi) = -A_-(\xi) = -A_+^*(\xi^*) \tag{6b}$$

Equation (6a) describes a direct or reflected local plane wave, or one that has undergone an even number of refractions at turning points. Equation (6b) describes a local plane wave with an odd number of turning point encounters.

III. RECOVERY OF THE TRANSIENT FIELD

1. Cagniard-DeHoop Method

For the canonical problem of reflection in a homogeneous half space, where the spectral representation has exactly the form in (2), a method for direct inversion of a ray integral into the time domain was proposed by Cagniard[12] and reformulated subsequently by DeHoop.[13] This method was applied by Felsen[14] to a wide class of diffraction problems and modified to allow for complex pole singularities in the spectral plane. A generalization to accommodate reflection in an inhomogeneous medium, by using a WKB approximation for the integrand of (1) which is thereby reduced to the form in (2), has also been carried out.[15]

In the Cagniard-DeHoop method, the integration contours $C_\pm$ of (2) are transformed into contours $C_\pm^t$ in the complex ξ plane, on which $L(\xi)$ is real. Changing the integration variable from ξ to t via

$$L(\xi) = vt, \tag{7}$$

one obtains from (2)

$$\hat{G}_\pm(\omega \gtrless 0) = (-ik)^{m+1} \int_{C_\pm^t} iA_\pm(d\xi/dt)e^{i\omega t}\, dt, \quad \frac{d\xi}{dt} = \frac{v}{L'(\xi)} \tag{8}$$

It follows from (4) that each of the integration contours $C_\mp^t$ in the complex ξ plane is symmetric with respect to the real axis. Then if $C_\pm^t = C_-^t$ and if $A_+ = A_-$, one may recognize (8) as a Fourier time transform from which the transient response G(t) can be recovered by inspection (recall that $\hat{G}_+(\omega)$ and $\hat{G}_-(\omega)$ should be Fourier transforms, with $\omega \gtrless 0$, of the same transient function G(t)):

$$G(t) = \left(\frac{d}{vdt}\right)^{m+1} \mathrm{Im}\left\{\frac{-2vA_+(\check{\xi})}{L'(\check{\xi})}\right\} \tag{9}$$

where $\check{\xi}(t)$ is the solution of (7) in the lower half of the ξ plane.

The above derivation of the Cagniard-DeHoop inversion illuminates its difficulties and why it cannot be applied to a general ray integral. The two requirements, which led from (8) to (9), restrict its applicability to a special class of problems. For example, one may show that the requirements are not satisfied, and the method can therefore not be applied, to ray species with a caustic (see Fig. 2), where the phase function $L(\xi)$ contains more than one stationary point as defined in (5). Also, it cannot be applied to turning rays whose $A_\pm$ functions satisfy (6b) rather than (6a).

2. Chapman Method[16]

Instead of inverting by inspection as in the Cagniard-DeHoop method, one may employ an alternative procedure that is capable of treating the non-dispersive general ray integral in (2) without the limitations inherent in the Cagniard-DeHoop approach. In this procedure, introduced by Chapman[16] and presented in this section, the order of the ω and ξ integrations is interchanged, with the restriction that both integration variables are kept real. By the method proposed in the next section, the ξ integration is carried out in the complex plane.

When the ω and ξ integrations are performed along the real axis, the sum of the Fourier transforms of $\hat{G}_+(\omega)$ and $\hat{G}_-(\omega)$ in (2) may be written as

$$G(t) = \frac{1}{2\pi}\int_{-\infty}^{\infty} d\omega \int_{-\infty}^{\infty} d\xi \, \exp[-i\omega(t - L(\xi)/v)] \cdot (-ik)^{m+1}\{i \, \mathrm{sgn}(\omega) \mathrm{Re}\, A_+(\xi) - \mathrm{Im}\, A_+(\xi)\} \tag{10}$$

where $\mathrm{sgn}(\omega) = \pm 1$ for $\omega \gtrless 0$, respectively. The two terms in the braces $\{...\}$ are a consequence of (4) and of the direction of $C_\pm$ in (2). After inverting the order of the ω and ξ integrations in (10), the ω-integration for the second term in the braces leads to the Dirac delta function $\delta[t - L(\xi)/v]$ whereas for the first term in the braces, one requires the Fourier transform of the $\mathrm{sgn}(\omega)$ function. As a result, one obtains:

$$G(t) = \left(\frac{d}{v\,dt}\right)^{m+1} \left\{ -\mathcal{H} \sum_{\bar{\xi}(t)} \frac{v \mathrm{Re} A_+(\bar{\xi})}{|L'(\bar{\xi})|} - \sum_{\bar{\xi}(t)} \frac{v \mathrm{Im} A_+(\bar{\xi})}{|L'(\bar{\xi})|} \right\} \tag{11}$$

where $\bar{\xi}(t)$ are the real solutions of (7) (i.e., in the visible spectrum range where L is real), and $\mathcal{H}$ is the Hilbert transform

$$\mathcal{H}(f) = \frac{P}{\pi}\int \frac{f(t')}{t'-t}\, dt' \tag{11a}$$

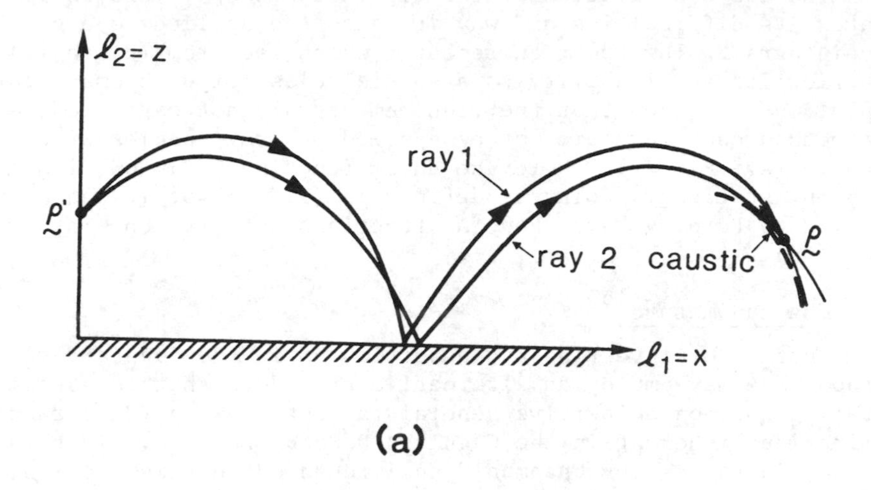

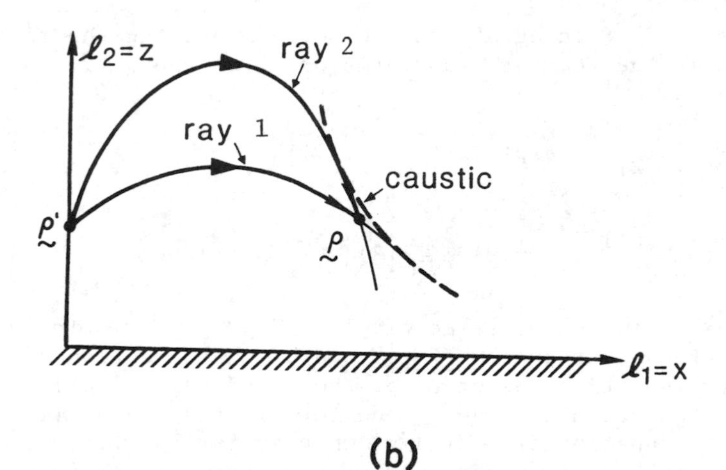

Fig. 2. Typical ray configurations, with caustics, in a vertically inhomogeneous medium.

(a) rays with reflection and refraction.
(b) rays with refraction only.

Ray 2 has, while ray 1 (which arrives at the observer before ray 2) has not, touched the caustic before reaching the observer.

with P denoting the principal value of the integral. In deriving (11), a contribution from the non-visible spectral range $|\xi| > \xi_c$, wherein $L(\xi)$ is complex, has been neglected. This contribution is, however, included in reference 16.

The second term in (11) is similar to (9). The field at t is contributed directly by the real plane wave $\bar{\xi}(t)$ whose wavefront arrives at ρ' at the time corresponding to (5) and (7). However, for plane wave species whose amplitude function A_+ is real in the visible spectrum (thereby yielding the first term in (11)), the Hilbert transform reveals no direct relation between the plane wave $\bar{\xi}(t)$ and the field at t. This obscures a physical interpretation for such wave species. A simple example is the canonical problem of reflection in a homogeneous half space, which has a simple closed form solution in (9) via the Cagniard-DeHoop method. Since, in this problem, $A_\pm$ are real in the visible spectrum, the solution via the Chapman route is given by the first term in (11) which contains the Hilbert transform. The contribution of this term is non-causal but the final result agrees (when the integration over the non-visible spectrum, not included in (11), is taken into account as well) with the exact result from (9) [ref. 16, Sec. 3.10]. This example highlights the complications introduced by the Hilbert transform for the class of ray species with $A_\pm$ real in the visible spectrum. The half-space problem is more easily doable via the Cagniard-DeHoop method, but as noted previously (Sec. III.1), this method is not applicable to other classes of problems. In the next section we shall present a generalized "Cagniard-type" solution (i.e. of the form in (9)) that removes these limitations and contains within it either the conventional Cagniard form, when that applies, or the Chapman form.

3. Transient Spectral Formulation

In this formulation, we follow Chapman by performing first the ω-integration. However, since we deal here with analytic functions in the complex ξ plane, we may not combine the $\hat{G}_\pm$ functions of (2) as in (10). Instead we shall Fourier invert $\hat{G}_+(\omega)$ and $\hat{G}_-(\omega)$ separately to obtain

$$G(t) = G_+(t) + G_-(t) \tag{12}$$

with

$$G_\pm(t) = \left(\frac{d}{vdt}\right)^{m+1} \frac{(\pm 1)}{2\pi} \int_{C_\pm} \frac{vA_\pm(\xi)}{vt - L(\xi)} d\xi, \quad L(\xi) \gtrless 0 \text{ on } C_\pm \tag{13}$$

The main feature of this representation is that the nature of G(t) is determined by the singularities of the transient spectral integrals, this facilitates a direct physical interpretation. The singularities consist of those in $A_\pm$ and $L_\pm$, and of the poles generated by the roots of (7). By appropriate contour deformation, one may reduce (13) either to a Chapman or to a generalized Cagniard formulation, whichever is more convenient. This avoids the Hilbert transform in (11) when it is not needed, and relates the complex spectrum directly to the time domain.

In the following, we shall concentrate on the poles due to the denominator of (13), i.e., the roots of (7). As a function of t, these poles lie on trajectories in the complex ξ plane, for example, those in Fig. 3, where the arrows indicate the direction of increasing t. For a certain time interval, poles may be found on the real axis, in the visible spectrum $-\xi_c < \xi < \xi_c$ where L is real. The pole trajectories leave the real axis at the stationary points $\xi_i < \xi_c$ (cf. (5)). The direction of the trajectories in the vicinity of the stationary points depends on the sign of the first non-vanishing derivative there. For example, if ξ_i is a simple stationary point of $L(\xi)$, with a non-vanishing second derivative (i.e., it corresponds to a conventional ray away from a caustic), than for $L''(\xi_i) > 0$, the poles move toward ξ_i on complex trajectories as t approaches $t_i = L(\xi_i)/v$, the arrival time of the wavefront along the ray. After that time, the poles move away from ξ_i on the real axis. The opposite holds when $L''(\xi_i) < 0$.

Figure 3 illustrates these features for wave species with two stationary points, corresponding to observation points in the lit region near a caustic, as in Fig. 2. The stationary point ξ_1 describes the ray which has not yet touched the caustic. It may be verified that its wavefront arrives first since $L(\xi_1) < L(\xi_2)$; furthermore, $L''(\xi_{1,2}) \gtrless 0$. As t approaches t_1, two poles, which lie on the complex branch associated with ξ_1 in Fig. 3, approach that point; in addition, there may be another pole on the right hand side of ξ_2. In the time interval between t_1 and t_2, there are three poles on the real axis. As t approaches t_2, two of the poles approach ξ_2, and for $t > t_2$, they move away from t_2 on the complex trajectories associated with it.

Other singularities of the integrand are the branch points of L and $A_\pm$ at the end points $\xi = \pm\xi_c$ of the visible spectrum. In addition, the amplitude functions $A_\pm$ may have (head (lateral)wave generating) branch point singularities in the visible spectrum, or (surface and leaky wave) pole singularities due to energy trapping properties of the boundaries.[14] Such singularities can be accounted for in a straightforward manner but we shall not be concerned with them here.

The evaluation of the integrals (13) is performed by contour deformation around the singularities of the integrands. One notes two important special cases corresponding to (6a) and (6b) respectively. In the first case, the pole contributions of the real roots of (7) (for the time interval when they exist) cancel in the overall result, while from the complex roots one obtains

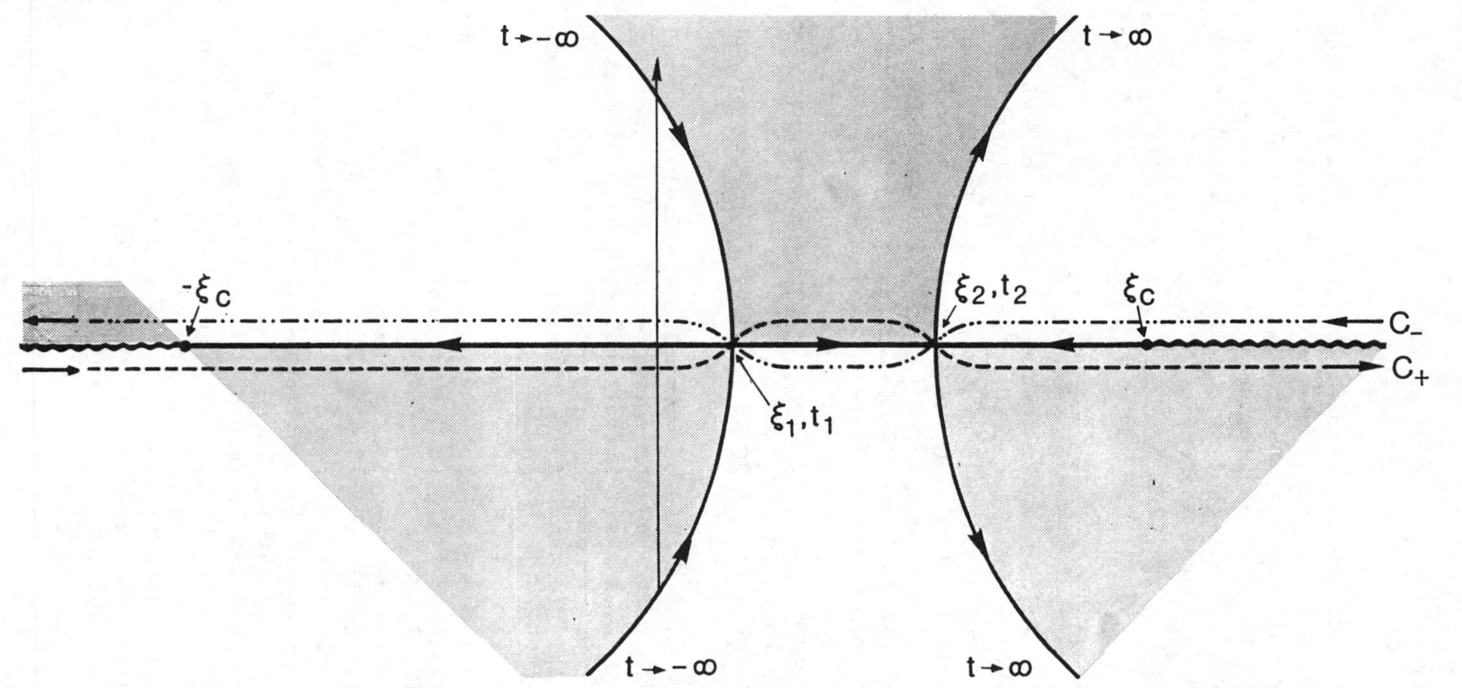

Fig. 3. Complex ξ plane map for local plane wave phase function L with two stationary points ξ_1 and ξ_2 descriptive of rays 1 and 2 in Fig. 2. The wavefront along ray 1, which has not touched the caustic, arrives first at t_1, whereas the wavefront along ray 2, which has touched the caustic, arrives at $t_2 > t_1$. Note that $L''(\xi_{1,2}) \gtrless 0$. The heavy lines are trajectories of the roots of (7) (poles of the transient spectral integral (13)) and time increases in the direction of the arrows. The integration contours $C_\pm$ in (2) and (13) (-- → and ← ·· – ··) are deformed into regions with $\mathrm{Im}\,L \gtrless 0$, the shaded region being the one with $\mathrm{Im}\,L > 0$. The branch points $\pm\xi_c$ are the end points of the visible spectrum and the branch cuts (〰) are drawn along the non-visible spectrum interval. For the configuration in Fig. 2, one may show that the amplitude functions $A_\pm$ are either real or imaginary in the visible spectrum.

$$G(t) = \left(\frac{d}{vdt}\right)^{m+1} \sum_{\check{\xi}(t)} \operatorname{Im} \frac{-2vA_{+}(\check{\xi})}{L'(\check{\xi})} \tag{14a}$$

This formula is similar to the Cagniard expression (9) except that here we allow for multiple stationary points (thereby multiple complex trajectories and multiple $\check{\xi}(t)$) as in Fig. 2, a case which cannot be accommodated by the Cagniard method. If, on the other hand, the amplitude functions satisfy (6b), the pole contributions from the complex roots of (7) cancel and one obtains

$$G(t) = \left(\frac{d}{vdt}\right)^{m+1} \sum_{\bar{\xi}(t)} \frac{viA_{+}(\bar{\xi})}{|L'(\bar{\xi})|} \tag{14b}$$

(cf. the second term in (11)) plus an integral over the non-visible spectrum $|\xi| > \xi_c$, which is neglected here as in (11). Since (6a) and (6b) describe, respectively, cases wherein $A_{\pm}$ are real or imaginary in the visible spectrum, one notes that (14a) and (14b) correspond to the first and second terms in (11), when (10) is specialized to either case (6a) or case (6b), respectively. Thus we see that wave species of type (6a) are naturally described by the complex roots of (7) as in (14a), rather than by the first term (with the Hilbert transform) in (11) plus the integration over the non-visible spectrum. Wave species of type (6b) are naturally described by the real roots of (7) as in (14b) or as in the second term in (11).

As an example, Figs. 4(a) and (b) show qualitatively the transient signals which correspond to the L functions of Fig. 3 for the two cases where $A_{\pm}$ satisfy either (6a) or (6b), respectively. Figs. 2(a) and (b) schematize possible ray species which give rise to these cases. In Fig. 4(a) the ray contribution for $t < t_1$ comes from poles on the complex trajectories in Fig. 3 associated with ξ_1, whereas the contribution for $t > t_2$ comes from poles on the complex trajectories associated with ξ_2. The contribution in Fig. 4(b) comes from poles on the real trajectories in Fig. 3. The field near $t = t_1$ in Fig. 4(b) is due to the two poles moving away from ξ_1, while the rising field as $t \to t_2$ is due to the two poles moving toward ξ_2. Finally, it is instructive to note how the first term in (11) is applied to the case in Fig. 4(a). Here, one would have to Hilbert transform the real pole contributions whose waveform would look like that in Fig. 4(b). Thus, in addition to the fact that there is no direct and easily interpretable relation between the pole contributions and the final signal, the Hilbert transform also generates spurious fields which can, however, be cancelled by including the integration over the non-visible spectrum (cf. the example of reflection in a homogeneous half space, discussed earlier [ref. 16, Sec. 3.10]).

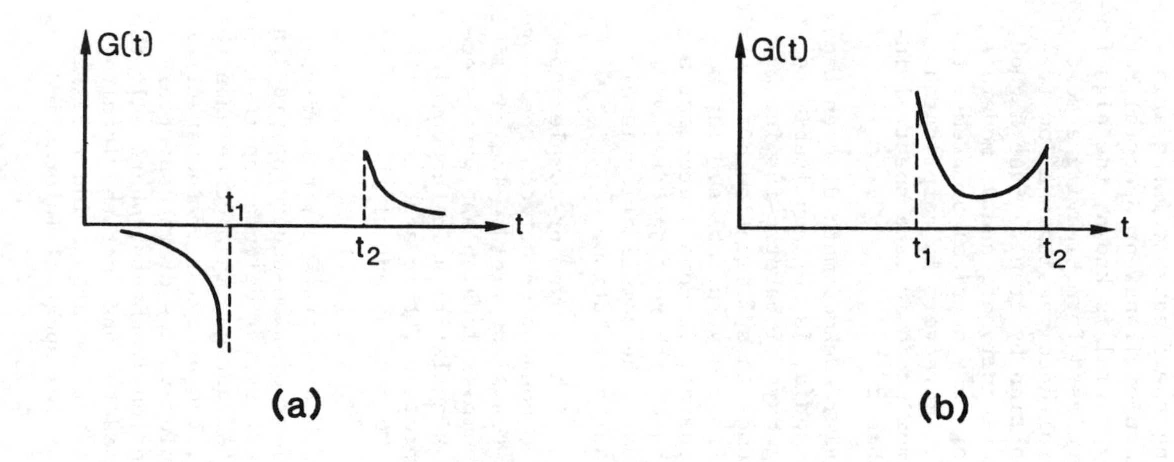

Fig. 4. Transient response for ray species with two rays whose L function is characterized as in Fig. 3.

(a) Case (6a), with transient response (14a) (complex pole trajectories in Fig. 3).

(b) Case (6b), with transient response (14b) (real pole trajectories in Fig. 3).

IV. SUMMARY

High-frequency approximated spectral integral representations of ray fields are useful for evaluating transient ray phenomena beyond the wavefront region. The nondispersive nature of these approximations makes them amenable to direct inversion into the time domain, either by inspection after a canonical transformation (Cagniard-DeHoop method) or by performing the frequency integration before the spectral integration (Chapman or Transient Spectral Methods). Whenever the exact spectral integral is known, the high frequency behavior may be extracted in closed form, leaving a more rapidly convergent correction integral encompassing the low frequency range. The high frequency portion by itself gives a good description of the ray field in the vicinity of, and at moderate distances behind, its wavefront. For many physical problems, this is the domain of principal interest. Moreover, in complicated configurations, the high-frequency approximation of the spectral integral is often the only one available.[8-10]

Explicit inversion via the Cagniard-DeHoop method, even though it can be extended to inhomogeneous media, is not applicable to general ray species, such as turning rays or caustic-forming multiple rays. The Chapman method overcomes this limitation but restricts consideration to real spectral and frequency variables. The resulting inversion formula, though general, is complicated for a certain class of problems, namely those with real spectral wave amplitudes $A_{\pm}$ in the visible spectrum. The complication involves a Hilbert transform over the spectral contribution.

The Transient Spectral Method, which allows for complex spectral variables, provides the analytic connection between the above alternatives for description of the transient field. The ray field may be expressed by either real or complex local plane wave spectrum contributions which arise from singularities in the complex spectral integral. In the other approaches, this aspect is either submerged or is absent entirely due to confinement of the spectrum variables to the real axis.

While the Transient Spectral Method offers a direct and unified spectral approach to a broad class of time-dependent propagation and scattering problems, its utility for calculating the transient field remains to be explored. Advantages may accrue when the complex spectrum is available, i.e., the behavior of the spectral functions is known for real and complex values of the spectral wavenumbers. However, in complicated environments, where only the <u>real</u> ray fields are known and are used to construct the local spectrum,[8] the ability to perform the analytic continuation requires further study. In the absence of complex spectral information, use of the Chapman procedure is appropriate.

ACKNOWLEDGEMENT

This summary is based on work sponsored by the Office of Naval Research under ContractNo. N00014-79-C-0013, by the Joint Services Electronics Program under Contract No. F-49620-82-C-004, and by the National Science Foundation under Grant No. EAR-8213147.

REFERENCES

1. Keller, J.B., "Geometrical Theory of Diffraction," J. Opt. Soc. Am. 52, 116-130, 1962.

2. Kouyoumjian, R.G., "The Geometrical Theory of Diffraction and its Application," in Numerical and Asymptotic Techniques in Electromagnetics, R. Mittra ed., pp. 166-213, Springer-Verlag, New York, 1975.

3. Lee, S.W., "Uniform Asymptotic Theory (UAT) of Electromagnetic Edge Diffraction: A review," in Electromagnetic Scattering, P.L.E. Uslenghi ed., pp. 67-120, Academic Press, New York, 1978.

4. Kouyoumjian, R.G. and P.H. Pathak, "A Uniform Geometrical Theory of Diffraction (UTD) for an Edge in a Perfectly Conducting Surface," Proc. IEEE 62, pp. 1448-1461, 1973.

5. Mittra, R., Y. Rahmat-Samii and W.L. Ko, "Spectral Theory of Diffraction (STD)," Appl. Phys. 10, pp. 1-13, 1976.

6. Friedlander, P.G., Sound Pulses, Cambridge University Press, 1958.

7. Felsen, L.B. and N. Marcuvitz, Radiation and Scattering of Waves, Prentice-Hall, Englewood Cliffs, NJ, 1973. Sec.3.3.

8a. Frazer, L.N. and R.A. Phinney, "The Theory of Finite Frequency Body Wave Synthetic Seismograms in Inhomogeneous Elastic Media," Geophys. J. 63, pp. 691-713, 1980.

8b. Sinton, J.B. and L.N. Frazer, "A Method for the Computation of Finite Frequency Body Wave Synthetic Seismograms in Laterally Varying Media," Geophys.J.Roy.Astr. Soc. 71, pp. 37-55, 1982.

9. Chapman, C.H. and R. Drummond, "Body-waves in Inhomogeneous Media Using Maslov Asymptotic Theory," Bull. Seism. Soc. Am. 72, pp. 277-317, 1982.

10. Arnold, J.M. and L.B. Felsen, "Rays and Local Modes in a Wedge-shaped Ocean," J. Acoust. Soc. Am., Vol. 70, No. 4, pp. 1105-1119, 1983.

11. Felsen, L.B., "Hybrid Ray-mode Fields in Inhomogeneous Waveguides and Ducts," J. Acoust. Soc. Am., 69, pp. 238-250, 1981.

12. Cagniard, L., <u>Reflection and Refraction of Progressive Seismic Waves,</u> translated from the 1939 French monograph by Flinn and Dix, McGraw-Hill, New York, 1962.

13. DeHoop, A.T., "A Modification of Cagniard's Method for Solving Seismic Pulse Problems," Appl. Sci. Res., B8, pp. 349-356, 1960.

14. Felsen, L.B., "Transient Solutions for a Class of Diffraction Problems," Quart. Appl. Math., 23, pp. 151-169, 1965.

15. Chen, P. and Y.H. Pao, "The Diffraction of a Sound Pulse by Circular Cylinder," J. Math. Phys., 18, pp. 2397-2405, 1977. By conformal mapping the cylindrical boundary can be transformed into a plane boundary with an inhomogeneous medium.

16. Chapman, C.H., "A New Method for Computing Synthetic Seismograms," Geophys. J., 54, pp. 481-518, 1978.

TIME-DOMAIN WEYL PLANE-WAVE REPRESENTATION FOR WAVE FUNCTIONS

Edward F. Kuester* and Anton G. Tijhuis†

*Department of Electrical and Computer Engineering
University of Colorado
Boulder, Colorado 80309 USA

†Laboratory of Electromagnetic Research
Delft University of Technology
P.O. Box 5031
2600 GA Delft THE NETHERLANDS

ABSTRACT

Use of double or triple Fourier transforms (one in the frequency domain and the rest over spatial wavenumbers) in solving transient propagation problems has usually proceeded by carrying out all calculations in the frequency domain, and finally going over to the time-domain using FFT or some similar algorithm. An idea first suggested explicitly by Chapman (though hinted at earlier by several authors) is to carry out the frequency transform(s) first, and analyze the problem in terms of a spectrum of time-domain plane waves. In this paper, we concentrate on the numerical aspects of this technique, and specifically on the application of the method to the transient scattering of the fields from a line source by a layered, lossy dielectric slab.

1. INTRODUCTION

Efficient computation of transient electromagnetic fields in the presence of dispersive dielectric media is desirable for many purposes. Notable applications are to problems of shielding and lightning protection, as well as to methods of remote sensing for the determination of the electromagnetic parameters of inaccessible regions.

For plane-stratified regions, one can hope to find numerical solutions when a plane wave is incident (1), and for nondispersive, stepwise constant profiles, the fields due to line and point sources can be found analytically using the Cagniard-de Hoop method and its generalizations (2). Otherwise, one must use methods applicable to very general sources and stratifications, such as finite-difference solutions to Maxwell's equations (3). Because they are so general, these methods cannot take advantage of simplifications due to planar stratification and computational efficiency may not be the best. Moreover, no insight into the relationship of the fields due to incident plane waves with those due to more general sources is gained.

In this paper we present a formalism by means of which the time-domain problem for fields of an arbitrary two-dimensional source in the presence of a plane-stratified dielectric and conducting region can be reduced to the plane-wave problem studied in (1). This is accomplished by means of a time-domain plane-wave spectrum representation for the fields first introduced by Poritzky (4) and Papadopoulos (5), ane examined recently in connection with scattering from layered media in geophysical problems by Chapman (6) and Phinney et al. (7). We can then solve the problem for incident transient plane waves from various angles by the numerical methods of (1), and the remaining calculation is a straightforward quadrature to obtain the field of the finite source by composing the corresponding plane-wave spectra.

2. TIME-DOMAIN PLANE-WAVE EXPANSIONS IN UNBOUNDED FREE SPACE

Before considering the scattering problem, let us obtain a time-domain plane-wave (or Weyl) expansion for the fields of given sources in free space.

Consider the pulsed line source of electric current whose volume current density is

$$j_z^{(0)}(x,y,t) = i(t)\delta(x)\delta(y) , \tag{1}$$

situated at $x=0$, $y=0$ in free space (Fig. 1). Here $i(t)$ is the current flowing in the source, while δ denotes the Dirac δ-function. From Maxwell's equations, a TE-polarized, z-independent field will result, whose only nonvanishing components are $e_z^{(0)}$, $h_x^{(0)}$ and $h_y^{(0)}$, where

$$\left(\nabla^2 - \frac{1}{c^2}\frac{\partial^2}{\partial t^2}\right)e_z^{(0)} = \mu_0 i'(t)\delta(x)\delta(y) , \tag{2}$$

Here $\nabla^2 = \partial^2/\partial x^2 + \partial^2/\partial y^2$ is the two-dimensional Laplacian,

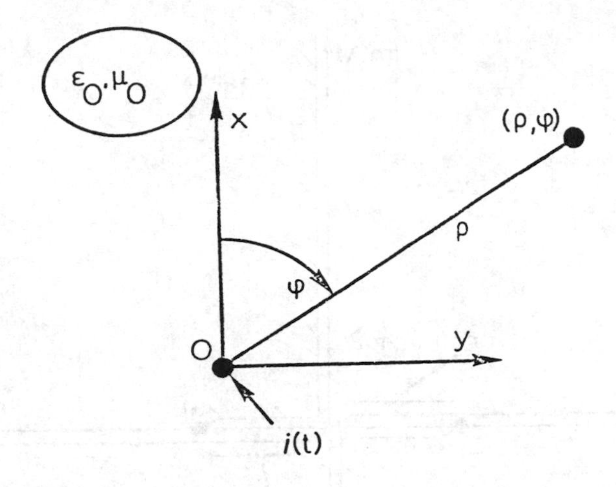

Fig. 1: Pulsed line source of current in free space.

$c = (\mu_0\varepsilon_0)^{-\frac{1}{2}}$ is the speed of light *in vacuo*, and ε_0 and μ_0 are respectively the permittivity and permeability of free space. Throughout this work, we will use lower-case letters to denote time-dependent fields or sources, and capitals to denote the corresponding Fourier transforms (frequency-domain quantities) according to

$$F(\omega) = \int_{-\infty}^{\infty} f(t)e^{i\omega t}dt \tag{3}$$

Taking the Fourier transform of both sides of eqn. 2 leads to an equation which is readily solved to give

$$E_z^{(0)} = \frac{i\omega\mu_0 I(\omega)}{4\pi}\int_{-\infty}^{\infty} \frac{e^{i\beta y-|x|(\beta^2-\omega^2/c^2)^{\frac{1}{2}}}}{(\beta^2-\omega^2/c^2)} d\beta \tag{4}$$

where the square root is taken to have nonnegative real part on the path of integration.

By the change of variable $\beta = |\omega|\lambda/c$, we can rewrite $E_z^{(0)}$ as

$$E_z^{(0)} = \frac{i\omega\mu_o I(\omega)}{4\pi}\int_{-\infty}^{\infty} \frac{e^{i|\omega|(\lambda y + iu_1|x|)/c}}{u_1} d\lambda \tag{5}$$

where $u_1 = (\lambda^2-1)^{\frac{1}{2}}$, and $\mathrm{Re}(u_1) \geq 0$ defines a "proper" Riemann sheet of the complex λ-plane in which the path of integration for eqn. 5 must lie (Fig. 2).

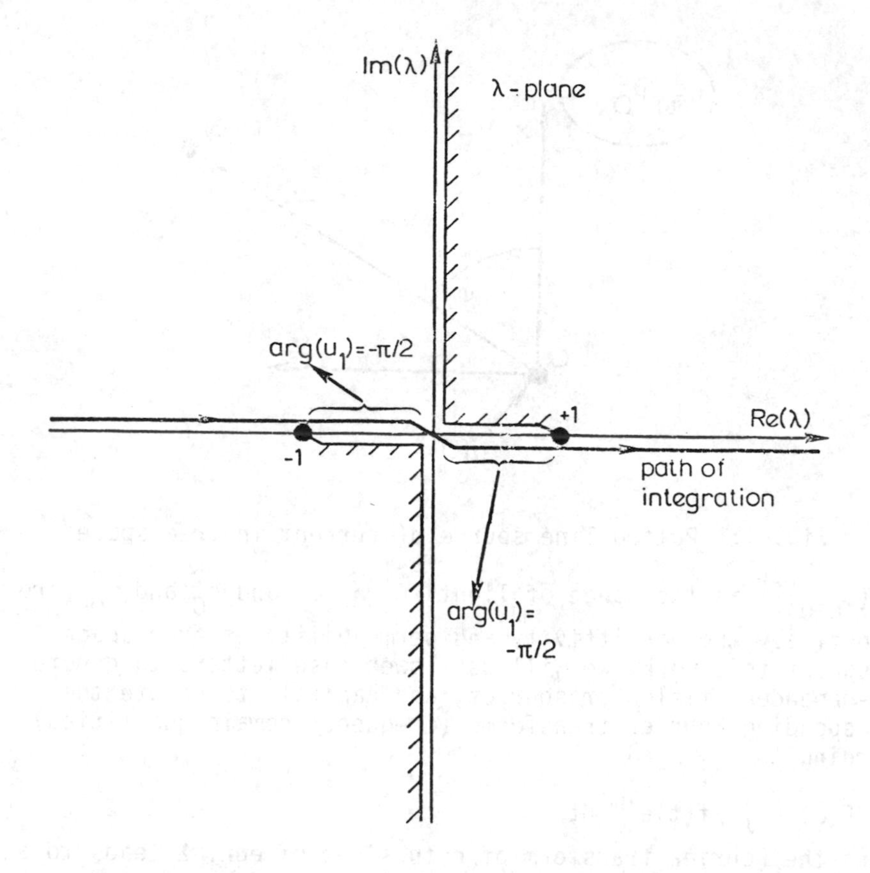

Fig. 2: Branch cuts and path of integration for eqn. 5.

Now, one means of evaluating the inverse Fourier transform of a function $F(\omega)$ is by using its analytic signal $f_1(\tau)$ (8):

$$f_1(\tau) = \frac{1}{\pi}\int_0^\infty F(\omega)\, e^{-i\omega\tau} d\omega \qquad [\mathrm{Im}(\tau) < 0] \tag{6}$$

whence

$$\begin{aligned} f(t) &= \mathrm{Re}\{f_1(t - i0)\} \\ &= \mathrm{Re}\{f_1(t)\} \end{aligned} \tag{7}$$

We will hereafter omit the implied limit of $\mathrm{Im}(\tau) \to 0^-$ as in eqn.7 for brevity. We can now proceed to the crucial step of using eqns. 6 and 7 with eqn. 5 to recover $e_z^{(0)}$ but interchanging the order of the λ and ω integrations so that the ω-integral is performed first. The result of the ω-integral, by eqn. 6, can be identified with the analytic signal $i_1'(\tau)$ corresponding to $i'(t)$, with argument $\tau = t - [\lambda y + iu_1|x|]/c$:

$$e_z^{(0)}(x,y,t) = -\frac{\mu_0}{4\pi}\,\mathrm{Re}\int_{-\infty}^{\infty} i_1'(t-\tfrac{1}{c}[\lambda y + iu_1|x|])\,\frac{d\lambda}{u_1} \tag{8}$$

In fact, the analytic signal corresponding to $e_z^{(0)}$ will be given by

$$e_{z1}^{(0)}(x,y,\tau) = -\frac{\mu_0}{4\pi}\int_{-\infty}^{\infty} i_1'(\tau-\tfrac{1}{c}[\lambda y + iu_1|x|])\,\frac{d\lambda}{u_1} \tag{9}$$

When the line source of eqn. 1 is replaced by a more general current distribution $j_z(x,y,t)$, we can obtain the corresponding transient field e_z by convolution with eqn. 8 in the space variables x and y (Duhamel's theorem). The result is

$$\begin{aligned} e_z(x,y,t) = -\frac{\mu_0}{4\pi}\,\mathrm{Re}\int_{-\infty}^{\infty}\Big\{&\frac{\partial}{\partial t} i_1^+(t-\tfrac{1}{c}[\lambda y + iu_1x],\lambda,x) \\ &+ \frac{\partial}{\partial t} i_1^-(t-\tfrac{1}{c}[\lambda y - iu_1x],\lambda,x)\Big\}\frac{d\lambda}{u_1} \end{aligned} \tag{10}$$

where

$$\begin{aligned} i_1^+(t,\lambda,x) &= \int_{-\infty}^{\infty} dy' \int_{-\infty}^{x} dx' j_{z1}(x',y',t+\tfrac{1}{c}[\lambda y'+iu_1x']) \\ i_1^-(t,\lambda,x) &= \int_{-\infty}^{\infty} dy' \int_{x}^{\infty} dx' j_{z1}(x',y',t+\tfrac{1}{c}[\lambda y'-iu_1x']) \end{aligned} \tag{11}$$

are, respectively, contributions from those parts of the sources lying below x (propagating upwards) and from those parts lying above x (and propagating downwards) and j_{z1} is the analytic signal corresponding to j_z.

3. A MODIFIED RADON TRANSFORM

The representations of eqns. 9 and 10 are actually equivalent (through some changes of variable) to a Radon transform representation (9); in fact, for an analytic signal $v_1(y,\tau)$ there is a second analytic signal satisfying

$$v_1(y,\tau) = \int_{-\infty}^{\infty} w_1(\lambda, \tau-\lambda y/c)\,d\lambda \tag{12}$$

$$w_1(\lambda,\tau) = \frac{i}{2\pi c}\,\frac{\partial}{\partial\tau}\int_{-\infty}^{\infty} v_1(y,\tau+\lambda y/c)\,dy \tag{13}$$

This transform pair allows us to deal directly with functions of λ rather than functions of the space variable y.

4. SCATTERING BY AN INHOMOGENEOUS SLAB

Suppose a given "external" current source distribution $j_z^e(x,y,t)$ produces a known incident field upon an inhomogeneous slab located in $0 < x < d$. The slab is characterized by the x-dependent quantities $\sigma(x)$ and $\varepsilon(x) = \varepsilon_0[1 + \chi(x)]$ as shown in Fig. 3. We wish to find the transient field scattered by this structure.

We can use the formalism developed in section 2 if we replace the slab by its equivalent conduction and polarization currents, now radiating in free space, but of course, dependent on the as yet unknown field $e_z(x,y,t)$. In terms of analytic signals,

$$j_{z1}(x,y,t) = j_{z1}^e(x,y,t) + \sigma(x)e_{z1}(x,y,t) + \varepsilon_0\chi(x)\frac{\partial}{\partial t} e_{z1}(x,y,t). \quad (14)$$

Putting this representation into eqns. 10 and 11, and taking the modified Radon transform (eqn. 13) of the result, we obtain an integro-differential equation for $f_1(x,\lambda,t)$, the modified Radon transform of e_{z1}:

$$\begin{aligned} f_1(x,\lambda,t) &= f_1^{inc}(x,\lambda,t) \\ &+ \frac{i\zeta_0}{2u_1} \int_0^d [\sigma(x')f_1(x',\lambda,t - \frac{iu_1}{c}|x-x'|) \\ &+ \varepsilon_0\chi(x')\frac{\partial}{\partial t} f_1(x',\lambda,t - \frac{iu_1}{c}|x - x'|)]dx' \end{aligned} \quad (15)$$

where $\zeta_0 = (\mu_0/\varepsilon_o)^{\frac{1}{2}}$,

$$f_1^{inc}(x,\lambda,t) = \frac{i}{2\pi c}\frac{\partial}{\partial t} \int_{-\infty}^{\infty} e_{z1}^{inc}(x,y,t + \lambda y/c)dy \quad (16)$$

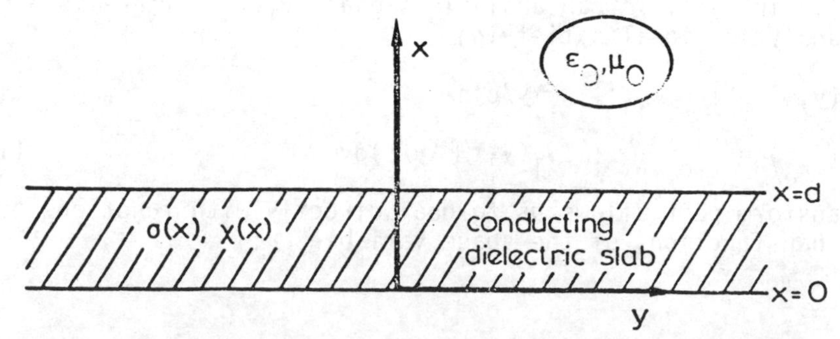

Fig. 3: Lossy, inhomogeneous dielectric slab.

if e_z^{inc} is the incident field produced by j_z^e, and e_{z1} is recovered from f_1 by the use of eqn. 12:

$$e_{z1}(x,y,t) = \int_{-\infty}^{\infty} f_1(x,\lambda,t-\lambda y/c)d\lambda \ , \tag{17}$$

Equation 15 is identical with the equation obtained by Tijhuis (1) for the scattering by an incident plane wave at an angle θ , if we put $\lambda = \sin\theta$ and $u_1 = -i\cos\theta$ therein. The method of solution used in (1) can be used here as well, although some numerical instability can be present when $|\lambda| > 1$ (complex angles of incidence). There are several possible ways to overcome this difficulty and these will be discussed elsewhere. In most cases, the contributions from the troublesome λ are not large, and can often be completely ignored.

5. CONCLUSION

As a sample of calculations performed in this way, we present a time history of an incident Gaussian pulse from a line source located along $y=0$ at a distance of 2d from a homogeneous lossless slab with $\varepsilon_r = 2.25$. The field is observed in the cut $y=0$. The multiple reflections of the pulse from the slab boundaries can be clearly observed as time evolves.

We have primarily been concerned with numerical applications of this representation, but it seems likely that it will be of value in analytical solutions as well, as Chapman (6) suggests. Development of analytical solutions for the integral equation 15 near $|\lambda| = 1$ and $|\lambda| \to \infty$ would be of considerable value in

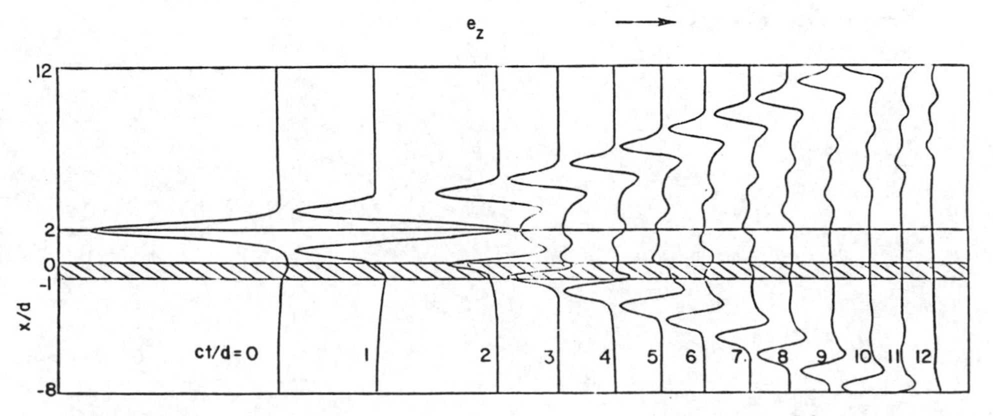

Fig. 4: Evolution of the field along $y = 0$ due to a line source with a Gaussian-pulse current.

improving the efficiency of numerical algorithms for this class of problems. Generalizations to cylindrical, spherical and three-dimensional planar geometries should also be possible. There appear to be many ways in which this representation can be used in practical problems.

REFERENCES

1. Tijhuis, A.G., Iterative Determination of Permittivity and Conductivity Profiles of a Dielectric Slab in the Time Domain, IEEE Trans. Antennas and Propagation 29(1981) 239-245.
2. de Hoop, A.T., Pulsed Radiation from a Line Source in a Two-Media Configuration, Radio Science 14(1979) 253-268.
3. Yee, K.S., Numerical Solution of Initial Boundary Value Problems Involving Maxwell's Equations in Isotropic Media, IEEE Trans. Antennas and Propagation 14(1966) 302-307.
4. Poritzky, H., Extension of Weyl's Integral for Harmonic Spherical Waves to Arbitrary Wave Shapes, Communications on Pure and Applied Mathematics 4(1951) 33-43.
5. Papadopoulos, M., The Reflection and Refraction of Point-Source Fields, Proceedings of the Royal Society of London A273 (1963) 198-221.
6. Chapman, C.H., A New Method for Computing Synthetic Seismograms, Geophysical Journal 54 (1978) 481-518.
7. Phinney, R.A., K.R. Chowdhury and L.N. Frazer, Transformation and Analysis of Record Sections, Journal of Geophysical Research 86(1981) 359-377.
8. Born, M. and E. Wolf, Principles of Optics (Oxford, Pergamon Press, 1975).
9. Deans, S.R., The Radon Transform and Some of Its Applications (New York, Wiley, 1983).

PART IV
NUMERICAL MODELING

NUMERICAL MODELS IN UNDERWATER ACOUSTICS

Finn B. Jensen

SACLANT ASW Research Centre
Viale San Bartolomeo 400
19026 La Spezia, Italy

ABSTRACT

The physics of sound propagation in the ocean is briefly reviewed. The mathematical foundation of the most widely used acoustic models (ray, mode, fast field, parabolic equation) is presented and the areas of applicability of the various models are indicated. A few numerical examples are included to show the consistency among the different computer models in overlapping regimes of validity. Finally, we show a series of computational examples that demonstrate the applicability of these models to a wide range of general wave-propagation problems.

INTRODUCTION

The modern era of underwater acoustics essentially dates back to the beginning of World War II, where considerable effort went into improving submarine detection by acoustic means. This effort has continued, promoted by naval interests in developing still better and more reliable sonar systems. To achieve optimum sonar design one needs to know how sound propagates in the ocean as a function of frequency for different source/receiver configurations and for different environmental conditions.

By now the theory of acoustic propagation is well developed, providing both a good general understanding and a detailed description of how sound travels in the ocean. The theoretical basis is the acoustic wave equation, which has to be solved with realistic boundary conditions at the sea surface and at the sea floor. This problem is generally too complex for applying analytical solutions, and hence we must resort to numerical methods. Several solution techniques (ray, mode, FFP, PE) have been introduced over the years, with the acoustic models increasing in complexity as computers became faster and more powerful. Ray theory was the only practical technique for solving propagation problems until the beginning of the 1970s. Then advances in computer technology made it possible to consider solving more complex equations, and, consequently, new techniques (normal mode, fast field, parabolic equation) came into extensive use during the last decade.

In this paper we outline the basics of sound propagation in the ocean, including important propagation and loss mechanisms, and a simplified environmental description for use in numerical models. We then proceed to outline the mathematical foundation of the most widely used numerical models, and we demonstrate the consistency and inter-relationship between the various models through a few numerical examples for deep and shallow water environments. We then indicate the areas of applicability of the various acoustic models taking into account both limitations in the underlying theory and the numerical efficiency of the actual computer codes. A demonstration of the wide range of applicability of these ocean-acoustic models is provided by a series of computational examples, where the models have been applied to some general wave propagation problems, including beam reflection at fluid/solid interfaces, propagation from ducts into free-space, up- and down-slope propagation involving mode coupling and mode cutoff.

2 SOUND PROPAGATION IN THE OCEAN

The goal of ocean acoustic modelling is to estimate the spatial properties of the sound pressure field as a function of source frequency. To clarify the complexity of the modelling problem, let us briefly review the environmental acoustics of the ocean. Figure 1 is a schematic of some important sound propagation paths; two possible sound-source locations are on the left and sound is propagating from left to right. Two dashed lines at 0 and 80 km range indicate two of the innumerable ways in which sound speed in the water can vary with depth from place to place (or from time to time). Lines A, B, C, and D represent four possible sound-propagation paths whose shapes are determined by the location of the source and the sound-speed structure over the extent of the propagation.

Path A from the shallow source is "surface-duct" propagation, because the sound-speed profile is such that the sound is trapped near the surface of the ocean. Paths B, C, and D are from the deeper source. Ray B, leaving the source at a small angle from

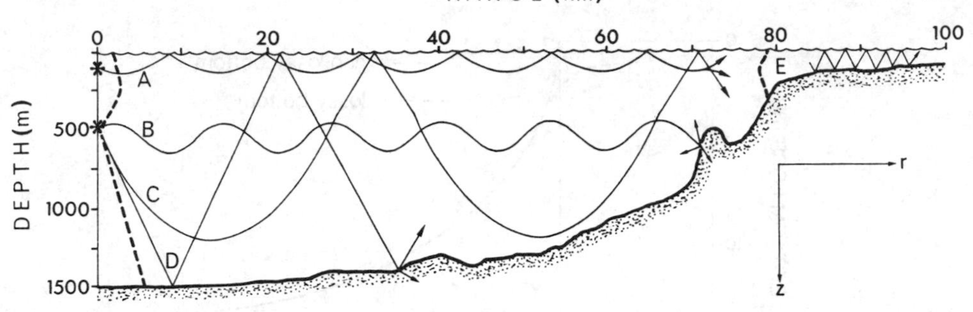

Fig. 1 Schematic of sound propagation in the ocean

the horizontal, tends to propagate in the "deep sound channel" without interacting with the boundaries (surface and bottom) of the ocean. At slightly steeper angles (path C) we have "convergence zone" propagation, which is a spatially periodic phenomenon of zones of high intensity near the surface. Here sound interacts with the ocean surface but not with the bottom. Path D is the "bottom-bounce path", which has a shorter cycle period than the convergence zone path. The right-hand side of Fig. 1 depicts propagation on the continental shelf (shallow water) where a complicated bottom structure combined with variable sound-speed profiles result in rather complicated propagation conditions not always suited for a ray representation.

Our ability to model acoustic propagation effectively in the ocean is determined by the accuracy with which acoustic loss mechanisms in the ocean environment are handled. Aside from geometrical spreading loss (spherical, cylindrical, etc.) the main loss mechanisms are volume absorption, bottom-reflection loss, surface and bottom scattering loss.

Volume absorption in sea water, caused by viscosity and chemical relaxation, increases with increasing frequency. This loss mechanism is the dominant attenuation factor associated with path B in Fig. 1, since this path does not interact with the boundaries. Because there is very little volume absorption at low frequencies, deep-sound-channel propagation has been observed to distances of many thousands of kilometres.

When sound interacts with the sea floor, the nature of the bottom becomes important. Figure 2 depicts simple bottom-loss curves, with zero dB loss indicating perfect reflection. For an ideal bottom without volume absorption (non-lossy) we still get severe reflection loss above a certain critical angle θ_c due to transmission into the bottom. For a real bottom with volume

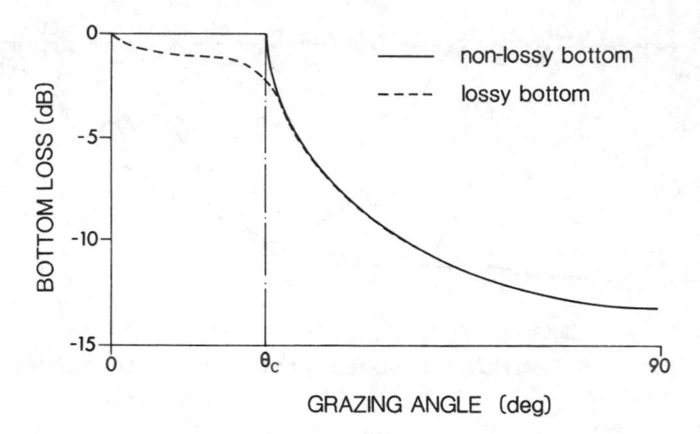

Fig. 2 Bottom loss versus grazing angle

absorption (lossy) we never get perfect reflection, even though the curves look similar. Path D in Fig. 1, the bottom-bounce path, often corresponds to angles near or above the critical angle; therefore after a few bounces it is highly attenuated. On the other hand, for shallower angles, many more bounces are possible; hence in shallow water (path E) most of the energy that propagates is close to the horizontal. In reality, much of the ocean bottom is layered and also supports shear waves; in this case bottom loss becomes a complicated function of frequency and grazing angle. The overall effect of bottom loss on sound propagation in the water column is an increasing loss with decreasing frequency.

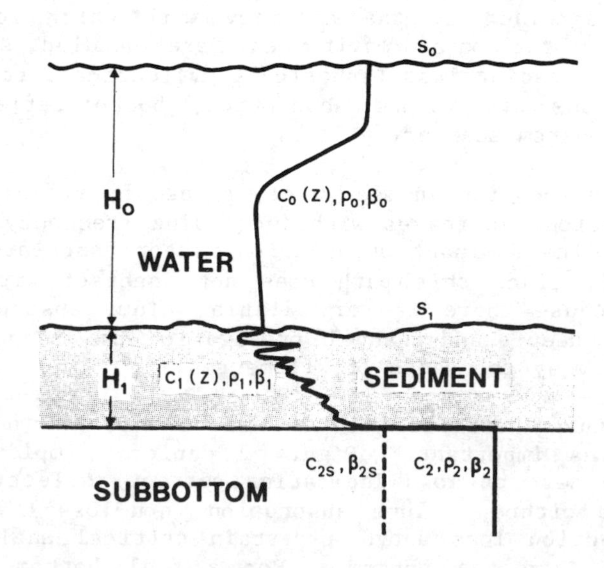

Fig. 3 Environmental input to ocean acoustic models

A rough sea surface or sea floor causes scattering of the incident sound. The result is a decay of the mean acoustic field in the water column as a function of range (scattering loss), with the scattered energy being lost to the ocean bottom through steep-angle propagation. The scattering loss increases with increasing frequency, and the propagation paths mainly affected are paths A and C (surface scattering loss) and paths D and E (surface and bottom scattering loss).

A consistent mathematical model of sound propagation in the ocean must contain the physics that governs the above-mentioned propagation and loss mechanisms. A summary of the environmental inputs needed for a realistic description of the ocean waveguide is given in Fig. 3. In this simplified model the ocean consists of a water column of depth H_o limited by a rough sea surface and a rough sea floor. The sound speed c_o in the water column may vary arbitrarily with depth, while density ρ_o and attenuation β_o are considered constant. Even though real ocean bottoms exhibit a complicated layering, we have found that a simple two-layered geoacoustic model generally provides the necessary degrees of freedom to accurately include bottom effects in numerical models for many ocean areas. Hence the bottom may consist of just a sediment layer of thickness H_1 and a semi-infinite subbottom. The model should allow for sound speed, density, and attenuation to vary arbitrarily with depth in the sediment layer, while the sub-bottom can be considered homogeneous. It is desirable that the model can handle shear-wave propagation in both bottom layers. Finally, in real ocean environments the parameters given in Fig. 3 may all vary with range.

3 MATHEMATICAL FOUNDATION OF OCEAN ACOUSTIC MODELS

We briefly present the mathematical foundations of the four models discussed in this paper. A more detailed description can be found in references <1-10>.

The starting point for all the models is the wave equation for a harmonic point source with time dependence exp(-iωt),

$$\nabla^2\phi(x,y,z) + \left[\frac{\omega}{c(x,y,z)}\right]^2 \phi(x,y,z) = -\delta(x-x_o)\delta(y-y_o)\delta(z-z_o) \tag{1}$$

$$\psi = \phi\exp(-i\omega t) \tag{2}$$

At any point (x,y,z) in the medium, the velocity potential ϕ satisfies Eq. (1) where c(x,y,z) is the sound speed of the medium and δ is the Dirac delta function. The source is at the coordinate (x_o, y_o, z_o) where z is the depth coordinate, which is taken to be positive in the downward direction from the ocean surface.

For the boundary condition at the ocean surface we take the density of air to be negligible compared with that of water; hence, the pressure must vanish at the ocean surface ("pressure-release surface"). At a boundary between two media such as the ocean and the ocean bottom, the balancing of forces at the interface require that physical quantities such as particle velocity and pressure be continuous across the boundary:

$$v_i = -\frac{\partial \phi}{\partial x_i} \; ; \; x_i = x, y, \text{ or } z \tag{3}$$

$$p = -i\omega\rho\phi \tag{4}$$

If the ocean bottom is treated as an elastic medium that can support shear motions, there is the additional boundary condition that tangential stress must be continuous. Since the water column cannot support shear waves, this requires that the tangential stress in the ocean bottom vanishes at the interface.

Four widely used solution techniques for Eq. (1) are schematically represented in Fig. 4. The derivation of the classical ray solution can be found in most text books on acoustics, as can the details of the well-established normal-mode solution. The fast-field technique is not yet in standard text books, but it is a powerful tool for solving propagation problems in stratified environments. The parabolic equation technique is a recent advent in acoustic modelling. This method is particularly suited for propagation in range-dependent environments.

We briefly describe the derivation of the above four solution techniques, starting with range-independent wave theory.

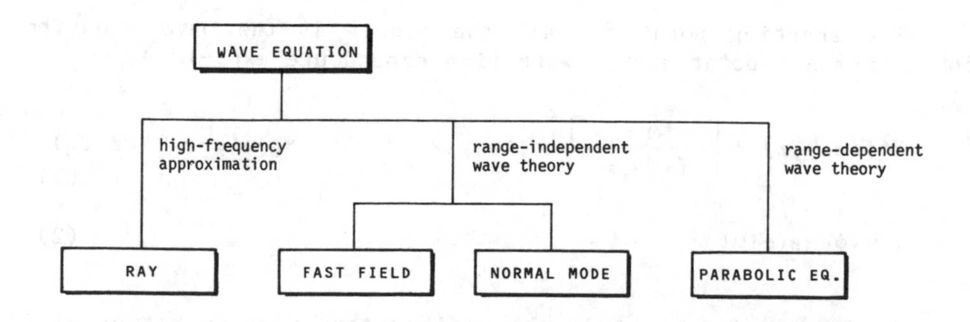

Fig. 4 Four techniques for solving the wave equation

3.1 Fast field solution

Here we are solving the wave equation for the case where the sound-speed profile is only a function of depth and the bottom is flat; this type of environment is often referred to as the horizontally stratified ocean. From Eq. (1) we therefore have that c(x,y,z) is simply c(z). Because the environment is independent of "r", the horizontal coordinates (x,y), one possible method of solving Eq. (1) is to Fourier decompose the acoustic field into an infinite set of horizontal waves:

$$\phi(x,y,z) = \frac{1}{2\pi} \int d^2\vec{\eta}\; u(\vec{\eta},z)\, e^{i\vec{\eta}\cdot\vec{r}} . \tag{5}$$

Substituting Eq. (5) into Eq. (1) we obtain the equation for $\phi(\eta_x,\eta_y;z)$,

$$\frac{\partial^2 u(\eta_x,\eta_y;z)}{dz^2} + [k^2(z) - \eta^2]u(\eta_x,\eta_y;z) = -\frac{1}{2\pi}\delta(z-z_o), \tag{6}$$

where $k(z) = \omega/c(z)$ and $\eta^2 = \eta_x^2 + \eta_y^2$, is the horizontal wavenumber of the individual plane waves.

Using polar coordinates we can rewrite Eq. (5) as

$$\phi(r,z) = \frac{1}{2\pi} \int_o^{2\pi} d\theta \int_o^{\infty} \eta\, d\eta\, u(\eta,z) e^{i\eta r\cos\theta} . \tag{7}$$

We now integrate over the azimuthal angle to obtain

$$\phi(r,z) = \int_o^{\infty} \eta\, d\eta\, u(\eta,z) J_o(\eta r), \tag{8}$$

where J_o is the zeroth order Bessel function. Using the relationship that

$$J_o(\eta r) = 1/2\, [H_o^{(1)}(\eta r) + H_o^{(2)}(\eta r)],$$

where the H's are Hankel functions and noting from Eq. (6) that $u(\eta,z)$ is even in η, we can rewrite Eq. (8) as

$$\phi(r,z) = 1/2 \int_{-\infty}^{\infty} \eta\, d\eta\, u(\eta,z)\, H_o^{(1)}(\eta r), \tag{9}$$

where now the integration over η is from $-\infty$ to ∞. For ranges greater than a few wavelengths from the source, the asymptotic form of the Hankel function can be used:

$$H_o^{(1)}(\eta r) \sim (2/\pi\eta r)^{1/2} \exp[i(\eta r - \pi/4)]$$

and, hence, Eq. (9) can be expressed as

$$\phi(r,z) = \frac{e^{-i\pi/4}}{\sqrt{2\pi}} \cdot \frac{1}{\sqrt{r}} \int_{-\infty}^{\infty} d\eta \sqrt{\eta} u(\eta,z) e^{i\eta r}, \tag{10}$$

where the factor $1/\sqrt{r}$ indicates cylindrical spreading.

Equation (10) can be numerically integrated to obtain the acoustic field at the range r and depth z. In order to do this we must solve Eq. (6) for many η's to have a sufficient set of u's as a function of η so that the integration over η in Eq. (10) can be performed. Given that u has been obtained numerically as a function of η, the integration can be done using an FFT algorithm. This total procedure is called the Fast Field Program (FFP) <11-13>, although most of the numerical effort goes into solving Eq. (6) for the many η's. For computation we discretize Eq. (10) by letting

$$\eta_m = \eta_o + m\Delta\eta;\ r_n = r_o + n\Delta r;\ (m,n) = 0,1,2\ldots,N-1, \tag{11}$$

with the additional relation

$$\Delta r \Delta\eta = \frac{2\pi}{N} \tag{12}$$

and N is an integral power of two. Note that the discretization relations of Eq. (11) restricts the solution to outgoing waves. Substituting Eq. (11) into Eq. (10) we obtain

$$\phi(r_n,z) = \Delta\eta \frac{e^{-i\pi/4}}{\sqrt{2\pi r}} e^{i\eta_o r_n} \sum_{M=0}^{N-1} X_m e^{i2\pi mn/N}, \tag{13}$$

and hence the input to the FFT is

$$X_m = \sqrt{\eta_m}\, u(\eta_m,z) e^{imr_o\Delta\eta}. \tag{14}$$

Equations (13) and (14) specify the numerical procedure to be employed in solving the wave equation using the FFP approach after Eq. (6) has been solved numerically for the complete set of u's as a function of η. As mentioned above, the main effort in the FFP approach is the numerical integration of Eq. (6) and not the final implementation of Eqs. (13) and (14), which is a simple FFT computation. Numerical procedures for integrating Eq. (6) are given in references <11-13>.

The first implementation of the FFP approach was done by DiNapoli <11> around 1970. His method is for a stratified fluid environment, and the numerical procedure is quite fast, since it uses recurrence relations of hypergeometric functions to solve Eq. (6) for all the wavenumbers (η's) rather than solve the equation for one η at the time. A more general solution technique was devised by Kutschale <12> for an arbitrary stratification of solid layers. This technique employs the Thomson-Haskell matrix method, and Eq. (6) is here solved separately for each wavenumber. This solution technique is computationally quite slow and there is no efficient way of doing calculations simultaneously for many sources and receivers. The most recent FFP technique was developed by Schmidt <13>. Again the solution is for an arbitrary stratification of homogeneous solid layers. Displacements and stresses are expressed in terms of three scalar potentials for each layer, as described in <14>. Boundary conditions are then matched at each interface yielding a linear system of equations in the Hankel transforms of the potentials. Equation (6) is again solved at discrete horizontal wavenumbers; with the coefficient matrices being of band form, the equations are solved very efficiently by gaussian elimination. This solution technique is considerably faster than the Kutschale technique, and it furthermore allows for an efficient evaluation of the acoustic field for many sources and receivers at a time. This, in turn, means that the model can be applied to a variety of new problems, including beam reflection problems as shown in Sect. 5.1.

The main advantage of the FFP is that it provides the full solution to sound propagation in a multilayered solid medium, and hence consitutes a benchmark against which other approximate solutions can be checked. Its main disadvantage is that the procedure is not easily automated.

3.2 Normal-mode solution

The alternative to a direct numerical integration of Eq. (6) is to expand u into a complete set of normal modes:

$$u(\eta_x,\eta_y;z) = \Sigma\ a_n(\eta_x,\eta_y)u_n(z), \qquad (15)$$

where the u_n's are the solutions to the eigenvalue equation

$$\frac{d^2u_n(z)}{dz^2} + [k^2(z) - k_n^2]u_n(z) = 0 \qquad (16)$$

that satisfies the above-mentioned boundary conditions. In addition, we require the $u_n(z)$ be bounded as $z \to \infty$. The normal modes $u_n(z)$ form a complete orthogonal set that satisfies the relation

$$\int_0^\infty \rho(z)u_n(z)u_m(z)dz = \delta_{nm}, \tag{17}$$

where the density $\rho(z)$ takes its appropriate value in each layer and δ_{nm} is the Kronecker-delta symbol. The spectrum of eigenvalues consists of a discrete part and a continuous part, the discrete eigenvalues occurring in the interval

$$\omega/c_2 < k_n < \max[\omega/c(z)], \tag{18}$$

where c_2 is the highest speed of the system. In the present treatment we consider only the discrete eigenvalues, since, in general, the continuous spectrum makes a negligible contribution beyond the nearfield of the source (and requires an FFP-type calculation in any event).

We now substitute Eq. (15) into Eq. (6), multiply the resulting equation by $\rho(z)u_m(z)$, and integrate over z from 0 to ∞, giving:

$$a_n = \frac{1}{2\pi}\,\frac{\rho(z_o)u_n(z_o)}{\eta^2-k_n^2} \tag{19}$$

Substituting Eqs. (15) and (19) back into Eq. (5) we obtain an integral representation of the velocity potential,

$$\phi(x,y,z) = \frac{\rho(z_o)}{(2\pi)^2}\int_{-\infty}^{\infty} d\eta_x \int_{-\infty}^{\infty} d\eta_y \sum_n \frac{u_n(z_o)u_n(z)}{\eta^2-k_n^2} \times \exp[i(\eta_x x + \eta_y y)]. \tag{20}$$

We evaluate the above integral by choosing a path about the poles so as to lead to an outgoing wave from the source point r = 0. Each integral in Eq. (20) is proportional to the two-dimensional plane-wave representation of the zero-order Hankel function of the first kind <15>, and therefore $\phi(x,y,z)$ can be expressed as:

$$\phi(r,z) = \frac{i}{4}\,\rho(z_o)\sum_n u_n(z_o)u_n(z)H_o^{(1)}(k_n r). \tag{21}$$

The asymptotic form of the Hankel function can then be used to obtain

$$\phi(r,z) = \frac{i\rho(z_o)}{(8\pi r)^{1/2}} e^{-i\pi/4} \sum_n \frac{u_n(z_o)u_n(z)}{k_n^{1/2}} e^{ik_n r}. \tag{22}$$

In addition to the decay of the field due to cylindrical spreading, other loss mechanisms such as volume attenuation in the water column and bottom are included in Eq. (22) because the eigenvalues, k_n, have positive imaginary parts <16>, thereby resulting in an exponential attenuation of each normal-mode term. Equation (22) gives us the important result that the field at a depth z is proportional to a sum of the products of normal modes evaluated at the source and the receiver depth. The normal modes are the "natural vibrations" of the system and if a point source is located at the null of a particular normal mode, that mode will not be excited. Similarly, if a point receiver is placed at the null of a particular mode, that mode contribution to the total field will not be sensed.

In analogy to the FFP procedure, the main numerical effort for the normal-mode procedure is the solution of the eigenvalue problem defined by Eq. (16) and the boundary conditions. There are many techniques to solve this equation <17-18> but they are mainly applicable to low-frequency or shallow-water propagation, where there is only a small number of modes <19>. However, there are also techniques to handle deep-water high-frequency propagation using normal modes <20>.

The advantages of the normal-mode procedure are, first, that once Eq. (16) is solved we have the solution for all source/receiver configurations, and, second, that the whole solution procedure can be highly automated. In addition, the normal-mode procedure can be easily extended to slightly range-dependent environments using the adiabatic approximation where mode coupling is neglected <21-26>; numerical methods for including mode-coupling effects are present areas of research <27-31>. The disadvantages of the normal-mode solution are that conventional procedures do not include nearfield effects (the exception is Stickler's work <32>, but even there the nearfield is evaluated with a procedure similar to the FFP approach) and there are restrictions on how one can treat shear propagation in the bottom.

Both FFP and normal modes are solutions of Eqs. (5) and (6). The difference between the two is that the normal-mode method restricts the integration to horizontal wavenumbers in the interval corresponding to the discrete portion of the spectrum defined by Ineq. (18). From Eq. (20), we see that the integrand has poles at $\eta = k_n$. Hence, the function u in Eq. (10) and X_m of Eqs. (13) and (14) should have poles at the same locations.

We shall demonstrate the inter-relationship between normal mode (NM) and fast field (FFP) solutions through a numerical example using the environment given in Fig. 5. The upward-refracting sound-speed profile defines a surface duct of thickness 1500 m. We consider propagation for source and receiver both a 500 m depth, and for a frequency of 25 Hz. Two plots of the FFP integrand are shown in Fig. 6. The horizontal axis is the wave-number (η) and the vertical axis is a normalized absolute value of the integrand amplitude. The vertical dashed line in Fig. 6a separates the discrete spectrum (defined by Ineq. 18) and the continuous spectrum. A blowup of the discrete spectrum is shown in Fig. 6b. In this particular case there are eleven propagating modes. The asterisks on the plot are the locations and the amplitudes of the modes from a NM calculation <18>. We see that the eigenvalues as calculated from the NM model (Table 1) coincide with the peaks (poles) in the FFP integrand, and also that the NM amplitudes correspond to the amplitudes of the peaks. (It turns out that this one-to-one correspondence in the amplitudes is because there is virtually no loss in this problem. Otherwise, the correspondence would not be as precise because loss shows up in the FFP calculation as widths in the peaks; nevertheless, for realistic losses the location of the poles would be the same).

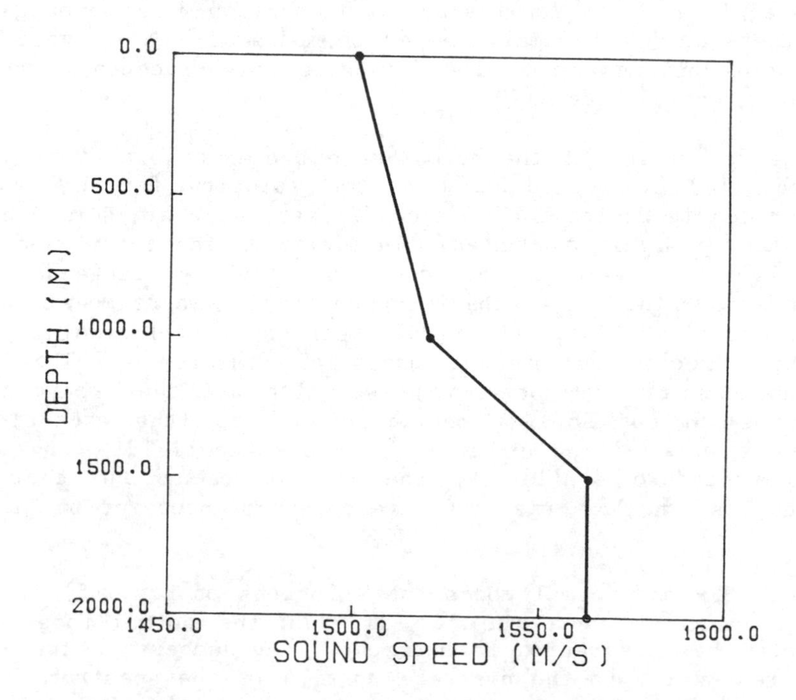

Fig. 5 Sound-speed profile for test problem

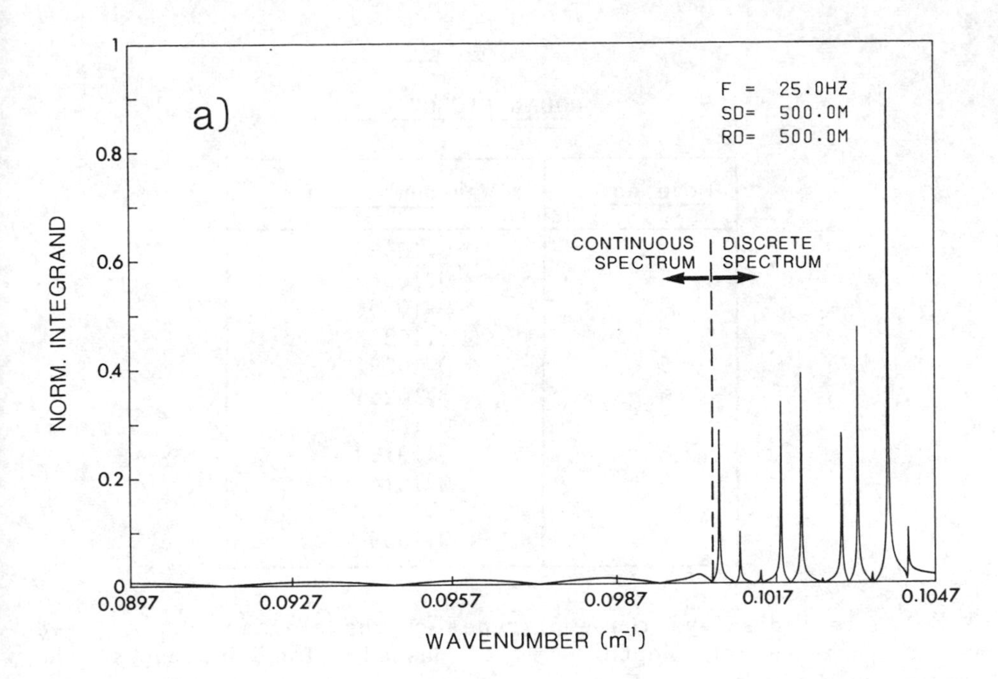

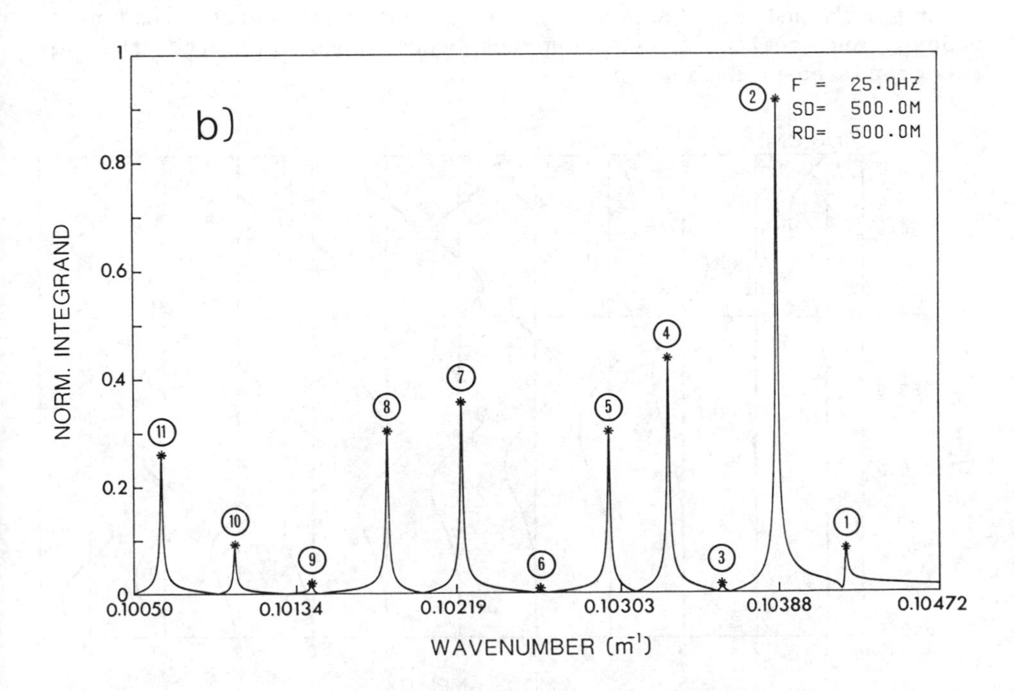

Fig. 6 FFP integrand for test problem
a) full spectrum, b) discrete spectrum

TABLE 1

MODAL EIGENVALUES

Mode no.	Wavenumber (m^{-1})
1	0.10423
2	0.10386
3	0.10356
4	0.10328
5	0.10297
6	0.10261
7	0.10221
8	0.10182
9	0.10142
10	0.10102
11	0.10064

Figure 7 displays the amplitudes of the eleven modes plotted as a function of depth. The dashed line indicates the source/receiver position. Notice the high excitation of the second mode and the low excitation of the third, sixth, and ninth modes, and notice the one-to-one correspondence with the FFP integrand shown in Fig. 6b.

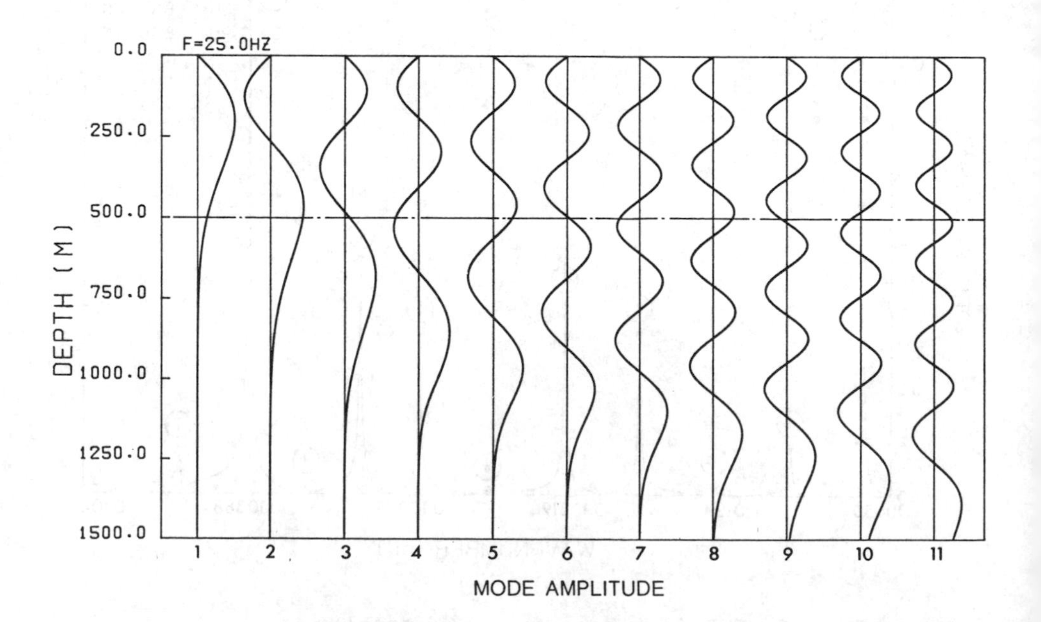

Fig. 7 Mode-amplitude functions for test problem

This particular example shows the consistency between FFP and NM calculations. In the next section we show the results on transmission loss when discrete and discrete-plus-continuous spectra are included and we compare them with calculations using the parabolic equation technique.

3.3 Parabolic equation

If the environment varies both in range and depth, the wave equation cannot be separated and therefore direct numerical integration is required. At present there are no practical methods to perform this direct integration of the three-dimensional wave equation, which is a boundary-value problem. An alternative approach is to derive an approximate wave equation that lends itself to practical numerical solution. We now outline the derivation of such an approximation, the Parabolic Equation (PE), which was introduced into underwater acoustics in 1973 by Tappert and Hardin <17,33>.

The velocity potential is decomposed as follows:

$$\phi = \psi(r,z) \cdot S(r), \tag{23}$$

and we substitute ϕ into Eq. (1) in a source-free region:

$$\psi \cdot \left[\frac{\partial^2 S}{\partial r^2} + \frac{1}{r}\frac{\partial S}{\partial r} \right] + S \cdot \left[\frac{\partial^2 \psi}{\partial r^2} + \frac{\partial^2 \psi}{\partial z^2} + \left(\frac{1}{r} + \frac{2}{S}\frac{\partial S}{\partial r}\right) \cdot \frac{\partial \psi}{\partial r} + k_o^2 n^2 \psi \right] = 0. \tag{24}$$

That is, we will eventually end up with an equation that allows a "marching-in-range" solution and we will have to initialize the solution in some way (see below). We use the notation that

$$k^2 = k_o^2 n^2, \tag{25}$$

where n is an "index of refraction" equal to c_o/c, where c_o is a reference speed.

Equation (24) may be separated into two differential equations by setting the terms in the first bracket equal to $-Sk_o^2$ and the terms in the second bracket equal to ψk_o^2, where k_o^2 is the separation constant. The functions S(r) and $\psi(r,z)$ then have to satisfy the following two equations:

$$\frac{\partial^2 S}{\partial r^2} + \frac{1}{r}\frac{\partial S}{\partial r} + k_o^2 S = 0 \tag{26}$$

$$\frac{\partial^2\psi}{\partial r^2} + \frac{\partial^2\psi}{\partial z^2} + \left(\frac{1}{r} + \frac{2}{S}\frac{\partial S}{\partial r}\right) \cdot \frac{\partial\psi}{\partial r} + k_o^2 n^2\psi - k_o^2\psi = 0. \tag{27}$$

The solution of Eq. (26) is the zeroth order Hankel function, $H_o^{(1)}(k_o r)$, whose asymptotic form has been given in Sect. 3.1. Substituting the asymptotic form of the Hankel function into Eq. (27) and making the paraxial approximation

$$\frac{\partial^2\psi}{\partial r^2} \ll 2k_o \frac{\partial\psi}{\partial r}, \tag{28}$$

we obtain

$$\frac{\partial^2\psi}{\partial z^2} + 2ik_o \frac{\partial\psi}{\partial r} + k_o^2(n^2 - 1)\,\psi = 0, \tag{29}$$

which is the parabolic wave equation.

The paraxial approximation is a narrow-angle approximation. It implies that the rapid range dependence of Eq. (23) is included in S(r), while ψ is a function varying more slowly in r. An approximation to solving Eq. (29) is to assume that n is not a function of the spatial variables but is a constant. It is shown elsewhere <33,34> that the error introduced can be made arbitrarily small by using numerical methods. With n a constant, we can fourier transform ψ with respect to z,

$$\psi(r,s) = \frac{1}{2\pi} \int_{-\infty}^{\infty} \psi(r,z)e^{-isz}dz, \tag{30}$$

which together with Eq. (29) gives

$$-s^2\psi + 2ik_o \frac{\partial\psi}{\partial r} + k_o^2(n^2 - 1)\,\psi = 0. \tag{31}$$

Equation (31) is a first-order differential equation with constant coefficients and has the solution

$$\psi(r,s) = \psi(r_o,s) \cdot e^{-\frac{k_o^2(n^2-1) - s^2}{2ik_o}(r - r_o)} \tag{32}$$

where the initial condition at r_o must be specified. The field as a function of depth is the inverse transform of Eq. (30)

$$\psi(r,z) = \int_{-\infty}^{\infty} \psi(r_o,s) \cdot e^{\frac{ik_o}{2}(n^2-1)\Delta r} e^{-\frac{i\Delta r}{2k_o}s^2} \cdot e^{isz} ds \tag{33}$$

where $\Delta r = r - r_o$.

By introducing the symbol $\mathcal{F}$ for the fourier transform from the z-domain and $\mathcal{F}^{-1}$ as the inverse transform, Eq. (33) may be written as

$$\psi(r + \Delta r,z) = e^{\frac{ik_o}{2}(n^2-1)\Delta r} \cdot \mathcal{F}^{-1}\left\{ e^{-\frac{i\Delta r}{2k_o}s^2} \cdot \mathcal{F}\{\psi(r,z)\}\right\}. \tag{34}$$

Equation (34) is the so-called "split-step" marching solution of the parabolic equation. The fourier transforms are performed using an FFT. It is the solution for n constant, but the error introduced when n (profile and bathymetry) varies with range and depth can be made arbitrarily small by increasing the transform size and decreasing the range-step size <33,34>.

The parabolic equation is not a boundary-value equation as we have numerically formulated it above. We can include the surface boundary condition by taking an anti-symmetric FFT about the sea surface (z = 0). In practice this is performed by taking sine transforms. The boundary conditions in the bottom are simulated by including the discontinuity in velocity in the sound-speed profile. There are methods to also include the density discontinuity <33>. The radiation condition as z goes to infinity is simulated by requiring the field to exponentially tail off for large values of z beyond which there would not be any significant acoustic interaction.

As mentioned above, the PE method requires an initial starting solution. Two methods have been used for describing a point source. The first method is to initialize the field with a set of normal modes descriptive of the point source in the starting environment. This would not include the continuous portion of the spectrum (see Sect. 3.2), but for long-distance propagation this approximation is adequate. A second approach has proved to be simpler and as effective. The point source is approximated by two gaussians that are anti-symmetric about the sea surface, thereby automatically including the pressure-release boundary condition at the surface. Both starting techniques have been used in this paper. We will see that by using the gaussian starting field part of the continuous spectrum is included in the PE solution.

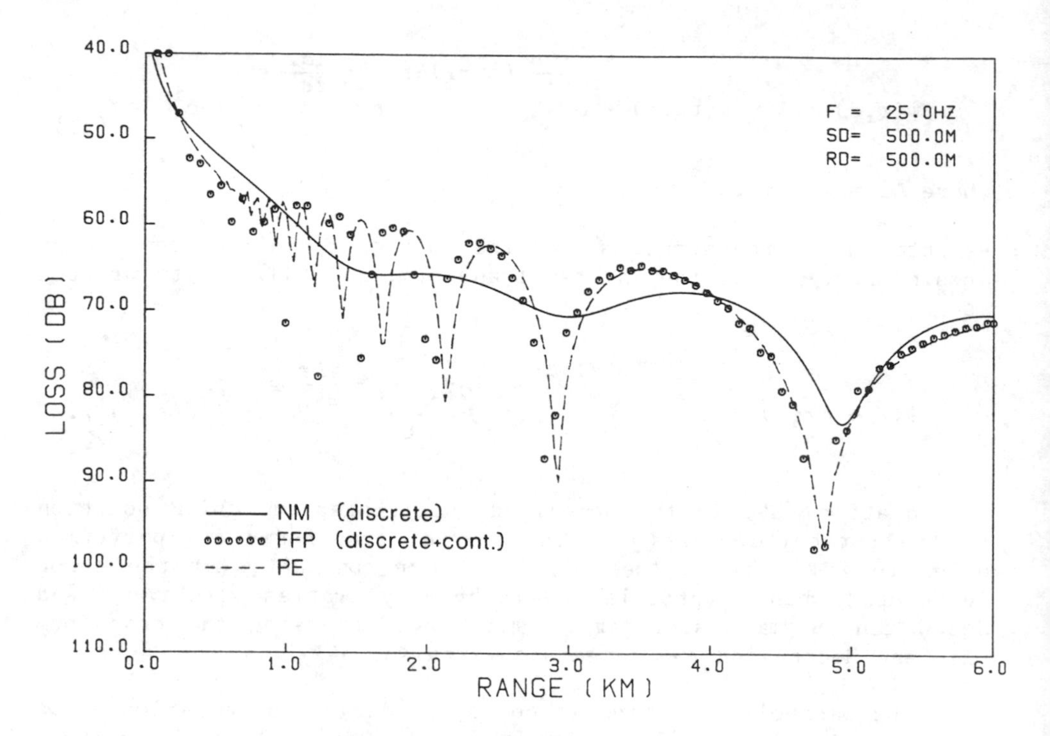

Fig. 8 Computed transmission losses for test problem

In Sect. 3.2 we compared FFP and NM, using a simple environment (Fig. 5) as an example. We now look at transmission loss from the point of view of discrete, discrete-plus-continuous, and PE, which does not obviously distinguish between regions in the spectrum. Figure 8 shows computed transmission loss from 0 to 6 km, defined as $TL = -20 \log (P/P_1)$, where P_1 is the pressure amplitude at 1 m distance from the source. Note that the PE tracks the FFP results in the nearfield indicating that at least part of the continuous portion of the spectrum is included in the PE calculation when using a gaussian starting field. We can also see how the three model results converge in the farfield; recall that the NM calculation does not include the full nearfield contribution. This particular example clearly illustrates the consistency and inter-relationship between the three models.

The advantages of the PE are that it handles a range-dependent environment and gives the acoustic field in the entire water column without additional computational effort. Its disadvantages are that the procedure is not easily automated, and it is practical only for low-frequency propagation since computation times increase with frequency squared. Moreover, there is no straightforward way of handling shear propagation in the bottom.

The parabolic wave equation as given by Eq. (29) is based on the paraxial approximation, and hence only propagation close to the horizontal (± 20°) is accurately handled. This angle limitation is of minor importance for a wide class of ocean acoustics problems. However in studying bottom-interacting propagation, the narrow-angle approximation becomes a serious limitation. It has recently been shown <35,36> that a slight modification to Eq. (20) can improve the angle coverage to ± 40°, yielding a modified parabolic equation, which, however, can no longer be solved by the split-step technique. Instead finite-difference solution techniques have been applied <35,37>, and a working computer code is already available <38>.

Numerical PE results given in this paper were all done with a model based on the standard parabolic equation technique as delineated above.

3.4 Ray theory

This paper is concerned mainly with wave theory; nevertheless, for completeness, we include a brief description of ray theory. In this case we assume a solution of Eq. (1) (with right-hand side equal to zero) as

$$\phi = \psi(x,y,z) \cdot e^{iS(x,y,z)}. \tag{35}$$

$S(x,y,z)$ is a phase function that includes rapid variations as a function of range, and $\psi(x,y,z)$ is a more slowly varying envelope function in which geometrical spreading and loss mechanisms are included (in the PE, S contains the cylindrical spreading factor). Substituting Eq. (35) into the wave equation and separating real and imaginary parts, we obtain

$$\frac{1}{\psi}\nabla^2\psi - (\nabla S)^2 + k^2 = 0, \quad 2(\nabla\psi \cdot \nabla S) + \psi\nabla^2 S = 0. \tag{36}$$

We now make the geometrical-acoustics approximation

$$\frac{1}{\psi}\nabla^2\psi \ll k^2, \tag{37}$$

that is, the amplitude of the phase function varies slowly in range with respect to wavelength. Substituting Eq. (37) in Eq. (36) gives the eikonal equation,

$$(\nabla S)^2 = k^2. \tag{38}$$

The trajectory of the rays is perpendicular to the surfaces of constant phase (wavefronts), S, and is expressed by

$$\frac{d}{d\ell}\left\{k\frac{d\vec{X}}{d\ell}\right\} = \nabla k, \tag{39}$$

where ℓ is the arc length along a ray and X is the coordinate. It can be shown that the direction of the average energy flow is along these trajectories and the amplitude of the field at any point can be obtained from the density of these rays; formally, having solved for S, the amplitude is obtained from solving the second part of Eq. (36). We also mention here that corrected ray theory assumes that ψ is a function of frequency and an expansion in powers of inverse frequency is made, the leading term being the infinite-frequency solution with the additional terms being corrections from the infinite-frequency solution.

The advantages of ray theory methods are that the computations are rapid and that ray traces give a very physical picture of the acoustic paths. The disadvantage is that ray-theory is an infinite-frequency approximation and therefore does not include diffraction and other wave effects. This shortcoming also prevents ray theory from adequately describing significant bottom interaction and low-frequency ducted propagation.

4 NUMERICAL MODELS: THEIR APPLICABILITY AND CONSISTENCY

The four acoustic models described in this paper are a representative subset of the many different propagation models in use in underwater acoustics today. The reason for developing new models is to obtain either more accurate solutions or faster solutions to specific problems. Each model has its area of applicability depending on the theoretical limitations in the model and on the numerical efficiency of the computer code.

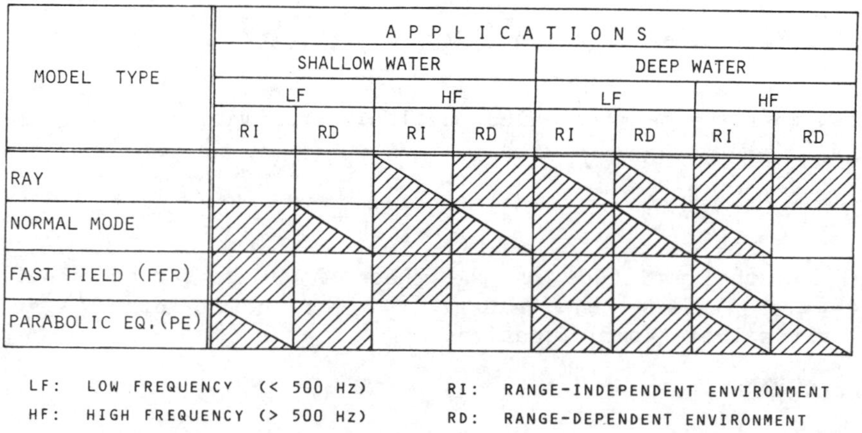

MODEL TYPE	APPLICATIONS							
	SHALLOW WATER				DEEP WATER			
	LF		HF		LF		HF	
	RI	RD	RI	RD	RI	RD	RI	RD
RAY								
NORMAL MODE								
FAST FIELD (FFP)								
PARABOLIC EQ. (PE)								

LF: LOW FREQUENCY (< 500 HZ) RI: RANGE-INDEPENDENT ENVIRONMENT
HF: HIGH FREQUENCY (> 500 HZ) RD: RANGE-DEPENDENT ENVIRONMENT

Fig. 9 Applicability of four propagation models

To indicate with some precision the type of ocean environment for which a given model should be used, we have classified environments according to water depth, frequency, and environmental complexity, as shown in Fig. 9. Here shallow water indicates all water depths for which sound interacts significantly with the ocean bottom. The separation frequency of 500 Hz between the low- and high-frequency regimes is arbitrarily chosen.

When indicating the applicability of a propagation model to a given type of environment we take into consideration limitations in the underlying theory. Ray models are applicable mainly to high-frequency propagation. Only ray and PE models accurately handle a range-dependent environment. The normal-mode model treats range dependence in the adiabatic approximation. When indicating a model's practicality we consider exclusively the computation time, which, of course, depends on the required accuracy. The computation time increases with both frequency and water depth for wave models (mode, FFP, PE), while the time is relatively independent of these parameters for the ray models. Likewise, computation time is proportional to the number of profiles in a range-dependent environment for both ray and mode models, while a PE model takes essentially the same time for range-dependent and range-independent environments.

Full-box shading in Fig. 9 means that a model is applicable as well as practical. On the other hand, if a box is only partially shaded, it means that the model is applicable with caution (theoretical limitations), or that computation times are excessive. The above judgements are, of course, relative. For instance, in our first evaluation some columns contained no applicable models. In these columns we therefore selected the model we felt was the most practical end denoted it by a fully shaded box. For a column where more than one box is fully shaded, the choice of model will depend on the actual models on hand, the running time, input/output options available, etc.

Since the various models are approximate solutions of the wave equation, it is valuable to check the validity of these approximations by doing an inter-model comparison for situations where all four models are considered applicable. Returning to Fig. 9, we note that all models should handle a range-independent shallow-water environment at around 500 Hz, even though the mode and FFP models are designated most applicable (fastest).

An example of a transmission loss computation for shallow water is given in Fig. 10. Here an isovelocity water column (1500 m/s) is 100 m deep and both source (SD) and receiver (RD) are at mid-depth. The bottom is considered homogeneous, with a sound speed of 1550 m/s, a density of 1.2 g/cm^3, and an attenuation of 1 dB/wavelength. We see from Fig. 10 that the three

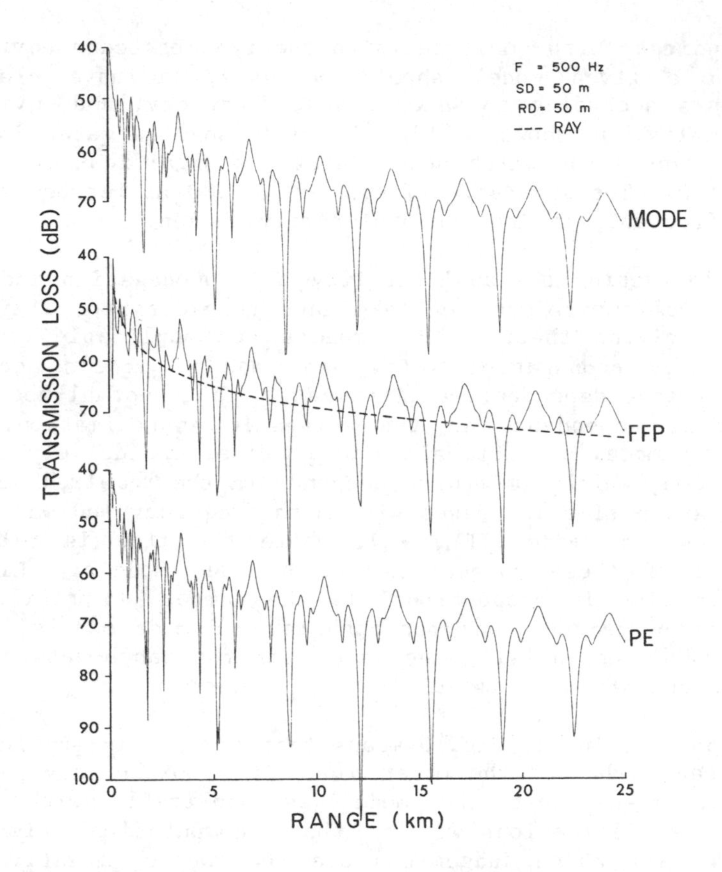

Fig. 10 Inter-model comparison for shallow-water environment

wave models (mode, FFP, PE) give virtually identical results both for level and for the multipath interference structure as a function of range. The ray model, though it cannot reproduce the interference pattern, does yield the same approximate level (dashed line).

Many consistency tests of acoustic models have to be carried out in order to check the complex computer programs within the framework of the theoretical limitations particular to each model <39>. A positive outcome of an inter-model comparison helps us to gain confidence in particular numerical models. However, we should remember that the final check on an acoustic model is a comparison with experimental data. This demonstrates whether or not the model includes all the physics necessary for explaining and understanding sound propagation in a real ocean environment. A series of model/data comparisons can be found in <40-43>.

5 SPECIAL MODEL APPLICATIONS

This section is dedicated to a study of some general wave-propagation problems, for which illustrative numerical solutions can be obtained in a straightforward manner using one of the aforementioned ocean-acoustic models. We shall first address the problem of reflection of a gaussian beam of arbitrary width at a fluid/solid interface near the Rayleigh angle, where a leaky surface wave is excited causing a complex reflection pattern with beam splitting and beam displacement. This is a well-researched problem in both optics and acoustics, which we can easily solve with the fast-field program. Next we study propagation in range-dependent waveguides using the parabolic equation technique. We shall address acoustic radiation from a duct into free space as well as mode coupling in tapered waveguides.

5.1 Beam reflection at fluid/solid interface

Bounded beam reflection near a critical angle is a subject that has received considerable attention in the past, both within the field of electromagnetics <44-48> and acoustics <49-55>. The phenomenon of interest has mainly been the displaced reflected beam, while the transmitted field has been studied in much less detail. There are basically two different wave phenomena that can account for the observed features of the reflected field. One is the excitation of a lateral wave at the interface when a beam of finite width is incident on the interface at grazing angles lower than the critical angle. The reflected field is then composed of contributions from both the specular reflection and the lateral wave field, causing an apparent lateral displacement of the reflected beam. In acoustics, the lateral wave phenomenon is associated with beam reflection at fluid/fluid interfaces.

The second phenomenon is associated with the excitation of a leaky surface wave, which again complicates the reflection pattern when added to the specularly reflected field. This phenomenon occurs in acoustics when a bounded beam is incident on a fluid/solid interface just below the shear critical angle.

Various aspects of bounded beam physics have been studied theoretically as well as experimentally in the past 35 years. The lateral displacement of a reflected light beam was first observed by Goos and Hänchen <44>, and the phenomenon is therefore often referred to as the Goos-Hänchen effect. Several theoretical papers have addressed the beam reflection problem, for instance <45-49>, though always with some theoretical limitations, such as beamwidth large compared to the wavelength, lossless media, parallel beams of particular shape (gaussian), etc.

We shall here apply the fast-field program (FFP) to numerically solve the reflection problem, and, as we shall see, this technique treats the complete problem without any of the above limitations. We are going to study reflection at a fluid/solid interface, where the reflection phenomenon is associated with the excitation of a leaky surface wave (Rayleigh wave). However, the FFP model could as well be applied to reflection and transmission for a solid plate in a liquid <52-53> or to the transmission through a fluid/fluid boundary <54-55>.

The FFP model <13> provides an exact numerical solution for the acoustic field generated by a point source in a multilayered fluid/solid environment (Sect. 3.1). Wave attenuation for both compressional and shear waves is included in the theory. We have modified the standard code to efficiently solve the system of equations for a number of point sources, equidistantly spaced and forming a vertical line array. The total acoustic field is found by superposition of the contributions from individual point sources. Hence, in this model the beam is generated by a vertical source array, and the beam direction is varied by appropriately phasing the source elements. By using a gaussian amplitude weighting across the array, a gaussian beam can be generated, propagating at any angle with respect to the horizontal. By varying the array distance from the interface and the number of source elements (half-wavelength spacing), a beam of arbitrary width can be generated, and we can obtain parallel, diverging, or converging beams, as we wish. Hence, the model is very general in concept, and should handle the reflection problem accurately for any beamwidth and for multilayered systems (plates) as well as the reflection and transmission at a single fluid/solid interface. The solution is for a plane geometry, i.e. a two-dimensional beam.

We first consider computational results for a water/steel interface. Information on material parameters were taken from Breazeale et al <51>, and the FFP results will be compared with the theory of Bertoni and Tamir <47>, here named the BT theory. Results for reflection at the Rayleigh angle (59.35°) are given in Fig. 11 for two different beamwidths.

Computations were done at 20 kHz, and the acoustic field is displayed as iso-loss contours given in arbitrary decibels (low values correspond to high intensity). We are considering incoming parallel gaussian beams, where the beamwidth is 10 and 64 wavelengths respectively. Note in Fig. 11a that the narrow beam, when reflected, is being split up in two beams separated by a "null strip". The beam to the left is displaced slightly backwards, while the rightmost beam is displaced forwards 18 wavelengths. This behavior is in complete agreement with the BT theory. With increasing beamwidth, the energy shifts to the forwardly displaced beam, and we end up (Fig. 11b) with a single-beam reflection with a displacement of 50 wavelengths. The general behavior seen in these examples is predicted by Bertoni and Tamir, and the measured

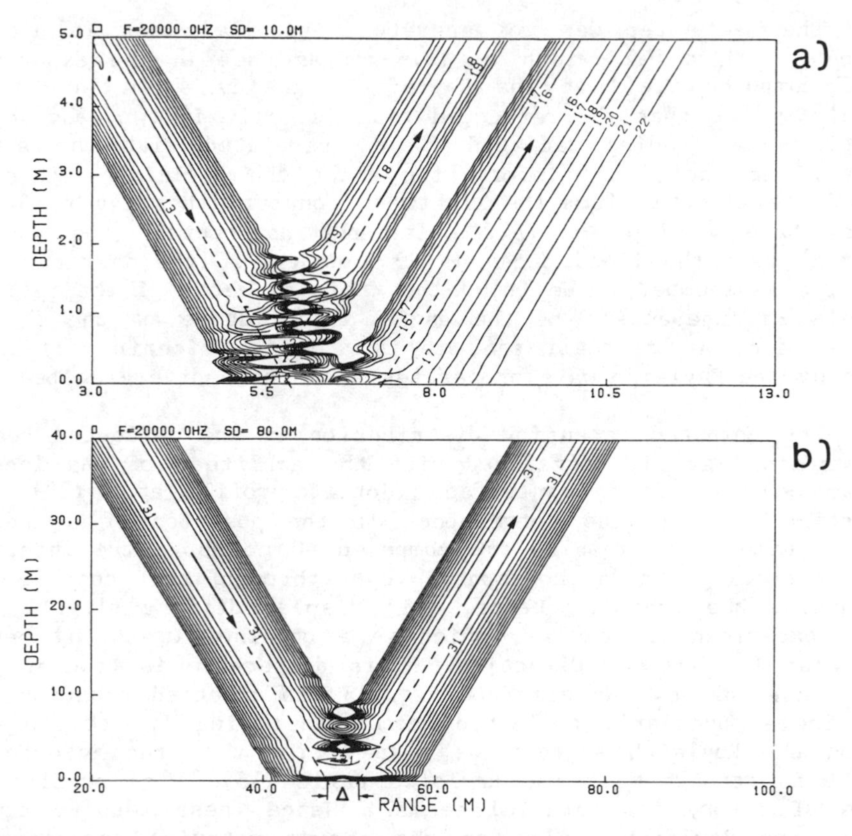

Fig. 11 Beam reflection at water/steel interface at the Rayleigh angle.
a) narrow beam of width 10λ,
b) broad beam of width 64λ.

displacement versus beamwidth gives data points that fall exactly on the displacement curve computed from the BT theory.

A more challenging investigation was to explain the disagreement between theory (BT) and experimental results for aluminium oxide, as reported by Breazeale et al <51>. We proceed to calculate the Rayleigh-wave properties for a water/aluminium-oxide interface using the material parameters given in <51>. For a water speed of 1490 m/s, a frequency of 2 MHz, and with realistic attenuation coefficients in both media, we found the leaky surface wave to have a phase velocity of 5825.6 m/s corresponding to an angle of 75.18°. The leakiness of the surface wave is given by the imaginary part of the wave number, which is calculated to be 23.3 m^{-1}. These Rayleigh-wave properties were determined from a separate computer code that finds the poles for the complex reflection coefficient.

The next step was to generate a parallel beam (plane wave front) with a half-width of 16.6 mm as used in the experiment. The computed field at the Rayleigh angle is shown in Fig. 12. Again we see that the reflected beam is split in two, and we also notice the leading radiated field associated with the surface wave. In fact, full information about the Rayleigh wave can be read off directly from the rightmost contour lines (46 to 54 dB). They have a slope of 75.2° with the horizontal (the Rayleigh angle), and the field decay is 23.3 Nepers/m (the imaginary part of the wavenumber). We found that this property of the reflected field is independent of the angle of incidence of the incoming beam, and, as we shall see, it is a valid criterion for determining the Rayleigh-wave properties also for a diverging beam.

The computed intensity distribution in the reflected beam is given in Fig. 13, normalized with the amplitude of the incoming beam measured at the interface (dotted profile at 75.2°). The vertical dashed line corresponds to the position of a specular reflection. The results are computed 40 cm above the interface, corresponding to a horizontal cut through the contour plot (outside the frame). Hence, this display differs slightly from the experimental results, which were obtained measuring perpendicular to the beam direction. This difference is however minor at these angles. We see from Fig. 13 the expected behavior, i.e. a single specularly reflected beam when moving 1.5 to 2.0° away from the Rayleigh angle. We also notice that the interference null is strongest at the Rayleigh angle (75.2°) as predicted by the BT theory for parallel beams. Hence these results confirm that the Rayleigh angle for the chosen material parameters is 75.2°.

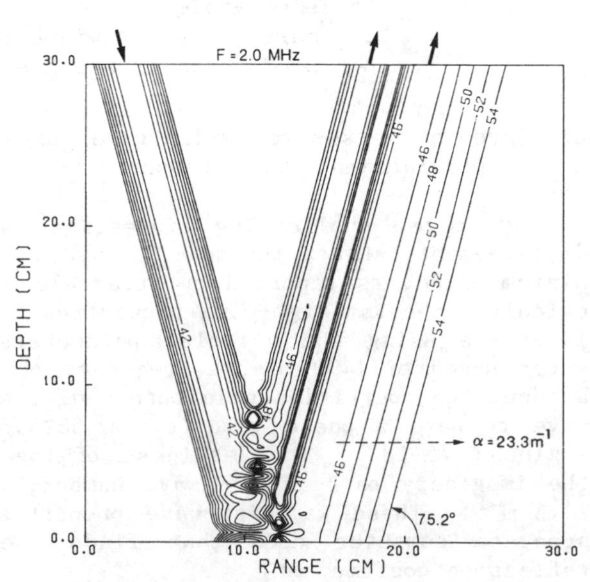

Fig. 12
Reflection of parallel beam at water/aluminium-oxide interface

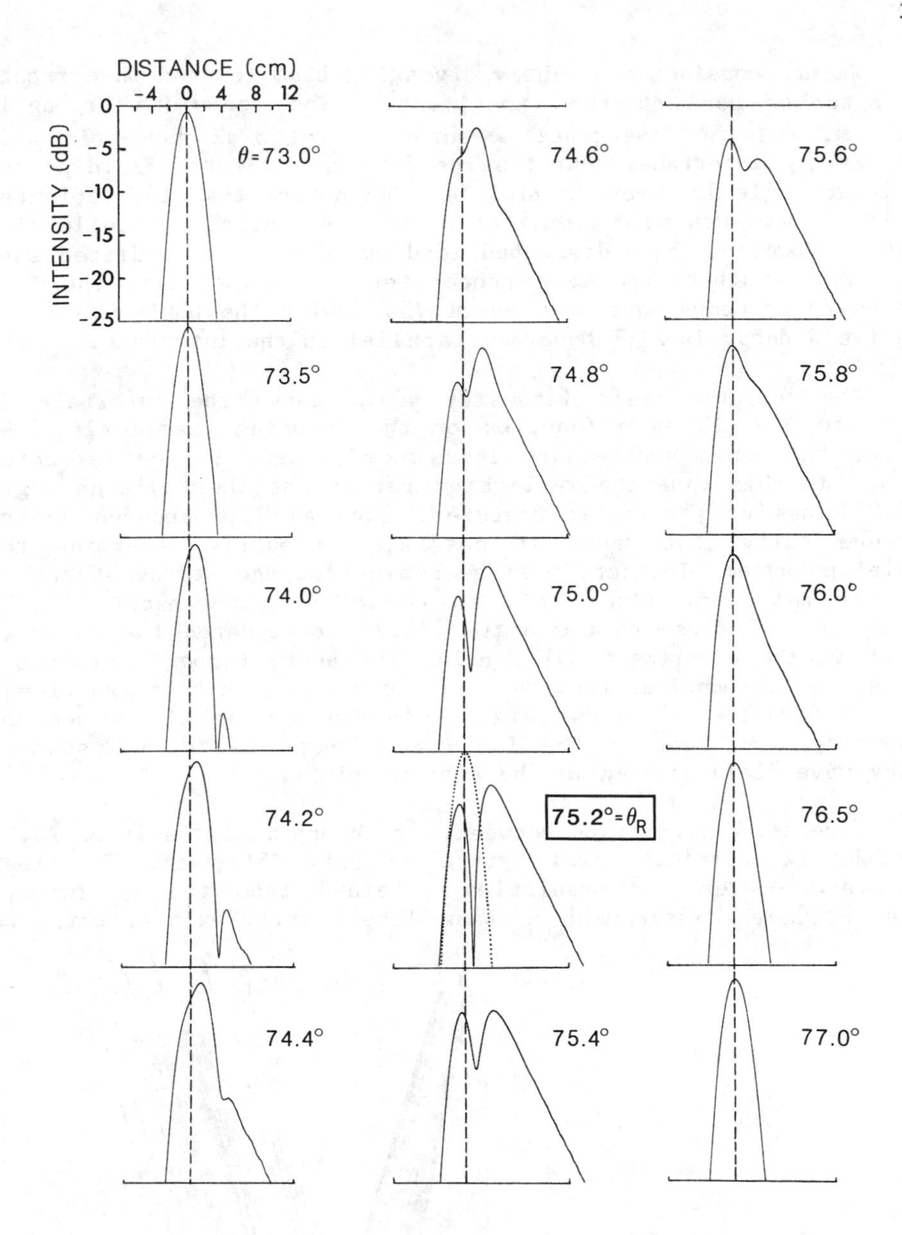

Fig. 13 Reflectivity pattern versus grazing angle for parallel beam incident on a water/aluminium-oxide interface

We now consider a slightly diverging beam (curved wave front) as actually used in the experiment. The beamwidth is again 16.6 mm, and the divergence measured at the 3 dB down-points is 2.7° at the interface. Our results for the reflected field at the Rayleigh angle is given in Fig. 14. We notice that the reflected field is now much more complicated, with essentially 3 reflection peaks. However, the undisturbed leading edge of the radiated surface wave exhibits the same properties as before, i.e. the iso-intensity contours have a slope of 75.2° with the horizontal, and the field decay is 23.3 Nepers/m parallel to the interface.

The computed field intensity 40 cm above the interface is given in Fig. 15 as a function of the incoming beam angle. We notice the many interference lobes now present in the reflected beam. In this case the reflection pattern at the Rayleigh angle (75.2°) has no particular features, such as a pronounced interference null, that makes it possible to easily determine the Rayleigh angle. In fact, when searching for the strongest interference null, one finds this to occur at approximately 75.8°, which is very close to the angle (75.7°) designated the Rayleigh angle in the experiment <51>, using the above (wrong) criterion. Hence, we may conclude from this set of curves, that an experimental verification of the Rayleigh angle can most easily be done by measuring the slope of the isoloss contours in the undisturbed leaky wave field as seen on the contour plots.

A detailed comparison between our theoretical result at 75.8° and the experimental result given in Ref. <51>, shows a clear improvement over the prediction obtained from the BT theory. However, some disagreement on peak levels still exists, which we

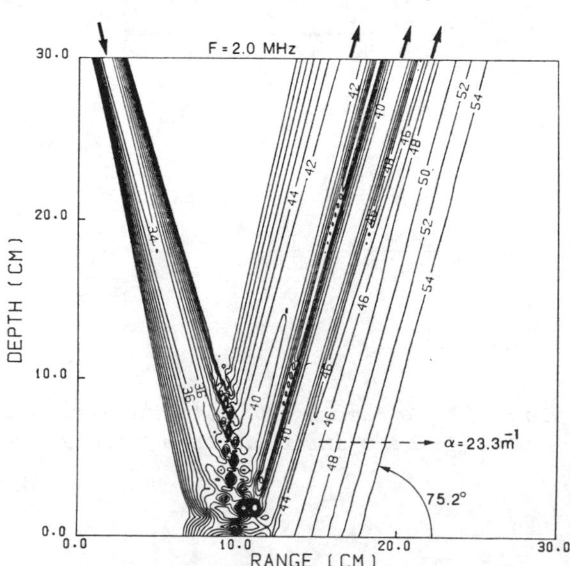

Fig. 14
Reflection of diverging beam at water/aluminium-oxide interface

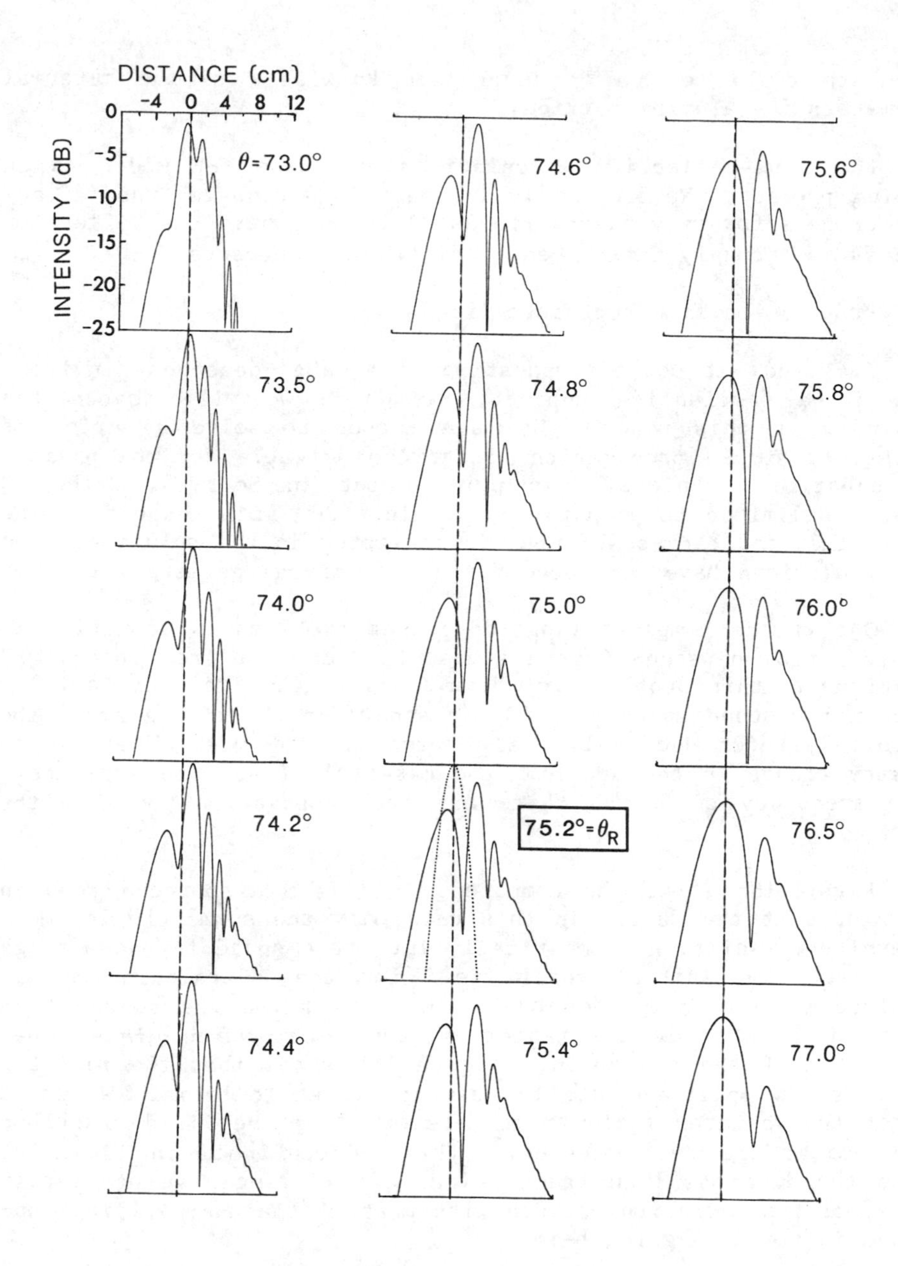

Fig. 15 Reflectivity pattern versus grazing angle for diverging beam incident on a water/aluminium-oxide interface

speculate could be due to inaccurate knowledge of the material parameters for aluminium oxide.

The beam reflection calculations with the FFP model is an ongoing project. We are mainly interested in studying the reflection process for very narrow (focused) beams, where the reflection process is strongly influenced by diffraction effects.

5.2 Propagation from duct into free space

The study of sound propagation in a range-dependent environment is a fascinating subject, to which we will devote the remainder of this paper. We have chosen to solve a series of relatively simple propagation situations suitable for the parabolic equation technique. As pointed out in Sect. 3.3, the PE method is limited to propagation within ± 20° with respect to the horizontal, and back-scattering is neglected in the solution. The PE calculations have been done for a cylindrical geometry.

One of the simplest range-dependent problems in acoustics is the radiation of sound from a symmetric duct into free space. PE solutions to this problem are shown in Fig. 16. The duct is 100 m wide with a sound speed of 1500 m/s and a density of 1 g/cm^3. The infinitely thick duct wall has a speed of 1550 m/s. There is no density change in the problem, and material losses are neglected. At a frequency of 50 Hz, there are two propagation modes in the duct.

Figure 16a shows the computed field from a source placed in the middle of the duct. In this case only the symmetric 1st mode is excited, radiating symmetrically into free space beyond a range of 2 km. The initial modal field for the PE calculation was supplied by a normal-mode model. Next we moved the source to a depth of 27 m below the center of the duct, which gives equal excitation of the two modes. Figure 16b shows that the radiated field is now split up into two almost symmetric beams. We would expect the radiated field to be determined by the field distribution across the duct opening. This is confirmed in Fig. 16c, where the duct has been truncated at 1.5 km range; we now obtain an asymmetric radiation pattern with most of the energy being contained in the down-going beam.

The above numerical results provide considerable physical insight into the duct radiation problem. The examples were chosen so as to facilitate a physical interpretation of the contour plots. However, more complex situations could easily be investigated.

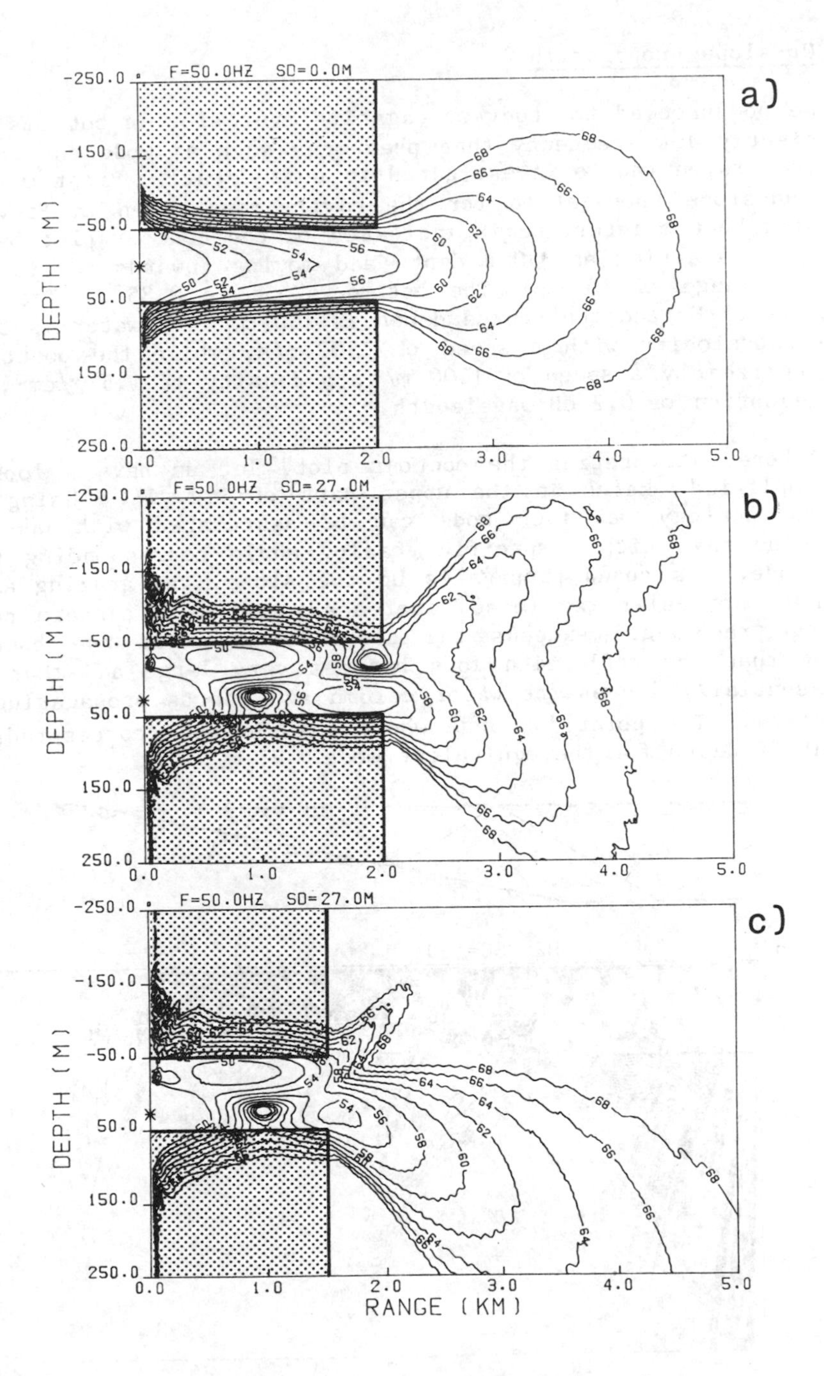

Fig. 16 Sound radiation from duct into free space
a) one symmetric mode excited in duct,
b) two modes excited in duct,
c) as b) but for a shorter duct.

5.3 Up-slope propagation

We now proceed to study propagation over sloping bottoms at a sufficiently low frequency that phenomena such as mode cutoff and mode conversion can be investigated in some detail. First we consider up-slope propagation for the environment given in Fig. 17. The water/bottom interface is indicated on the contour plot by the heavy line starting at 350 m depth and moving towards the surface beyond a range of 10 km. The bottom slope is 0.85°. The frequency is 25 Hz and the source depth is 150 m. The water is taken to be isovelocity with a speed of 1500 m/s, while the bottom is characterized by a speed of 1600 m/s, a density of 1.5 g/cm^3, and an attenuation of 0.2 dB/wavelength.

Before interpreting the contour plot, let us have a look at the simplified sketch in the upper part of Fig. 17. Using the ray/mode analogy, a given mode can be associated with up- and down-going rays with a specific grazing angle corresponding to a given mode. As sound propagates up the slope, the grazing angle for that particular ray (mode) increases, and at a certain point in range the angle exceeds the critical angle at the bottom, meaning that the reflection loss becomes very large and that the ray essentially leaves the water column and starts propagating in the bottom. The point in range where this happens corresponds to the cutoff depth for the equivalent mode.

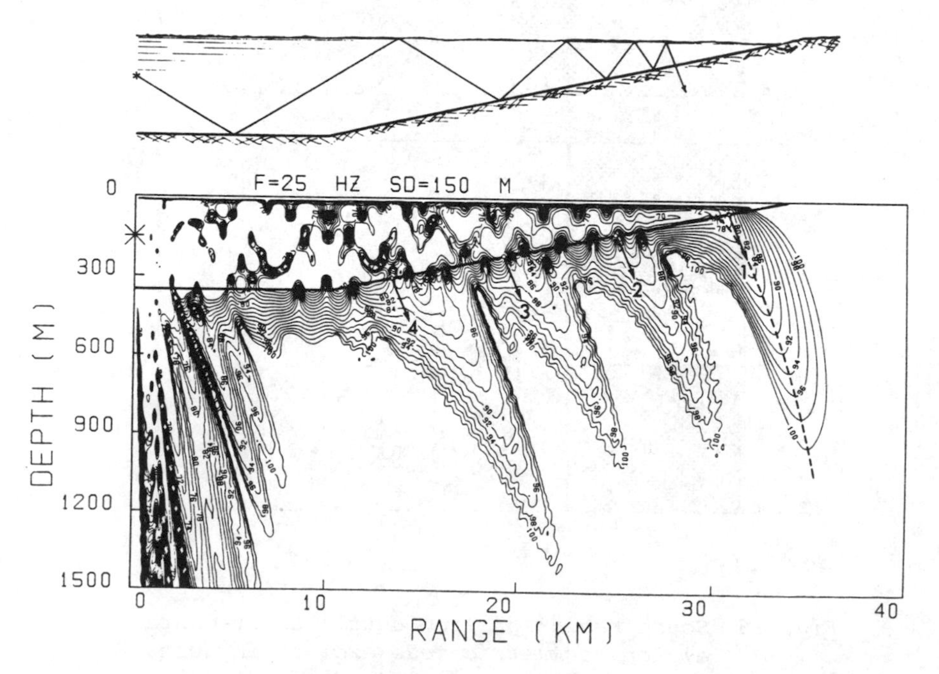

Fig. 17 Up-slope propagation showing discrete mode cutoffs

To emphasize the main features in Fig. 17, we have chosen to display contour levels between 70 and 100 dB in 2 dB intervals. Thus high-intensity regions (loss < 70 dB) are given as blank areas within the wedge, while low-intensity regions (loss > 100 dB) are given as blank areas in the bottom. The PE solution was here started off by a gaussian initial field, and there are four propagating modes. The high intensity in the bottom at short ranges (< 10 km) corresponds to the radiation of "continuous modes" into the bottom. As sound propagates up the slope we see four well-defined beams in the bottom, one corresponding to each of the four modes. This phenomenon of energy leaking out of the propagation channel as discrete beams has been confirmed experimentally <56>, and a detailed study of this phenomenon using the PE method has been reported elsewhere <57>.

This particular example of mode coupling where discrete modes trapped in the water column couple into "continuous" modes propagating in the bottom, has received much attention recently; several theoretical papers <31> and <58-60> have appeared offering solutions to the wedge-propagation problem.

We now proceed to study the problem of mode coupling within the wedge itself. Strong coupling can be achieved either by increasing the bottom slope or by increasing the frequency. It is the latter case that will be considered here. Figure 18 shows PE calculations for a simple wedge problem. The initial water depth is 100 m and the bottom slope is 2°. The water is taken to be isovelocity with a speed of 1500 m/s, while the bottom has a speed of 1550 m/s. There is no density change in the problem, and material losses are neglected. The source depth is 50 m, and the initial field for the PE calculation was supplied by a normal-mode model. The two plots are for source frequencies of 50 and 500 Hz, respectively. In both cases, only the first mode was propagated up the slope.

We notice in Fig. 18a that no mode coupling takes place at a frequency of 50 Hz. The contour lines are smooth, indicating that the local mode at range zero adapts to the changing water depth until it reaches the cutoff depth, where the energy radiates into the bottom. This is an example where propagation is well described by adiabatic mode theory <21-26>.

Strong mode coupling occurs when we increase the frequency to 500 Hz (Fig. 18b). The field within the wedge becomes complicated, and so does the radiation pattern into the bottom. We have done no attempt to analyse this complex mode-coupling problem in detail, but the PE technique could certainly be used for such a study.

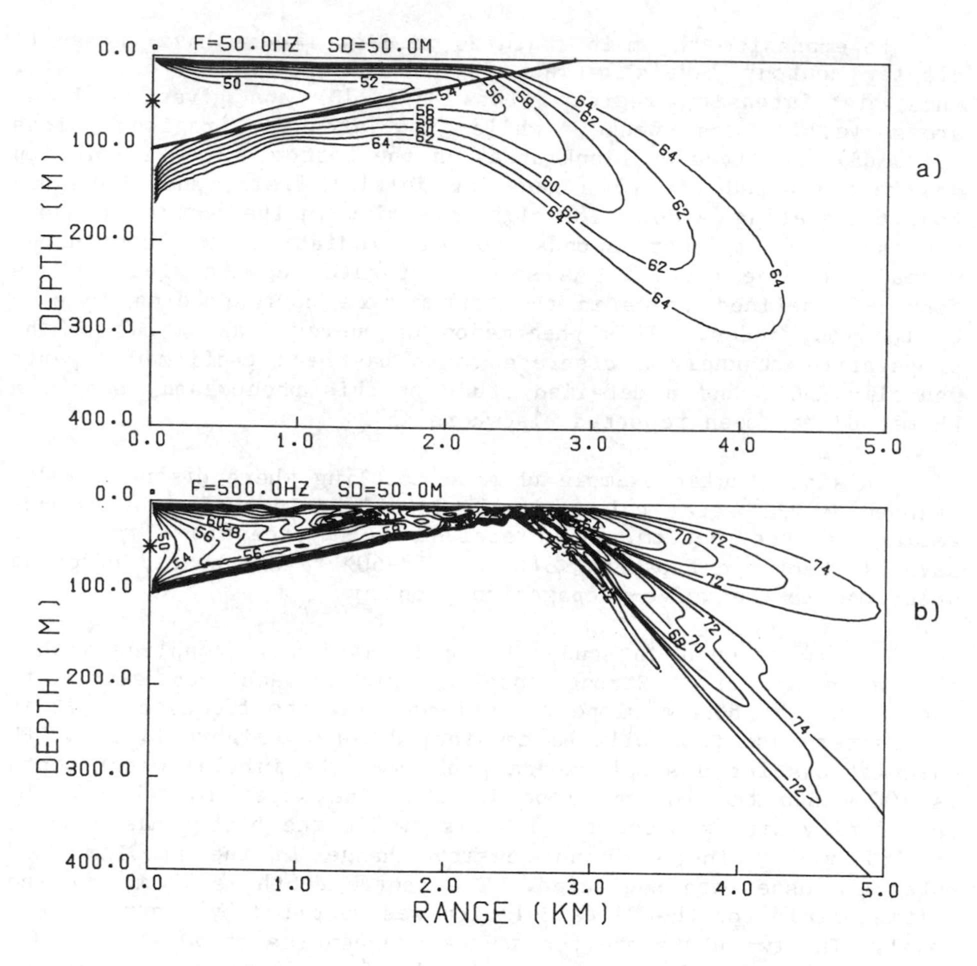

Fig. 18 Up-slope propagation over a constant 2° slope
a) 50 Hz, no mode coupling,
b) 500 Hz, strong mode coupling.

5.4 Down-slope propagation

We now consider the problem of down-slope propagation as illustrated in Fig. 19. The initial water depth is 50 m and the bottom slope is 5°. The water column is isovelocity with a speed of 1500 m/s, while the bottom has a speed of 1600 m/s. The density ratio between bottom and water is 1.5, and a wave attenuation of 0.5 dB/wavelength has been included in the bottom. The two contour plots are for source frequencies of 25 and 500 Hz, respectively. The initial fields for the PE calculations were supplied by a normal-mode model, and, in both cases, only the first mode was propagated down the slope.

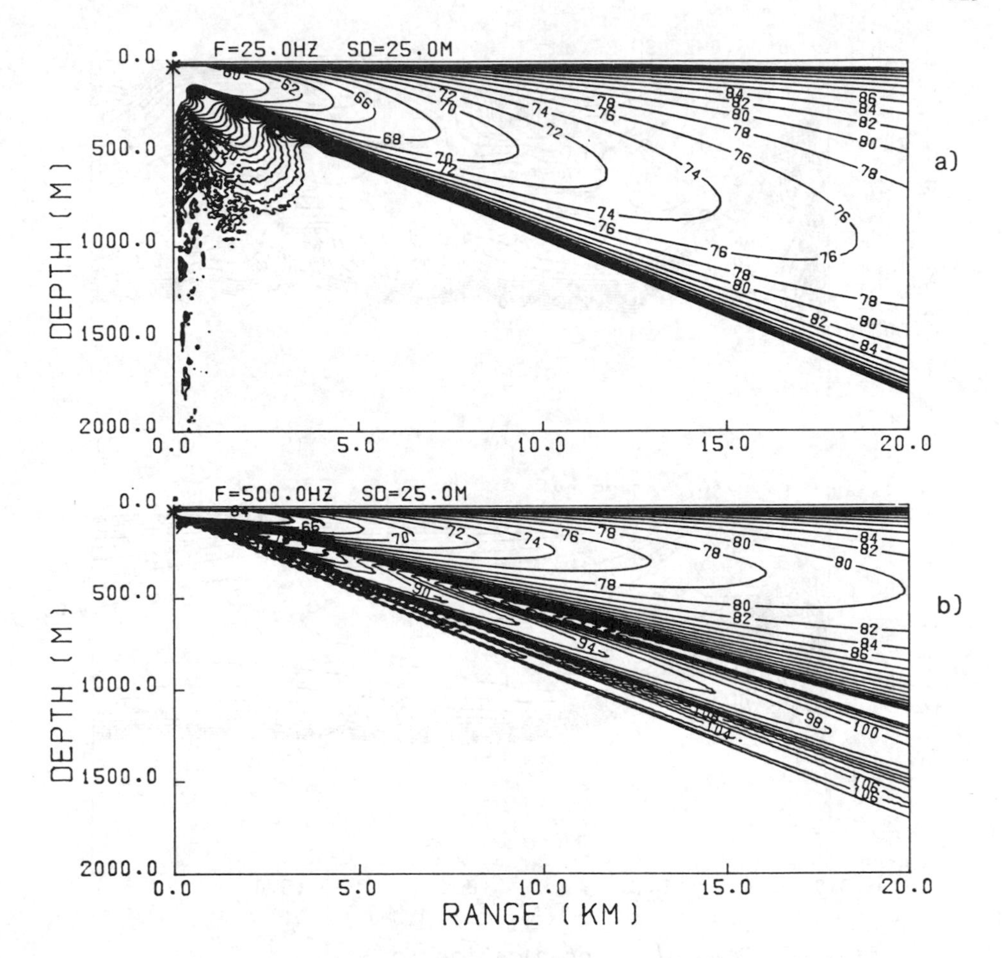

Fig. 19 Down-slope propagation over a constant 5° slope
a) 25 Hz, no mode coupling,
b) 500 Hz, strong mode coupling.

Fig. 19a shows that some energy propagates straight into the bottom at short ranges (coupling into the continuous spectrum). However, beyond the nearfield, propagation within the wedge is clearly adiabatic with the one propagating mode adapting well to the changing water depth. At range 20 km the energy is entirely contained in the local first mode, even though as many as 21 modes can exist in a water depth of 1800 m.

By increasing the frequency to 500 Hz (Fig. 19b), strong mode coupling occurs within the wedge. At long ranges, two interference nulls are present in the energy distribution over depth indicating that the energy is now partitioned among a few lower-order modes.

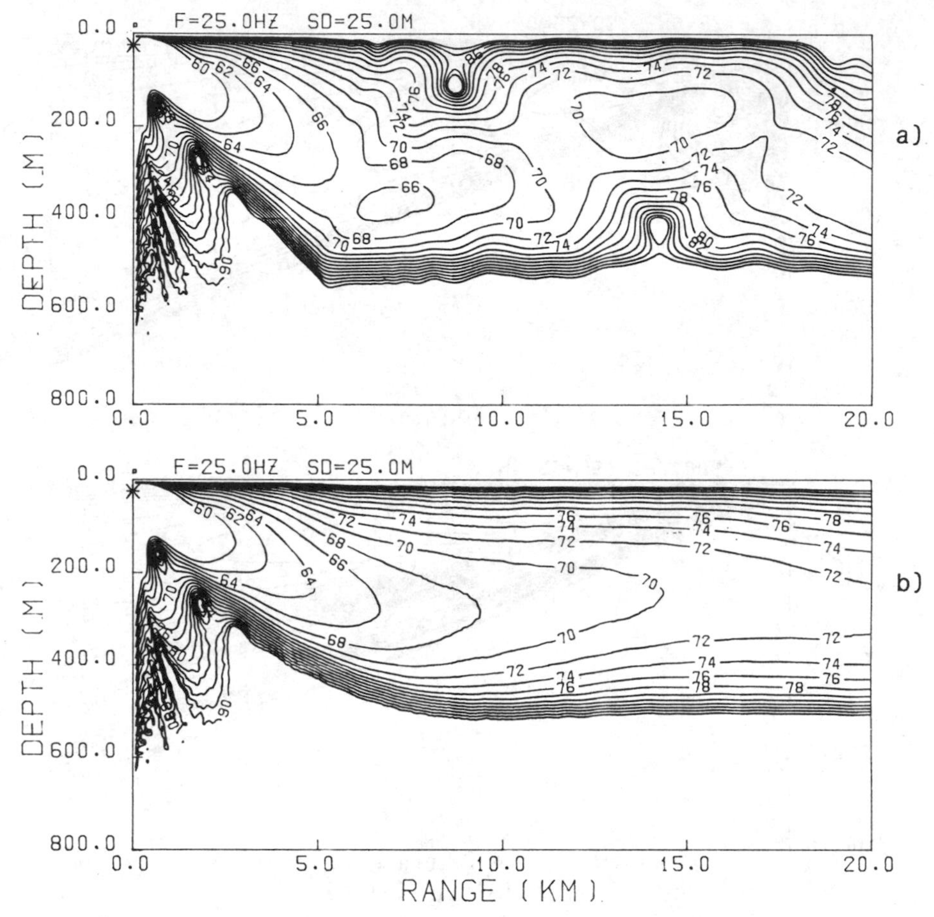

Fig. 20 Down-slope propagation for:
a) a 5° slope followed by a flat bottom, and
b) a smoothly changing bottom slope.

There are several ways to increase the degree of coupling between modes for down-slope situations. Besides the increased coupling with frequency and slope angle, coupling can also be caused by abrupt changes in bottom slope. This is shown in Fig. 20a, which is the same environment as in Fig. 19a, except that the bottom now is flat beyond 5 km. The abrupt change in bottom slope at 5 km results in a complicated contour pattern indicating interference between modes and, hence, the initial first mode has generated (or coupled into) higher modes after the vertex. In this case as many as six modes can exist in the flat part beyond 5 km. That mode coupling is associated with the sharp change of the vertex is shown in Fig. 20b where we gradually go from a 5° slope to a flat bottom. We see that the propagating mode now adapts to the changing water depth exhibiting an "adiabatic" behavior.

6 SUMMARY AND CONCLUSIONS

We have briefly presented an overview over the most commonly used propagation models in underwater acoustics pointed out their areas of applicability, and demonstrated their ability to accurately describe acoustic propagation in complicated ocean environments. It has also been shown that a good agreement between theory and experimental data can be obtained only by including such features as bottom layering, bottom rigidity (shear), scattering at rough boundaries, range-varying environments, etc. Hence the full complexity of a real ocean environment must be considered in the numerical models for accurately predicting the propagation conditions in a given area for a broad range of source frequencies. Finally, the general applicability of the numerical models has been demonstrated by applying the models to some basic wave-propagation problems not tractable by analytical methods.

7 ACKNOWLEDGMENT

This review paper on numerical techniques in underwater acoustics is largely based on material prepared in collaboration with Dr. William A. Kuperman, US Naval Ocean Research and Development Activity, during his 5 years stay at SACLANTCEN.

REFERENCES

1. Officer, C.B. Introduction to the Theory of Sound Transmission. (McGraw-Hill, New York, 1958).
2. Tolstoy, I. and Clay, C.S. Ocean Acoustics: Theory and Experiment in Underwater Sound. (McGraw-Hill, New York, 1966).
3. Tucker, D.G. and Gazey, B.K. Applied Underwater Acoustics. (Pergamon, Oxford, 1966).
4. Tolstoy, I. Wave Propagation. (McGraw-Hill, New York, 1973).
5. Urick, R.J. Principles of Underwater Sound. (McGraw-Hill, New York, 1975).
6. Clay, C.S. and Medwin, H. Acoustical Oceanography. (Wiley, New York, 1977).
7. Keller, J.B. and Papadakis, J.S. (eds.). Wave Propagation and Underwater Acoustics. (Springer-Verlag, Berlin, 1977).
8. De Santo, J.A. (ed.). Ocean Acoustics. (Springer-Verlag, Berlin, 1979).
9. Brekhovskikh, L.M. Waves in Layered Media. (Academic, New York, 1980).
10. Brekhovskikh, L.M. and Lysanov, Yu. Fundamentals of Ocean Acoustics. (Springer-Verlag, Berlin, 1982).

11. DiNapoli, F.R. and Deavenport, R.L. "Theoretical and numerical Green's function field solution in plane multilayered medium", J. Acoust. Soc. Am. 67, 92-105, (1980).
12. Kutschale, H.W. "Rapid computation by wave theory of propagation loss in the Arctic Ocean." Rpt. CU-8-73, Columbia University, Palisades, N.Y. (1973).
13. Schmidt, H. "Modelling of pulse propagation in layered media using a new fast field program." (These proceedings).
14. Ewing, W.M., Jardetzky, W.C. and Press, F. Elastic Waves in Layered Media. (McGraw-Hill, New York, 1957), Chapter 4.
15. Morse, P.M. and Feshbach, H. Methods of Theoretical Physics. (McGraw-Hill, New York, 1953), p. 823.
16. Ingenito, F., Ferris, R., Kuperman, W.A. and Wolf, S.N. "Shallow-water acoustics, summary report, (first phase)". Rpt. 8179, US Naval Research Laboratory, Washington, D.C. (1978).
17. Spofford, C.W. "A synopsis of the AESD workshop on acoustic propagation modelling by non-ray-tracing techniques". Rpt. TN-73-05, Office of Naval Research, Washington, D.C. (1973).
18. Jensen, F.B. and Ferla, M.C. "SNAP: the SACLANTCEN normal-mode acoustic propagation model." Rpt. SM-121, SACLANT ASW Research Centre, La Spezia, Italy (1979).
19. Jensen, F.B. and Kuperman, W.A. "Environmental acoustic modelling at SACLANTCEN". Rpt. SR-34, SACLANT ASW Research Centre, La Spezia, Italy (1979).
20. Ferla, M.C., Jensen, F.B. and Kuperman, W.A. "High-frequency normal-mode calculations in deep water", J. Acoust. Soc. Am. 72, 505-509 (1982).
21. Pierce, A.D. "Extension of the method of normal modes to sound propagation in an almost stratified medium", J. Acoust. Soc. Am. 37, 19-27, (1965).
22. Milder, D.M. "Ray and wave invariants for SOFAR channel propagation", J. Acoust. Soc. Am. 46, 1259-1263 (1969).
23. Williams, A.O. "Normal-mode methods in propagation of underwater sound", in Underwater Acoustics, edited by R.W.B. Stephens (Wiley, London 1970), pp. 23-56.
24. Nagl, A., Uberall, H., Haug, A.J. and Zarur, G.L. "Adiabatic mode theory of underwater sound propagation in range-dependent environment", J. Acoust. Soc. Am. 63, 739-749 (1978).
25. Rutherford, S.R. and Hawker, K.E. "An examination of the influence of range dependence of the ocean bottom on the adiabatic approximation", J. Acoust. Soc. Am. 66, 1145-1151 (1979).
26. Rutherford, S.R. "An examination of multipath processes in a range-dependent ocean environment within the context of adiabatic mode theory", J. Acoust. Soc. Am. 66, 1482-1486 (1979).

27. Chwieroth, F.S., Nagl, A., Uberall, H., Graves, R.D. and Zarur, G.L. "Mode coupling in a sound channel with range-dependent parabolic velocity profile", J. Acoust. Soc. Am. 64, 1105-1112 (1978).
28. Thompson, I.J. "Mixing of normal modes in a range-dependent model ocean", J. Acoust. Soc. Am. 69, 1280-1289 (1981).
29. McDaniel, S.T. "Comparison of coupled-mode theory with the small-waveheight approximation for sea-surface scattering", J. Acoust. Soc. Am. 70, 535-540 (1981).
30. Rutherford, S.R. and Hawker, K.E. "Consistent coupled mode theory of sound propagation for a class of nonseparable problems", J. Acoust. Soc. Am. 70, 554-564 (1981).
31. Evans, R.B. "A coupled mode solution for acoustic propagation in a waveguide with stepwise depth variation of a penetrable bottom," J. Acoust. Soc. Am. 74, 188-195 (1983).
32. Stickler, D.C. "Normal-mode program with both the discrete and branch line contributions", J. Acoust. Soc. Am. 57, 856-861 (1975).
33. Tappert, F.D. "The parabolic approximation method", in Wave Propagation and Underwater Acoustics, edited by J.B. Keller and J.S. Papadakis (Springer-Verlag, Berlin, 1977), pp. 224-287.
34. Jensen, F.B. and Krol, H.R. "The use of the parabolic equation method in sound propagation modelling". Rpt. SM-72, SACLANT ASW Research Centre, La Spezia, Italy, (1975).
35. Davis, J.A., White, D. and Cavanagh, R.C. "NORDA parabolic equation workshop". Rpt. TN-143, Naval Ocean Research and Development Activity, NSTL Station, MS (1982).
36. Lee, D. and Gilbert, K.E. "Recent progress in modeling bottom interacting sound propagation with parabolic equations", in Oceans 82 Conference Record (MTS-IEEE, Washington, D.C., 1982), pp. 172-177.
37. Lee, D., Botseas, G. and Papadakis, J.S. "Finite-difference solution to the parabolic wave equation", J. Acoust. Soc. Am. 70, 795-800 (1981).
38. Lee, D. and Botseas, G. "IFD: an implicit finite-difference computer model for solving the parabolic equation." Rpt. TR-6659, Naval Underwater Systems Center, New London, CT (1982).
39. Jensen, F.B. and Kuperman, W.A. "Consistency tests of acoustic propagation models". Rpt. SM-157, SACLANT ASW Research Centre, La Spezia, Italy (1982).
40. Jensen, F.B. "Sound propagation in shallow water: a detailed description of the acoustic field close to surface and bottom", J. Acoust. Soc. Am. 70, 1397-1406 (1981).
41. Ferla, M.C., Dreini, G., Jensen, F.B. and Kuperman, W.A. "Broadband model/data comparisons for acoustic propagation in coastal waters", in Bottom-Interacting Ocean Acoustics, edited by W.A. Kuperman and F.B. Jensen (Plenum, New York, 1980), pp. 577-592.

42. Jensen, F.B. and Kuperman, W.A. "Optimum frequency of propagation in shallow water environments", J. Acoust. Soc. Am. 73, 813-819 (1983).
43. Jensen, F.B. "Sound propagation over a seamount", in Proceedings of the 11th International Congress on Acoustics, (Paris, 1983).
44. Goos, F. and Hänchen, H. "A new and fundamental experiment on total reflection" (in German), Ann. Phys. (Leipzig) 1, 333-346 (1947).
45. Lotsch, H.K.V. "Beam displacement at total reflection: the Goos-Hänchen effect", Optik (Stuttgart) 32, 116-137 (1970); 32, 189-204 (1970); 32, 299-319 (1971); 32, 553-569 (1971)
46. Tamir, T. and Bertoni, H.L. "Lateral displacement of optical beams at multilayered and periodic structures", J. Opt. Soc. Am. 61, 1397-1413 (1971).
47. Bertoni, H.L. and Tamir, T. "Unified theory of Rayleigh-angle phenomena for acoustic beams at liquid-solid interfaces", Appl. Phys. 2, 157-172 (1973).
48. Shin, S.Y. and Felsen, L.B. "Lateral shifts of totally reflected gaussian beams", Radio Science 12, 551-564 (1977)
49. Ngoc, T.D.K. and Mayer, W.G. "Numerical integration method for reflected beam profiles near Rayleigh angle", J. Acoust. Soc. Am. 67, 1149-1152 (1980).
50. Neubauer, W.G. "Ultrasonic reflection of a bounded beam at Rayleigh and critical angles for a plane liquid-solid interface", J. Appl. Phys. 44, 48-55 (1973).
51. Breazeale, M.A., Adler, L. and Scott, G.W. "Interaction of ultrasonic waves incident at the Rayleigh angle onto a liquid-solid interface", J. Appl. Phys. 48, 530-537 (1977).
52. Pitts, L.E., Plona, T.J., and Mayer, W.G. "Theory of nonspecular reflection effects for an ultrasonic beam incident on a solid plate in a liquid", IEEE Trans. Sonics Ultrasonics SU-24, 101-109 (1977).
53. Claeys, J.M. and Leroy, O. "Reflection and transmission of bounded beams on half-spaces and through plates", J. Acoust. Soc. Am. 72, 585-590 (1982).
54. Muir, T.G., Horton, C.W. and Thompson, L.A. "The penetration of highly directional acoustic beams into sediments", J. Sound Vibr. 64, 539-551, (1979).
55. Tjøtta, J.N. and Tjøtta, S. . "Theoretical study of the penetration of highly directional acoustic beams into sediments, J. Acoust. Soc. Am. 69, 998-1008 (1981).
56. Coppens, A.B. and Sanders, J.V. "Propagation of sound from a fluid wedge into a fast fluid bottom", in Bottom-Interacting Ocean Acoustics, edited by W.A. Kuperman and F.B. Jensen (Plenum Press, New York, 1980), pp. 439-450.
57. Jensen, F.B. and Kuperman, W.A. "Sound propagation in a wedge-shaped ocean with a penetrable bottom", J. Acoust. Soc. Am. 67, 1564-1566 (1980).

58. Pierce, A.D. "Guided mode disappearance during upslope propagation in variable depth shallow water overlying a fluid bottom", J. Acoust. Soc. Am. 72, 523-531 (1982).
59. Arnold, J.M. and Felsen, L.B. "Rays and local modes in a wedge-shaped ocean", J. Acoust. Soc. Am. 73, 1105-1119 (1983).
60. Kamel, A. and Felsen, L.B. "Spectral theory of sound propagation in an ocean channel with weakly sloping bottom", J. Acoust. Soc. Am. 73, 1120-1130 (1983).

MODELLING OF PULSE PROPAGATION IN LAYERED MEDIA USING A NEW FAST FIELD PROGRAM

Henrik Schmidt

SACLANT ASW Research Centre
Viale San Bartolomeo 400
19026 La Spezia, Italy

ABSTRACT

Numerical modelling of acoustic pulse propagation in stratified solid media is often found to be impractical due to the comprehensive calculations involved. Here a new numerical model, of the fast field type, is presented. Instead, of using the Thomson-Haskell matrix method, as done in earlier models of the same type, the depth-separated wave equation is solved by a numerical technique very similar to that used in finite element programs. The speed improvement has been considerable, especially in cases with many sources and receivers. The model has been implemented on an FPS164 array processor and used for analysis of seismic pulse propagation in a shallow water environment and reflection of pulsed ultrasonic beams from a fluid-solid interface.

INTRODUCTION

A number of numerical models are available for investigating sound propagation in layered media. Each model is based on a set of assumptions and approximations and dedicated to specific applications. In relation to underwater acoustics four types of models are of interest, known as ray, normal-mode, parabolic equation, and fast field models. Jensen <1> has reviewed and classified these models and here we will concentrate on models of the last type, the fast field programs (FFP).

These models yield an exact solution to the depth-separated wave equation for horizontally stratified environments. Depending on the geometry (plane or cylindrical) the field is decomposed into plane or conical waves. The field corresponding to each horizontal wavenumber is then found by matching of boundary conditions at each interface, and the total field is found by superposition. In the case of cylindrical geometry the superposition is given by a Hankel transform integral. By replacing the Hankel function with its large argument approximation, this is changed to a Fourier integral, which after truncation can be evaluated by means of a fast Fourier technique. This approximation of the Hankel transform was originally introduced by Marsh <2> and is usually called the fast field technique.

The major part of the numerical effort is related to the solution of the depth-separated wave equation. DiNapoli <3> introduced an FFP-code that performed the solution very efficiently by means of recurrence relations for the hypergeometric functions. However, his approach allows only for fluid layers, and in such cases other models will often be more convenient, e.g. normal-mode models.

The main advantage of FFP-models are their ability to treat problems involving solid layers <1>. The first model to include this feature was introduced by Kutschale <4>. As shear properties are very important for sound propagation in shallow water <5>, this model has been used extensively for such problems during the last decade. Kutschale solved the depth-separated wave equation using a Green's-function approach based on the Thomson-Haskell matrix method. However, this method allows for only one source and receiver at a time, and even in such cases the computations are rather extensive, often the modelling of pulse propagation becomes impractical since this has to be carried out by means of Fourier synthesis involving many frequency components.

Here a new FFP-code, called SAFARI, is introduced. Instead of using the Thomson-Haskell method, a system of linear equations in the unknown potentials is set up and solved at each horizontal wavenumber. When the layer series is given, it is possible from the boundary conditions to determine *a priori* a mapping between the equations to be satisfied at each interface and a global set of equations. A significant number of wavenumber-independent expressions can then be evaluated once. The mapping technique is very similar to the one used in finite-element programs.

The source contributions are mapped into the righthand side of the global equations, and there is no theoretical limit for the number of sources.

The receiver depth is not included in the system of equations, and any number of receivers can therefore be treated with only one solution.

Even with one source/receiver combination, the calculation speed has been improved by up to one order of magnitude. This gain in speed and performance has made the FFP-technique a more realistic alternative to other numerical techniques, and in areas where this technique is the only possibility, the range of solvable problems has been increased considerably.

In the following the model and its mathematical background is described, and two examples of its use are given. First it is used to clarify some of the basic properties of the slow seismic interface waves that can be observed in shallow water environments. Synthetic seismograms are produced and compared qualitatively to experimental data. Then the model is used to determine the reflection of a pulsed beam from a water/aluminium-oxide interface.

1 THE MATHEMATICAL MODEL

The mathematical model is based on the assumption that the water column and the bottom consist of a series of range-independent layers. All materials are considered to be homogeneous and isotropic elastic continua with Lamé constants λ_n and μ_n and density ρ_n. The subscript refers to layer number n. The damping mechanisms are assumed to be linear viscoelastic.

A cylindrical coordinate system $\{r,\theta,z\}$ is introduced with the z-axis going through the source and being positive downwards (Fig. 1). The representation of the cylindrical displacement components $\{u,v,w\}$ in terms of scalar potentials and the subsequent expression of these as Hankel transforms closely follows the presentation given by Schmidt and Krenk <6>; hence, only an outline will be given here. If body forces are neglected, the displacement equation of motion will be satisfied if the displacement components in layer n are expressed in terms of three scalar potentials $\{\Phi_n,\Psi_n,\Lambda_n\}$ as

$$
\begin{aligned}
u\Big|_n &= \frac{\partial \Phi_n}{\partial r} + \frac{1}{r}\frac{\partial \Psi_n}{\partial \theta} + \frac{\partial^2 \Lambda_n}{\partial r \partial z} \\
v\Big|_n &= \frac{1}{r}\frac{\partial \Phi_n}{\partial \theta} - \frac{\partial \Psi_n}{\partial r} + \frac{1}{r}\frac{\partial^2 \Lambda_n}{\partial \theta \partial z} \qquad (1) \\
w\Big|_n &= \frac{\partial \phi_n}{\partial z} - \left(\frac{1}{r}\frac{\delta}{\partial r}\, r\, \frac{\delta}{\partial r} + \frac{1}{r^2}\frac{\partial^2}{\partial \theta^2}\right) \Lambda_n ,
\end{aligned}
$$

where the potentials satisfy the wave equations:

$$\left(\nabla^2 - \frac{1}{C_{Ln}^2}\frac{\partial^2}{\partial t^2}\right)\Phi_n = 0 \tag{2}$$

$$\left(\nabla^2 - \frac{1}{C_{Tn}^2}\frac{\partial^2}{\partial t^2}\right)\left(\Psi_n, \Lambda_n\right) = 0, \tag{3}$$

in which C_L and C_T are the velocities of the compressional and shear waves, respectively:

$$C_{Ln}^2 = \frac{\lambda_n + 2\mu_n}{\rho_n} \tag{4}$$

$$C_{Tn}^2 = \frac{\mu_n}{\rho_n} . \tag{5}$$

In the present case the field is axisymmetric due to the positioning of the source on the axis, and the angular displacement v vanishes everywhere. It is then clear from Eq. 1 that the potential Ψ_n must be constant and can be excluded.

In the following, only vibrations with angular frequency ω will be considered; displacements, stresses and potentials can then be expressed in complex form with the common factor $\exp(i\omega t)$. This factor will not be included in the following. The viscoelastic damping can now be accounted for by allowing the Lamé constant to be complex. After use of the Hankel transform on the wave equations, the following integral representations are obtained for the potentials:

$$\Phi_n(r,z) = \int_0^\infty \left\{A_n^-(s)e^{-z\alpha_n(s)} + A_n^+(s)e^{z\alpha_n(s)}\right\} J_o(rs)s\, ds \tag{6}$$

$$\Lambda_n(r,z) = \int_0^\infty \left\{B_n^-(s)e^{-z\beta_n(s)} + B_n^+(s)e^{z\beta_n(s)}\right\} J_o(rs)\; ds, \tag{7}$$

where

J_m is the Bessel function of the first kind and order m,

A_n^-, A_n^+, B_n^- and B_n^+ are arbitrary functions in the horizontal wavenumber S.

$\alpha_n(s)$, $\beta_n(s)$ are defined as

$$\alpha_n(s) = \begin{cases} \sqrt{(s^2-h_n^2)}, & s > \mathrm{Re}\{h_n\} \\ i\sqrt{(h_n^2-s^2)}, & s \leq \mathrm{Re}\{h_n\} \end{cases} \tag{8}$$

$$\beta_n(s) = \begin{cases} \sqrt{(s^2-k_n^2)}, & s > \mathrm{Re}\{k_n\} \\ i\sqrt{(k_n^2-s^2)}, & s \leq \mathrm{Re}\{k_n\}. \end{cases} \tag{9}$$

The wavenumbers h_n and k_n for the compressional and shear waves, respectively, are defined by

$$h_n^2 = \left(\frac{\omega}{C_{Ln}}\right)^2 = \frac{\omega^2\rho_n}{\lambda_n+2\mu_n} \tag{10}$$

$$k_n^2 = \left(\frac{\omega}{C_{Tn}}\right)^2 = \frac{\omega^2\rho_n}{\mu_n} \tag{11}$$

If Eqs. 6 and 7 are inserted into Eq. 1, the following expressions are obtained for the particle displacements:

$$w(r,z)\Big|_n = \int_0^\infty \{-\alpha_n A_n^- e^{-z\alpha_n} + \alpha_n A_n^+ e^{z\alpha_n} + sB_n^- e^{-z\beta_n} + s B_n^+ e^{z\beta_n}\} \, s J_0(rs)ds, \tag{12}$$

$$u(r,z)\Big|_n = \int_0^\infty \{-s A_n^- e^{-z\alpha_n} - s A_n^+ e^{z\alpha_n} + \beta_n B_n^- e^{-z\beta_n} - \beta_n B_n^+ e^{z\beta_n}\} \, s J_1(rs)ds. \tag{13}$$

The stress components involved in the boundary conditions follow from Hooke's law:

$$\sigma_{zz}(r,z)\Big|_n = \mu_n \int_0^\infty \{(2s^2-k^2)(A_n^- e^{-z\alpha_n} + A_n^+ e^{z\alpha_n}) + 2s\beta_n(-B_n^- e^{-z\beta_n} + B_n^+ e^{z\beta_n})\} \, s J_0(rs)ds, \tag{14}$$

$$\sigma_{rz}(r,z)\Big|_n = \mu_n \int_0^\infty \{2s\alpha_n(A_n^- e^{-z\alpha_n} - A_n^+ e^{z\alpha_n})$$

$$-(2s^2-k^2)(B_n^- e^{-z\beta_n} + B_n^+ e^{z\beta_n})\} \; s\, J_1(rs)\, ds. \tag{15}$$

In the case of a fluid layer the shear stiffness μ_n vanishes, and only the potential Φ_n is involved. The displacements follow directly from Eqs. 12 and 13 by setting B^- and B^+ to zero. The shear stress is identically zero, whereas Eq. 14 has to be replaced by

$$\sigma_{zz}(r,z)\Big|_n = -\lambda_n h^2 \int_0^\infty \{A_n^- e^{-z\alpha_n} + A_n^+ e^{z\alpha_n}\} \; s\, J_o(rs)ds. \tag{16}$$

The source is assumed to be in layer number m at depth z_s. In the absence of boundaries, the field produced in layer m would be <7>:

$$\Phi_s(r,z) = \frac{iS_\omega}{4\pi} \int_0^\infty \frac{e^{-|z-z_s|\alpha_m}}{\alpha_m} \; s\, J_o(rs)ds \; , \tag{17}$$

$$\Lambda_s(r,z) \equiv 0 \; , \tag{18}$$

where S_ω is the source strength. If Eq. 17 is inserted into Eq. 1, expressions similar to Eqs. 12 and 13 are obtained for the displacements, and again Hooke's law yields expressions like Eqs. 14 and 15 for the stresses involved in the boundary conditions.

For each value of the range, r, the boundary conditions must be satisfied. In the upper and lower half-spaces the arbitrary functions, with superscript - and + respectively, must vanish due to the radiation condition. At each interface, w and σ_{zz} must be continuous, and at all solid/liquid interfaces the shear stresses must vanish. At solid/solid interfaces, w, u, σ_{zz} and σ_{rz} must be continuous. This yields a linear system of equations in the arbitrary functions, to be satisfied at each horizontal wavenumber:

$$C_{ij}(s) \cdot A_j(s) = R_i(s). \tag{19}$$

The vector $A_j(s)$ contains all the non-vanishing arbitrary functions, $C_{ij}(s)$ is the coefficient matrix, and $R_i(s)$ contains the contributions from the source. When the arbitrary functions are found, the field parameters at any depth and range can be obtained from the Hankel transforms (Eqs. 12 to 16) plus the source contributions (if the source and receiver are in the same layer).

An analytical solution of Eq. 19 is of course possible, leading to closed-form expressions for the arbitrary functions; but for more than a few layers this procedure would be inconvenient. Further, the Hankel transforms do not lead to closed-form solutions, but need numerical evaluation. Thus the most general way to proceed is to create a numerical model based directly on the system of equations (Eq. 19). Such a model is described in the next chapter.

2 THE NUMERICAL MODEL SAFARI

The numerical evaluation of the Hankel transform necessitates a truncation and a discretization in the horizontal wavenumber s. As can be observed from Eqs. 8 and 17, the source terms decay exponentially for s going towards infinity. As the source terms form the righthand side of Eq. 19 the arbitrary functions will behave in the same way. It is therefore possible to truncate the integration interval in accordance with any accuracy demands. The fast-field technique introduced by Marsh <7> can then be used to evaluate the Hankel transforms.

The Bessel functions are expressed in terms of Hankel functions

$$J_m(rs) = \tfrac{1}{2}\left(H_m^{(1)}(rs) + H_m^{(2)}(rs)\right) \tag{20}$$

and each integral is split into two. As only outgoing waves are considered, the integrals involving $H_m^{(1)}(rs)$ are neglected, and $H_m^{(2)}(rs)$ is replaced by its asymptotic form

$$H_m^{(2)}(rs) \underset{rs\to\infty}{\sim} \sqrt{\left(\frac{2}{\pi rs}\right)}\; e^{-i\left[rs-\left(m+\frac{1}{2}\right)\frac{\pi}{2}\right]} \tag{21}$$

The integration over the truncated interval can now be performed by means of the fast Fourier transform, and the actual field parameter is found at a number of ranges equal to the number of discrete wavenumbers considered.

DiNapoli and Deavenport <3> have compared Marsh's method to the technique introduced by Tsang, Brown, Kiang and Simmons <8>, which does not use the asymptotic form of the Hankel function. Significant differences were found only for very short ranges.

The kernels in the Hankel transforms are now needed only for a limited number of discrete values of s. Kutschale <4> used a Green's-function approach based on the Thomson-Haskell matrix method. However, his approach allows for only one source/receiver combination at a time.

In the SAFARI model Eq. 19 is set up and solved directly. When the layer series is given, it is possible to determine *a priori* a mapping between the equations to be satisfied at each interface and the global system of equations, Eq. 19. The unknown arbitrary functions are mapped into the vector A_j, the coefficients into the matrix C_{ij}, and the source contributions into the right side R_j of Eq. 19. If more than one source is present, the contributions are simply added.

The receiver depth is not included in Eq. 19, thus yielding the possibility of determining the field at any depth from Eqs. 12 to 16 with only one solution of Eq. 19.

The mapping between the local and the global system of equations is very similar to the technique used in finite element programs. By using this technique the computer code can become very efficient since a significant number of calculations can be done only once. Furthermore, the code will be straightforward to vectorize, thus making it well suited for implementation on an array processor.

The solution of Eq. 19 is performed by means of gaussian elimination with partial pivoting.

In cases with only one source/receiver combination, SAFARI is found to be an order of magnitude faster than Kutschale's model. In cases with several sources or receivers the speed gain is of course much bigger.

This surprisingly high speed gain is believed to be due to the mapping technique, which partly ensures that a large amount of unnecessary calculation is avoided, and partly makes efficient programming possible.

The implementation of SAFARI on an FPS164 attached processor has given at least one order of magnitude further, and a number of problems, which were impractical to treat numerically, can now be solved with an acceptable amount of computation time. A couple of examples are given in the following.

3 MODELLING OF SEISMIC INTERFACE WAVES

The importance of the shear properties of the sea-bed for acoustic wave propagation in shallow water is well established <5>. Unfortunately the shear parameters are very difficult to isolate experimentally. The shear parameters are, however, indirectly pre-

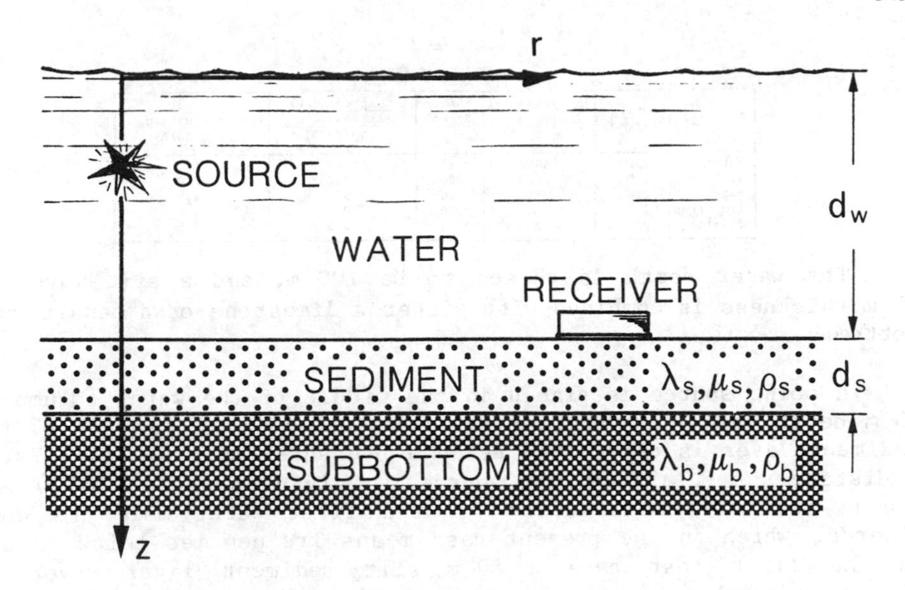

Fig. 1 Stratified shallow water environment

sent through the properties of the measurable seismic interface waves, and much effort has been put into an experimental investigation of these <9>, <10>.

Since no applicable inverse models are available the shear properties are determined by "trial-and-error" – methods using numerical propagation models <10>, <11>. Usually several parameters are unknown and, since the calculations needed for each combination are rather comprehensive, the determination of the shear parameters in this way can become very expensive in terms of calculation time. It is therefore important a priori to be able to determine approximate values directly from the experimental data.

With this in view, the SAFARI model has been used to clarify some of the propagation characteristics of the seismic interface waves for different shallow water environments. A detailed description of the investigation is given in <12>, and only a couple of examples will be given here.

In order not to obscure the basic principles, a simple 2-layered model was chosen for the sea-bed, Fig. 1. Below the water column of depth d_w, a single sediment layer of thickness d_s covers a half-space of rock or rock-like material. The bottom materials used in the examples and their assumed properties are listed in Table 1. The complex Lamé constants are not given explicitly in the table. Instead the compressional and shear velocities are shown, together with their respective dampings in dB per wavelength.

TABLE 1

MATERIAL PROPERTIES

Material	Density $\rho(g/cm^3)$	Compressional Speed $C_L(m/s)$	Shear Speed $C_T(m/s)$	Compressional attenuation $\gamma_L(dB/\Lambda)$	Shear attenuation $\gamma_T(dB/\Lambda)$
Water	1.0	1500	-	-	-
Silt	1.8	1600	200	1.0	2.0
Sand	2.0	1800	600	0.7	1.5
Limestone	2.2	2250	1000	0.4	1.0
Basalt	2.6	5250	2500	0.2	0.5

The water depth is chosen to be 100 m, and a silt layer of 50 m thickness is combined with either a limestone or a basalt sub-bottom.

A point source is placed in the middle of the water column at 50 m depth, and the vertical particle velocity at the top of the sediment layer is calculated using a pressure amplitude of 1 Pa at a distance of 1 m from the source. Only frequencies below and around the cut-off frequency for the first water mode are considered, which in the present case means frequencies below 10 Hz. In the first test case a 50 m silt sediment layer covers a limestone sub-bottom. Figure 2 shows the modulus of the integrand in the Hankel transform of the vertical velocity at the top of the sediment layer for a frequency of 1.5 Hz. No water modes are present at this frequency, but two mode-like peaks can be observed at wavenumbers corresponding to phase velocities of 300 m/s (peak '1') and 840m/s (peak '2'). These peaks correspond to interface waves, and will be denoted the first and second interface mode, respectively. The first interface mode is best excited (highest amplitude), but it also has the highest damping (widest peak).

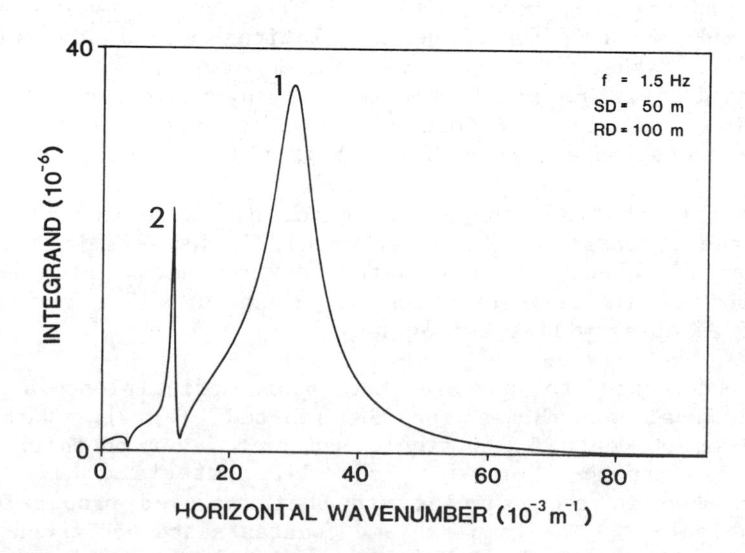

Fig. 2 Hankel transform of vertical particle velocity at 1.5 Hz. 50 m silt on limestone.

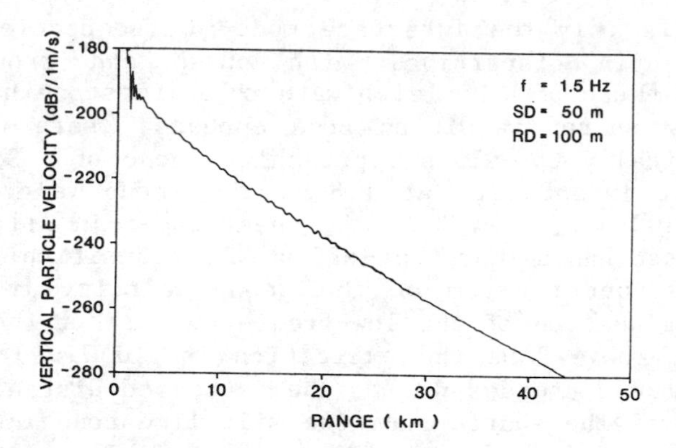

Fig. 3 Vertical particle velocity as function of range at 1.5 Hz. 50 m silt on limestone.

Figure 3 shows the corresponding vertical particle velocities at ranges up to 50 km. Due to the high damping of the first interface mode, its contribution is significant only for ranges shorter than 2 km, beyond which the second interface mode becomes dominant.

If the source is not of stationary type, but transient, the different velocities of the two interface modes will yield different arrival times, and the presence of the first interface mode will be measurable also at larger ranges. The phase and group velocities of the two interface modes have been calculated in the frequency band of interest, 0.1 to 10 Hz, with a resolution of 0.1 Hz. The results are shown in Fig. 4 together with an excitation measure, which somewhat arbitrarily has been chosen to be the particle velocity at 10 km range.

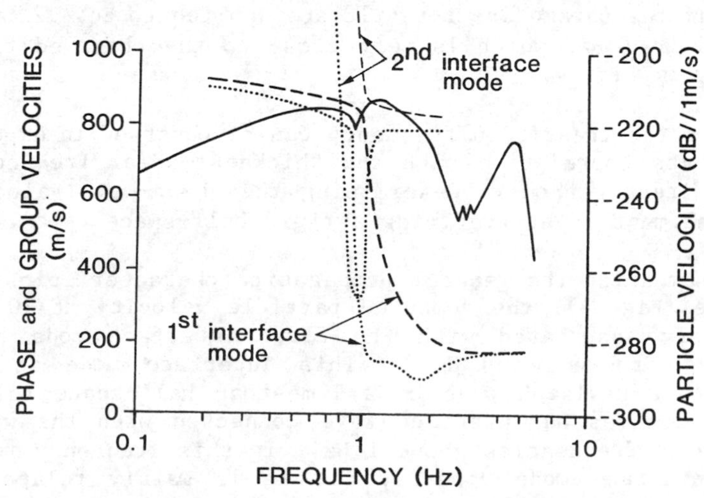

Fig. 4 Excitation and dispersion curves. 50 m silt on limestone.

There is only one interface mode at frequencies below 1 Hz. It is slightly dispersive, with phase and group velocities approaching those of a Rayleigh wave on a limestone half-space. At 1 Hz a very sharp transition zone appears, where the velocities drop dramatically to values approaching those of a Scholte wave at a water/silt interface. At 1.8 Hz the group velocity reaches a minimum of 100 m/s, i.e. half the shear speed in silt. The second interface mode has a sharp cut-off at the transition frequency, and after a distinct minimum of the group velocity it appears as a logical continuation of the low-frequency part of the first interface mode. Above 2 Hz the excitation (solid line) of the second interface mode decreases due to the increased distance in terms of wavelengths of the source from the silt/limestone interface, along which the second interface mode propagates. The sharp peak on the excitation curve around 4.5 Hz is the first propagating water mode.

The existence of mode transition zones is well known from the theory of vibration of elastic plates, Mindlin <13>, where they appear near the thickness-shear frequencies. These are the frequencies at which an infinite elastic plate can perform free shear vibrations with vanishing vertical displacements. Now consider the silt layer as an infinite elastic plate in welded contact with an infinite rigid half-space. The first thickness-shear frequency would then correspond to that of a free silt plate of the double thickness, Mindlin <13>:

$$f_{TS} = \frac{C_T}{4d_S} \qquad (22)$$

where C_T is the shear velocity and d_S is the thickness of the layer. If the parameters for silt are applied to Eq. 22, we obtain f_{TS} equal to 1 Hz, which is very close to the observed transition frequency in Fig. 4.

Since the transition frequency can be observed in experimental results, its correlation with the thickness-shear frequency could yield a direct method of determining the shear-wave velocity in a single sediment layer overlying a rigid half-space.

To summarize the general propagation characteristics for this test case (Fig. 4), the computed particle velocity at 10 km range is entirely associated with the first interface mode below the transition frequency (1 Hz). This interface mode is strongly related to a Rayleigh wave on a limestone half-space below 1 Hz, while it becomes an interface wave connected with the water/silt interface at frequencies above 1 Hz. In this frequency regime the second interface mode appears, and it is mainly related to the silt/limestone interface. Finally, the first water-borne mode appears at around 4.5 Hz.

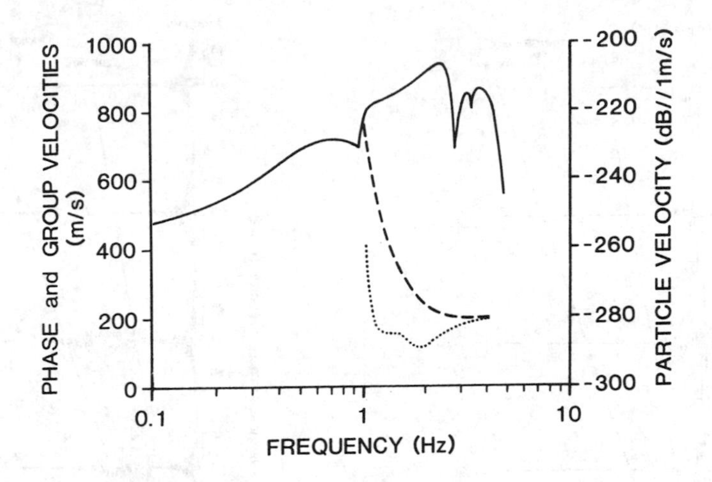

Fig. 5 Excitation and dispersion curves. 50 m silt on basalt.

A silt layer on basalt was also studied in order to analyze the effect of different sub-bottom materials. In this case (Fig. 5) the low-frequency part of the first interface mode is excited only slightly, but the transition zone can again be observed near the thickness-shear frequency of 1 Hz. Above 1 Hz the dispersion curves are very similar to those obtained earlier for a sub-bottom of limestone (Fig. 4).

Synthetic seismograms have been produced for the two test cases in order to illustrate the time-domain effect of the features described above. The source is assumed to be half a sine wave of 1.5 Hz sent through an ideal 0.5 to 3.2 Hz band-pass filter. This frequency range has been chosen as it contains frequencies on both sides of the transition frequency for a silt-sediment layer of 50 m thickness. The transfer functions were calculated to a resolution of 0.01 Hz and multiplied by the spectrum of the source. The time series were then created by means of the fast-Fourier transform at ranges of 1, 2, 3, 4 and 5 km.

Figure 6 shows the result for test case 1, i.e. a 50 m silt layer on a limestone half-space. The first weak arrival corresponds to a compressional wave in the limestone, whereas the first significant arrival corresponds to the Rayleigh wave velocity of the limestone (900 m/s). This arrival consists of the low-frequency parts of the first and second interface modes. A clear dispersion can be observed corresponding to the distinct minimum in the group velocity of the second interface mode in Fig. 4. The slow highly damped wave corresponding to the first interface mode above the transition frequency propagates with group velocities between 100 and 180 m/s, again in good agreement with Fig. 4. At ranges greater than 4 km this arrival has negligible amplitude.

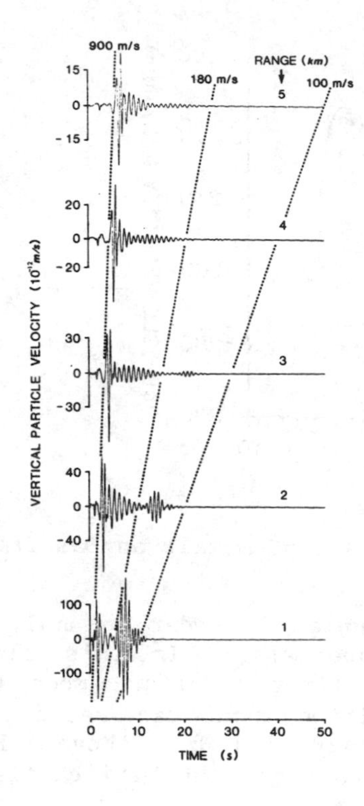

Fig. 6
Stacked synthetic seismograms.
50 m silt on limestone.

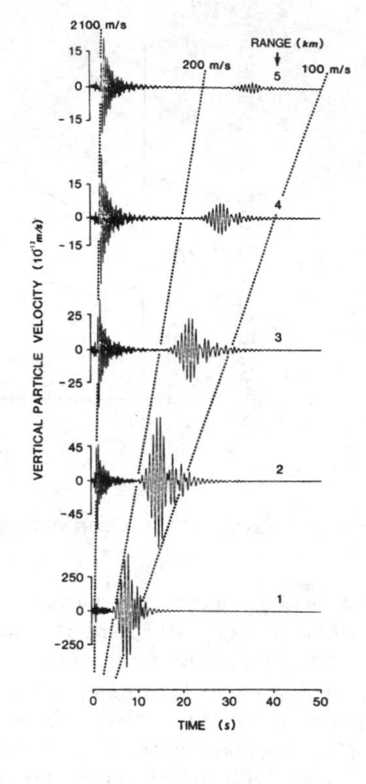

Fig. 7
Stacked synthetic seismograms.
50 m silt on basalt.

As described above, the low-frequency part of the first interface mode will not be significantly excited if the sub-bottom is basalt. The maximum excitation at 10 km range lies at 2.5 Hz (Fig. 5) and is due to a strong excitation of the second mode. These properties are also reflected in the synthetic seismograms (Fig. 7). The fastest arrival, apart from the weak compressional wave, has its major frequency content above 2 Hz and corresponds mainly to the high-frequency end of the second interface mode. The severe dispersion of the corresponding arrival in Fig. 6 is not present here, and the slow part of the first interface mode is well separated, travelling at group velocities between 100 and 200 m/s. The synthetic seismograms in Figs. 6 and 7 are, at least qualitatively, very similar to those observed during experiments, Fig. 8 <15>.

With 4096 sample points in the wavenumber space the calculation time on an FPS164 array processor, was 9 secnds for each frequency or 40 minutes in total for the synthetic seismograms in Figs. 6 or 7.

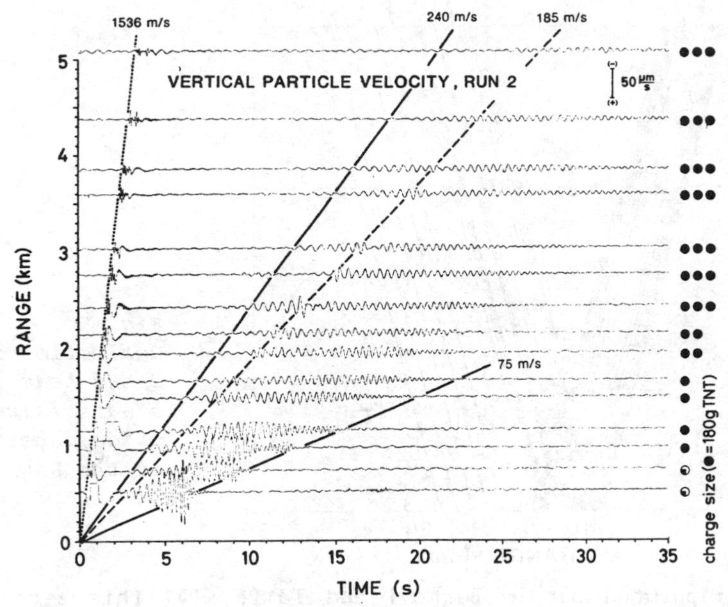

Fig. 8 Stacked experimental seismograms, <15>.

4 REFLECTION OF A PULSED BEAM AT A FLUID/SOLID INTERFACE

As mentioned above, the solution technique used in SAFARI yields the possibility of treating problems in which many sources and receivers are involved.

Jensen <1>, used the model to treat the problem of reflection of narrow ultrasonic beams at a water/aluminium-oxide interface at grazing angles near the Rayleigh angle, i.e. the angle where a leaky interface wave is excited. A monochromatic analysis was performed in order to clarify the influence of beam divergence on the reflection pattern, and excellent agreement was obtained with both experimental and theoretical results reported in the literature. In addition, some apparent discrepancies between experimental and theoretical results were resolved.

One of the results from <1> is shown in Fig. 9. A parallel gaussian beam is generated by a linear vertical source-array of 649 elements, spaced half a wavelength apart. The mid-point of the array is placed 40 cm above the interface, the frequency is 2 MHz, and the grazing angle of the beam is 75.2°, corresponding to the Rayleigh angle for water/aluminium-oxide. The splitting of the reflected beam is easily observed. The specularly reflected beam and the beam caused by the leaky interface wave are separated by a strip of low intensity.

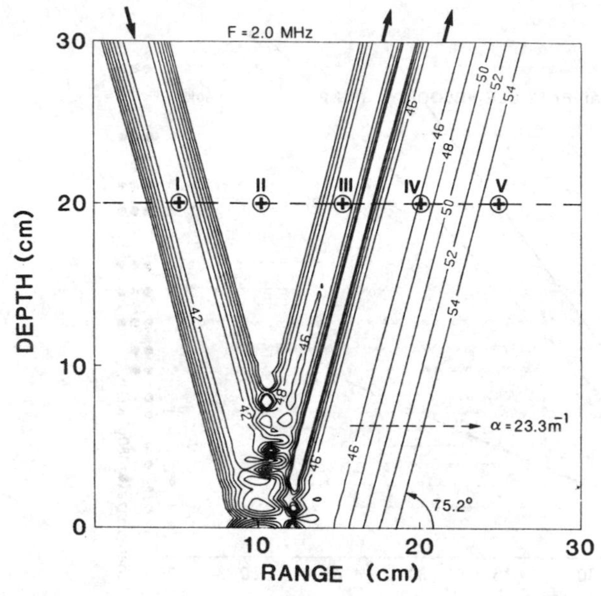

Fig. 9
Reflection of a 2 MHz beam from a water/aluminum-oxide interface at the Rayleigh angle.

As pointed out by Bertoni and Tamir <14> this zero-strip is due to destructive interference between the two parts of the reflected field. If the surfaces of equal phase are plane and perpendicular to the direction of propagation, which is the case for parallel beams, the two reflected beams thus have to be 180° out of phase.

This feature can of course be demonstrated by making a cut perpendicular to the propagation direction, but if the continuous beam is replaced by a pulsed beam, the phase shift will yield an amplitude inversion of the pulse in the two beams, independent of the position of the receivers within the beams. Furthermore, a pulse calculation will yield information on arrival times for the pulse at different receivers.

To demonstrate this the pulse version of SAFARI was used to calculate the received signals at five different receivers, all situated 20 cm above the interface as indicated in Fig. 9.

As pulse calculations in SAFARI are performed by means of discrete Fourier synthesis, it is very important to choose the time window correctly in order to avoid time-domain aliasing.

The approximate arrival times were therefore determined using a narrow-banded source pulse (1.8 to 2.2 MHz). The time window was chosen to be 0 to 1 ms, yielding frequency steps of 1 kHz. The calculated time series for the five receivers are shown in Fig. 10. The vertical scale is arbitrary, but is the same for all five receivers.

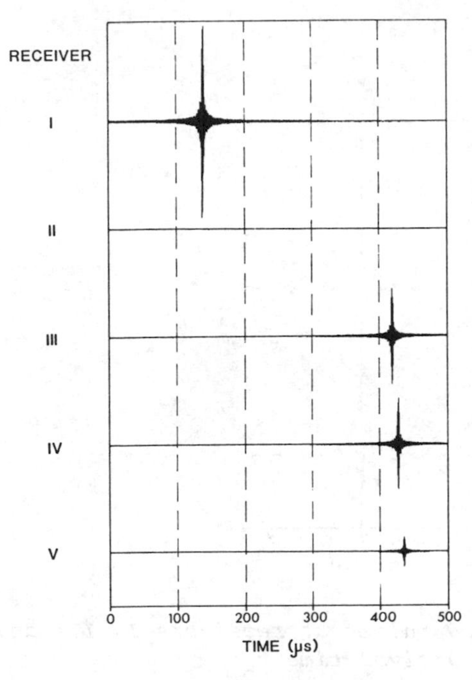

Fig. 10
Received narrow-band pulses at receivers I-V.

The pulse arrives at the reference receiver (I) at t ~ 140 μs. As could be expected from Fig. 9, no significant arrivals appear at receiver II, whereas the reflected pulses arrive at the three last receivers at times between 410 and 440 μs approximately.

A closer analysis is now possible using time windows of 125 μs width. In order to obtain short pulses, the bandwidth is increased to 1.5 MHz to 3.0 MHz. The results are shown for receiver I, III and IV in Fig. 11. The time scale shows the time relative to the expected arrival times. These are determined by geometrical means. A plane wave of grazing angle 75.2° that passes the reference receiver I at t = 139.30 μs will be reflected and pass receiver III at t = 416.00 μs. The calculated pulse at this receiver is seen to correspond exactly to this behaviour; the specularly reflected beam has no phase shift. The loss in amplitude is due to the fact that some of the energy is transferred into the leaky Rayleigh wave. The expected arrival at receiver IV is determined in a slightly different way. A ray at a grazing angle of 75.2° is assumed to pass the reference receiver at t = 139.30 μs. When it reaches the interface it travels at the speed of the Rayleigh wave (5825.55 m/s) along the interface, and finally it travels upwards at angle of 75.2° to receiver IV. The arrival time determined by these assumptions is 424.60 μs, again in excellent agreement with the numerical result in Fig. 11. The 180° phase shift is easily observed, thus verifying the explanation for the zero-strip given in <14>.

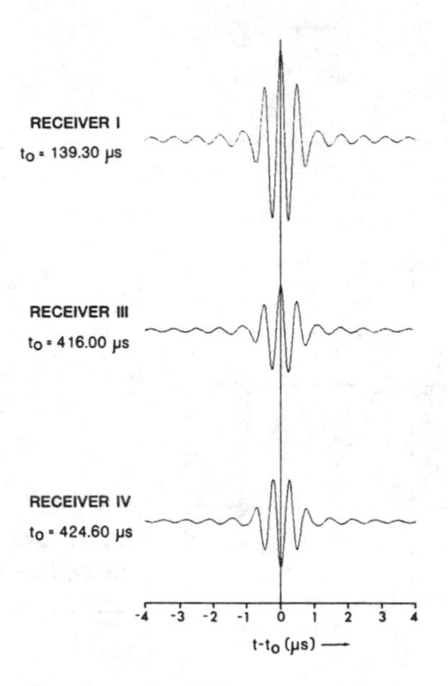

Fig. 11 Arrival of broad-band pulses at receivers I, III and IV related to expected arrival time t_o.

With 512 sampling points in the wavenumber space, the calculation time on the FPS164 was 8 seconds for each frequency, in total 40 minutes for Figs. 10 and 25 minutes for Fig. 11.

CONCLUSIONS

A new fast field program, SAFARI, has been developed. The program is meant for modelling sound and stress wave propagation in horizontally stratified fluid and solid media. By introducing a more efficient solution technique, the computational speed has been improved by an order of magnitude, in some cases even more, compared with earlier models of the same type. Furthermore, the SAFARI model is capable of treating several sources and receivers with only one solution of the wave equation, thus making it feasible to treat problems, in which the field is generated by vertical source arrays.

The model is basically monochromatic, but the increase in speed, together with the fact that the model is well suited for implementation on an array processor, has yielded the possibility of treating pulse propagation by means of Fourier synthesis with limited demands on computation time. This has been demonstrated by a couple of examples, but the range of problems that could be treated is of course much wider.

REFERENCES

1. Jensen, F.B. Numerical models in underwater acoustics (These proceedings).
2. Marsh, H.W., Schulkin, M. and Kneale, S.G. Scattering of underwater sound by a sea surface. Jnl Acoustical Society of America 33, 1961: 334-340.
3. DiNapoli, F.R. and Deavenport, R.L. Theoretical and numerical Green's function field solution in a plane multilayered medium. Jnl Acoustical Society of America 67, 1980: 92-105.
4. Kutschale, H.W. Rapid computations by wave theory of propagation loss in the Arctic Ocean, Rpt CU-8-73. Palisades, NY, Columbia University, 1973.
5. Akal, T. and Jensen, F.B. Effects of the sea-bed on acoustic propagation. In: PACE, N.G. ed. Acoustics and the Sea Bed, Proceedings of an Institute of Acoustics Underwater Acoustics Group Conference, Bath, U.K., University Press, 1983: pp 225-232.
6. Schmidt, H. and Krenk, S. Asymmetric vibrations of a circular plate on an elastic half-space. International Journal of Solids and Structures 18, 1982: 91-105.
7. Ewing, W.M., Jardetzky, W.C. and Press, F. Elastic Waves in Layered Media. New York, NY, McGraw-Hill, 1957.
8. Tsang, L., Brown, R., Kong, J.A., and Simmons, G. Numerical evaluation of electromagnetic fields due to dipole antennas in the presence of stratified media. Jnl Geophysical Research 79, 1974: 2077-2080.
9. Rauch, D. and Schmalfeldt, B. Ocean-bottom interface waves of the Stoneley/Scholte type: properties, observations and possible use. In: PACE, N.G. ed. Acoustics and the Sea Bed, Proceedings of an Institute of Acoustics Underwater Acoustics Group Conference, Bath, U.K., University Press, 1983: pp 307-316.
10. Holt, R.M., Hovem, J.M. and Syrstad, V. Shear modulus profiling of near-bottom sediments using boundary waves. In: PACE, N.G. ed. Acoustics and the Sea Bed, Proceedings of an Institute of Acoustics Underwater Acoustics Group Conference, Bath, U.K., University Press, 1983: 317-325.
11. Essen, H.-H. Model computations for low-frequency surface waves on marine sediments. In: KUPERMAN, W.A. and JENSEN, F.B. eds. Bottom-interacting Ocean Acoustics. Proceedings of a conference held at SACLANTCEN, La Spezia, Italy, 9-12 June 1980. New York, NY, Plenum, 1980: pp 113-118.
12. Schmidt, H. Excitation and propagation of interface waves in a stratified sea-bed. In: PACE, N.G. ed. Acoustics and the Sea Bed, Proceedings of an Institute of Acoustics Underwater Acoustics Group Conference, Bath, U.K., University Press, 1983: 327-334.

13. Mindlin, R.D. An Introduction to the Mathematical Theory of Vibration of Elastic Plates. Fort Monmouth, NY, US Army Signal Corps Engineering Laboratories, 1955.
14. Bertoni, H.L. and Tamir, T. Unified theory of Rayleigh-angle phenomena for acoustic beams at liquid-solid interfaces. Applied Physics 2, 1973: 157-172.
15. Barbagelata, A., Michelozzi, E., Rauch, D., & Schmalfeldt, B. Seismic sensing of extremely-low-frequency sounds in coastal waters. In: ICASSP 82, Proceedings of the IEEE International Conference on Acoustics, Speech and Signal Processing, Paris, France, 3-5 May 1982. Piscataway, NJ, IEEE Service Center, 1982: pp 1878-1881.

CALCULATION OF SYNTHETIC SEISMOGRAMS IN LAYERED MEDIA BY MEANS OF A MODIFIED PROPAGATOR MATRIX METHOD

Ghislain R. Franssens

University of Gent, Lab. for Electromagnetism and Acoustics, Sint Pietersnieuwstraat 41, B-9000 Gent, Belgium.

The calculation of synthetic seismograms in a staircase stratified elastic medium is investigated. A plane wave spectral representation is used for the displacement-stress field. The spectral field is computed with the aid of a hybrid formulation in terms of the propagator matrices and the matrices of second order minors of the propagator matrices (compound propagator matrices). This modified propagator matrix method avoids completely the numerical problems inherent in the classical Thomson-Haskell formulation when calculating P-SV waves in thick layered structures. Due to its matricial structure, the resulting algorithm is very well suited to code on a computer. In this paper a number of seismograms have been calculated for geometries encountered in seismic surveys of coal layers.

1 INTRODUCTION

We consider an elastic space consisting of a stack of n plane layers terminated on both ends by a semi infinite space, as shown in Fig. 1. Each material is assumed to be isotropic, homogeneous, loss free and perfectly elastic and can therefore be described by its volume density ρ and its shear and longitudinal velocities v_S and v_P. Hence the elastic parameters vary stepwize and in only one direction (z-axis). We address the problem of computing synthetic seismograms in this medium, generated by a line source parallel to the y-axis. The chosen geometry then yields a two-dimensional problem in the x,z-plane. In each material the wave motion is described by the linearized equations of motion, given in summation notation

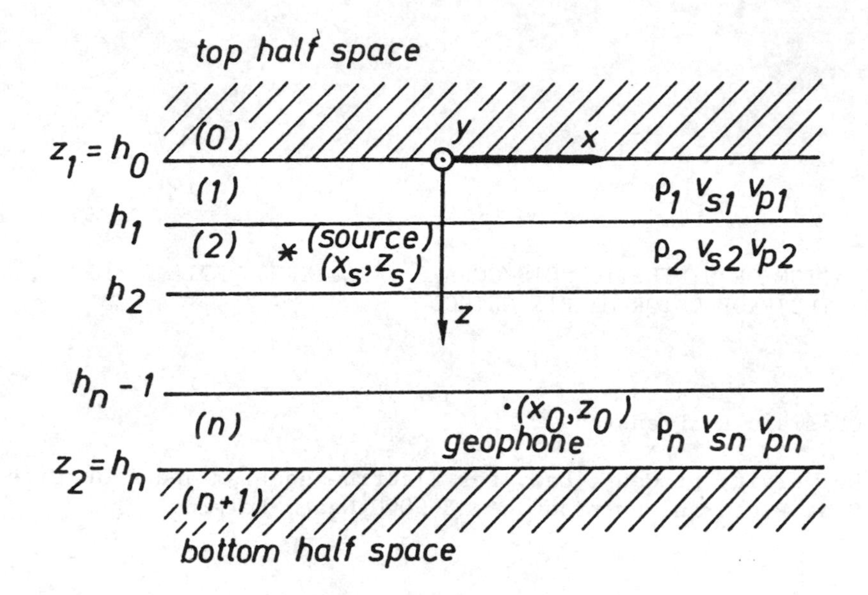

Fig. 1 The investigated stratification

by (1) :

$$\partial_j\, T_{ij} \;-\; \rho\partial_t^2\, u_i \;=\; s_i(t)\ \delta(x-x_s)\ \delta(z-z_s) \tag{1}$$

together with the constitutive equations :

$$T_{ij} = \frac{1}{2}\, C_{ijpq}\ (\partial_q u_p + \partial_p u_q)\ ,\quad (i,j,p,q=1,3), \tag{2}$$

Herein are :

u_i = the components of the displacement vector,

T_{ij} = the components of the stress tensor,

∂_j = the derivative with respect to the j-th axis,

∂_t^2 = the second derivative with respect to time,

$s_i(t)$ = the line source time functions,

C_{ijpq} = the fourth order stiffness tensor.

Due to the two-dimensionality of the problem and the simple form of the stiffness tensor for isotropic media, these equations split into two sets describing two uncoupled wave problems.
It is convenient to define the following two displacement-stress vectors, containing the displacements and the tractions normal to

the interfaces :

a) for P-SV waves :

$$B(x,z,t) = \left[u_1\ u_3\ T_{13}\ T_{33}\right]^T , \tag{3a}$$

b) for SH waves :

$$B(x,z,t) = \left[u_2\ T_{23}\right]^T . \tag{3b}$$

The scalar wave, with polarization along the y-axis, is called the Shear Horizontal wave type and the vectorial wave, with elliptic polarization in the x,z-plane, is called the comPressional Shear Vertical wave type. Both wave types can be handled separately. For media in welded contact the continuity of these vectors across the layer interfaces is assumed.
Let $\hat{B}(k,z,f)$ denote the two-dimensional Fourier transform of $B(x,z,t)$ so that :

$$B(x,z,t) = \frac{1}{2\pi} \int_{-\infty}^{+\infty} \exp(-j2\pi ft) \int_{-\infty}^{+\infty} \hat{B}(k,z,f)\exp(+jk(x-x_s))dkdf, \tag{4}$$

where f stands for the frequency and k for the wavenumber along the x-axis. The physical meaning of this representation is a decomposition of the displacement-stress field in an angular spectrum of plane waves. Such a representation is valid under very general conditions (2). Combining the equations (1)-(4) yields two uncoupled sets of first order ordinairy differential equations in z of the form :

$$\partial_z \hat{B}(k,z,f) = A(k,z,f)\hat{B}(k,z,f) + \hat{S}(f)\delta(z-z_s) . \tag{5}$$

The complex matrix A can be regarded as the system matrix and the source vector $\hat{S}$ contains the amplitude frequency spectra of the source time functions. A is a 2x2 matrix for SH waves and a 4x4 matrix for P-SV waves.
The boundary conditions are given by :

$$\lim_{z \to h_l^<} \hat{B} = \lim_{z \to h_l^>} \hat{B} , \qquad (l=0,1,\ldots n) \tag{6}$$

together with the radiation condition for $z \to \pm\infty$.

Before we continue to solve Eq. (5), let us first take a look at a summarized classification of a number of well known methods to compute synthetic seismograms, which make use of this double Fourier transform approach (Fig. 2).
A first distinction can be made depending on the order of integration while calculating Eq. (4). If the wavenumber integral is computed

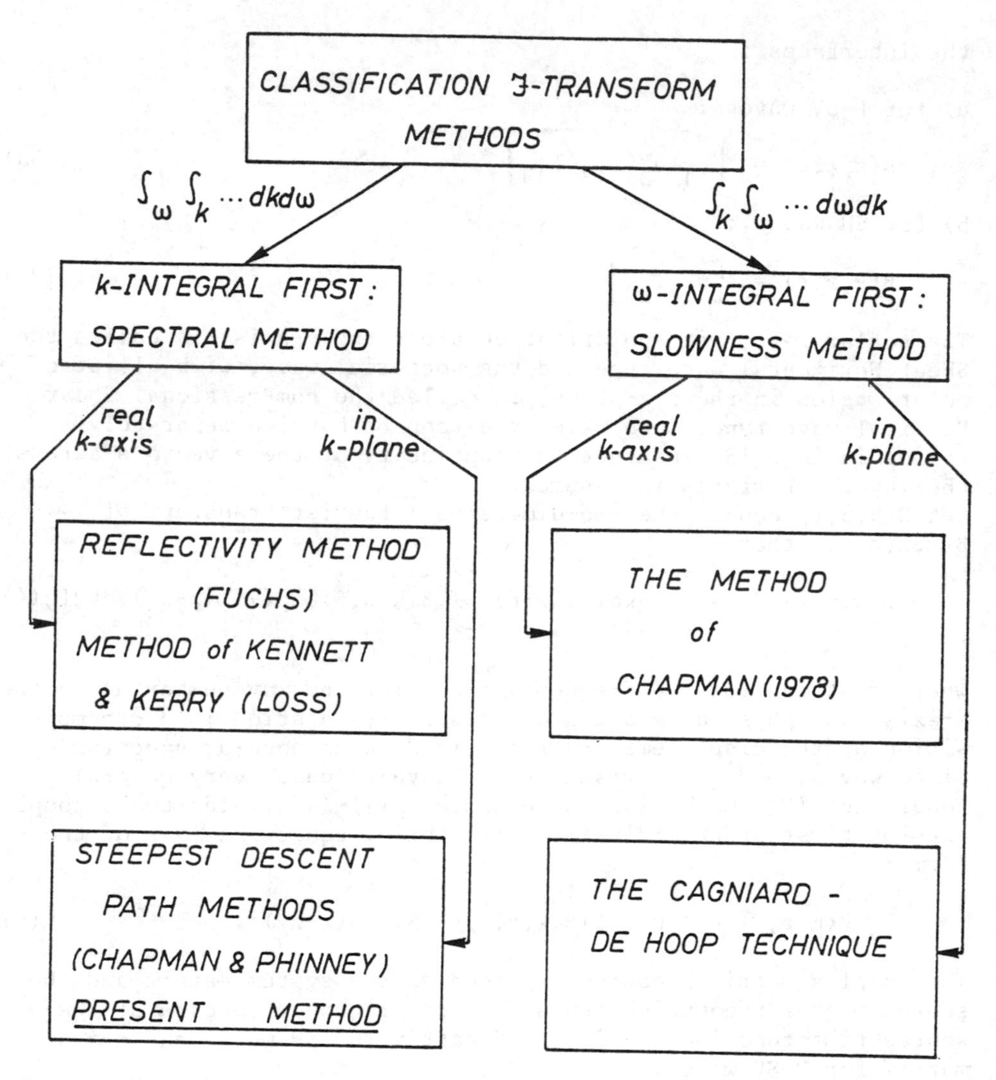

Fig. 2 A classification of methods to compute seismograms using a spectral decomposition of the wave field.

first we obtain the amplitude frequency spectrum of the medium response (for the given source and observer positions) as the intermediate result, and this approach is called the spectral method. Should we first evaluate the frequency integral, the intermediate result is the amplitude spatial spectrum, which is a function of the slowness 1/v, and this method is known as the slowness method.
A further separation can be made by considering the path of integration. If the integration of the wavenumber integral is performed entirely on the real k-axis, but stopped before the singularities of the integrand are encountered then we have the method of Fuchs,

also known as the reflectivity method (3). This is equivalent with considering only paraxial rays near the vertical direction (i. e. a spatial bandlimited spectrum). Therefore, this method is well suited to study the reflection and transmission properties of waves propagating across the layers. However, the effect of waveguiding in the layers can not be taken into account with this method since the poles of the integrand are neglected during the integration.
In the method of Kennett and Kerry (4) the integration is also performed on the real k-axis, but the poles on this axis are shifted off the real axis by giving the materials a slight loss.
Among the methods in which the integration is performed in the complex wavenumber plane, some authors distorce the path of integration to a path of steepest descent. An example of this is the work of Chapman and Phinney (5). In the present method the spatial Fourier integral is also computed along a path in the complex plane, so as to avoid the singularities, but this path is not a steepest descent path. When we consider the slowness method, we note the method of Chapman (6), where after performing the ω-integral, the evaluation of the k-integral takes place on the real axis.
Finally, we mention the Cagniard-de Hoop technique, which does not completely fits in this scheme because it starts from a Laplace-Fourier transform (7,8,9). The key of this method lies in finding a solution for the Laplace transform of the displacement field, with respect to time, precisely in the form of a forward Laplace transform. The solution is then found by inspection. The necessary manipulations, associated with this method, call for a well chosen path of integration in the complex slowness plane (10).
A further distinction can be made when one takes the method for integrating the system of first order differential equations into account (Fig. 3). When the elastic parameters vary continuously with depth, the integration of the differential equations can only be performed numerically, with for example the Runge-Kutta method (11,12). For a staircase stratification, as we have here, the differential equations can be integrated analytically. The importance of a staircase approximation in modeling a given continuously varying profile is due to a theorem by Volterra, concerning the Volterra product integral (13). The convergence of this integral expresses the convergence of the approximate solution to the exact one (1). The most straightforward and best known technique for solving the staircase stratification problem is the so called propagator matrix method. This method is due to Thomson and Haskell and developed in the fifties (14,15). However, this method possesses a numerical shortcoming so that it can not be applied for layered structures thick compared to the wavelength. To circumvent this difficulty various solutions have been presented. Among them we mention the method of Knopoff (16) and the method of Kennett and Kerry (4). In the latter case the system of first order differential equations is integrated in terms of the reflection and transmission properties of the regions above and below the source level. One particular approach is based on the use of the compound propagator

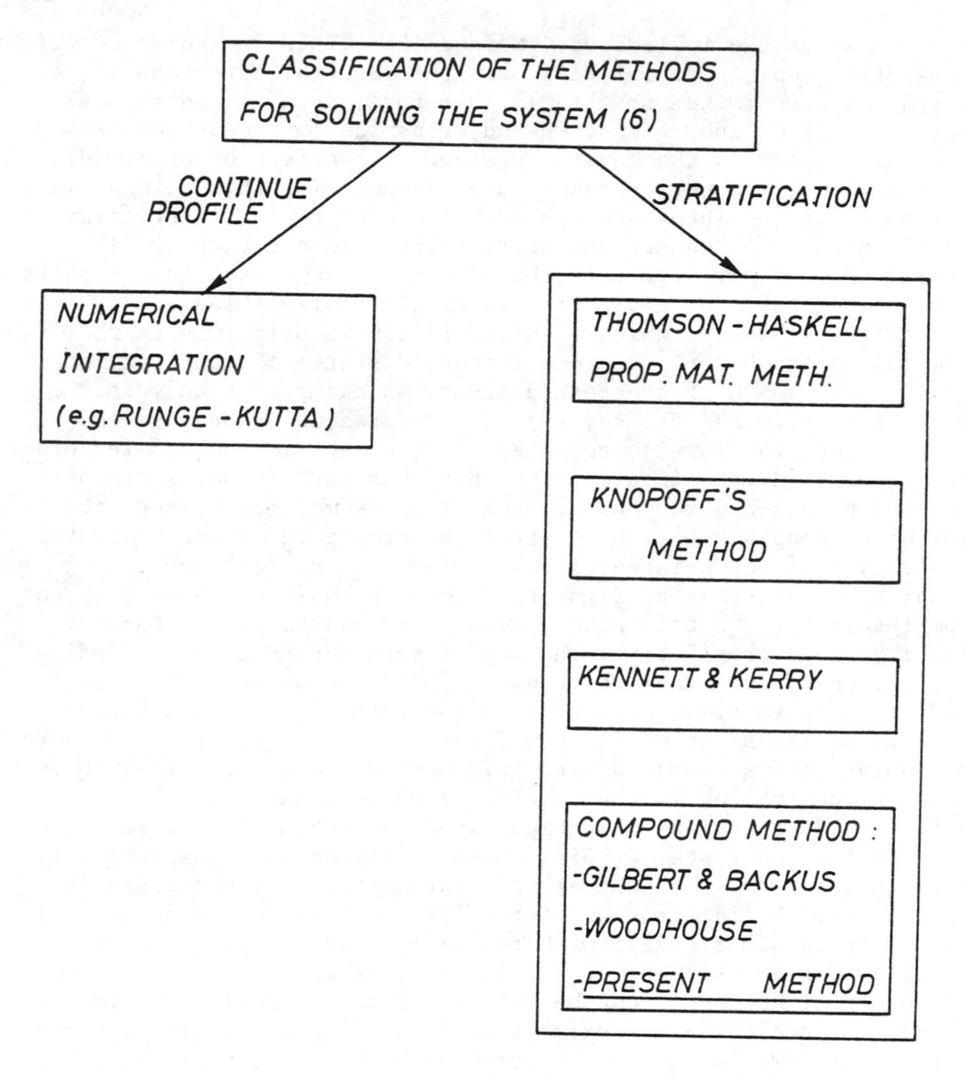

Fig. 3 A classification of methods to integrate Eq. (5).

matrix. This is the matrix of second order minors of the original propagator matrix (17). The general usefulness of this matrix was suggested by Gilbert and Backus (13). The recent work of Woodhouse (18) makes also use of this matrix. The present modified propagator matrix method combines the original propagator matrix and the compound matrix into an attractive computation scheme. The intention in developing this algorithm was to avoid the numerical instabilities of the classical method, without loosing the compactness of the propagator matrix method. This resulted in a general algorithm which is well suited to code on a computer. The mathematical derivation of this modified method can be found in (19).

2 THEORETICAL EXPRESSIONS

In this section we summarise the final expressions from which the integrand in Eq. (4) can be computed.
The displacement-stress vector $\hat{B}(k,z,f)$ in the k,f-domain is found as the product of a square matrix Ψ with the source vector $\hat{S}$:

$$\hat{B}(k,z_o,f) = \Psi(z_o,z_s^+)\hat{S} , \quad \Leftrightarrow \quad z_1<z_o<z_s<z_2 \quad , \tag{7}$$

$$\hat{B}(k,z_o,f) = \Psi(z_o,z_s^-)\hat{S} , \quad \Leftrightarrow \quad z_1<z_s<z_o<z_2 \quad . \tag{8}$$

Each element of this matrix is given as the ratio of two determinants :

$$\Psi_{ij}(z_o,z_s^+) = + \frac{\Delta_{ij}(z_o,z_s^+)}{\Delta} \quad , \tag{9}$$

$$\Psi_{ij}(z_o,z_s^-) = - \frac{\Delta_{ij}(z_o,z_s^-)}{\Delta} \quad . \tag{10}$$

These formulas are used if the observer point is located between the two half spaces. When the observation point lies in one of the half spaces, we first compute the $\hat{B}$-vector on the edge of that half space and propagate the displacement-stress vector in the half space with the propagator matrix containing only upward or downward propagating waves, as prescribed by the radiation condition. For $z_o<z_1$ we can write :

$$\begin{aligned} \hat{B}(z_1) &= \Psi(z_1,z_s^-)\hat{S} \\ \hat{B}(z_o) &= P_{\uparrow}(z_o,z_1)\hat{B}(z_1) \quad , \end{aligned} \tag{11}$$

and for $z_o>z_2$:

$$\begin{aligned} \hat{B}(z_2) &= \Psi(z_2,z_s^+)\hat{S} \\ \hat{B}(z_o) &= P_{\downarrow}(z_o,z_2)\hat{B}(z_2) \quad . \end{aligned} \tag{12}$$

If the source is located in one of the half spaces an additional layer, bounded by the source level, is added to the stack and we can apply again one of the above expressions.
The denominator determinant Δ can be written in the form :
a) for P-SV waves :

$$\Delta = \mathcal{C}^T(z_2)\,\mathcal{I}_{00}\,\mathcal{P}(z_2,z_1)\,\mathcal{C}(z_1) \tag{13a}$$

b) for SH waves :

$$\Delta = C^T(z_2)\,\mathcal{I}_{00}\,P(z_2,z_1)\,C(z_1) \tag{13b}$$

The determinants Δ_{ij} are given by :
a) for P-SV waves :

$$\Delta_{ij}(z_o,z_s^+) = \mathcal{G}^T(z_1)\,\mathcal{I}_{00}\,\mathcal{P}(z_1,z_s)(M^j)P(z_s,z_o)(M^i)^T\,\mathcal{P}(z_o,z_2)\,\mathcal{G}(z_2)$$

$$\Leftrightarrow \quad z_1<z_s<z_o<z_2 \qquad (14a)$$

$$\Delta_{ij}(z_o,z_s^-) = \mathcal{G}^T(z_2)\,\mathcal{I}_{00}\,\mathcal{P}(z_2,z_s)(M^j)P(z_s,z_o)(M^i)^T\,\mathcal{P}(z_o,z_1)\,\mathcal{G}(z_1)$$

$$\Leftrightarrow \quad z_1<z_o<z_s<z_2 \quad , \qquad (15a)$$

b) for SH waves :

$$\Delta_{ij}(z_o,z_s^+) = C^T(z_1)\,\mathcal{I}_{00}P(z_1,z_s)(N^j)(N^i)^T P(z_o,z_2)C(z_2)$$

$$\Leftrightarrow \quad z_1<z_s<z_o<z_2 \qquad (14b)$$

$$\Delta_{ij}(z_o,z_s^-) = C^T(z_2)\,\mathcal{I}_{00}P(z_2,z_s)(N^j)(N^i)^T P(z_o,z_1)C(z_1)$$

$$\Leftrightarrow \quad z_1<z_o<z_s<z_2 \qquad (15b)$$

For the SH case there are no compound matrices involved.
The assembly of the expressions for the determinants Δ_{ij} and Δ is illustrated in Fig. 4. The layered structure is shown again and for P-SV waves these determinants are computed as follows.
The Δ determinant is expressed as a (1 by 6) compound row vector, containing only information about the top half space, multiplicated with the set of compound matrices (with dimensions 6 by 6) for all the layers, and this product is terminated by the (6 by 1) compound column vector of the bottom half space. If one or both half spaces are replaced by a stress free surface, a rigid surface or a mass loaded surface, appropriate expressions for the compound boundary condition vectors $\mathcal{G}(z_1)$ and $\mathcal{G}(z_2)$ have to be inserted in these formulas.
For the determinants Δ_{ij} we start from the same row vector of the top half space and propagate down to the level of the source z_s. Here we multiply with the connection matrix M^j (with dimensions 6 by 4 and contains only ones and zeroes) and propagate further to the level of the observer point z_o with the propagator matrix itself (having dimensions 4 by 4). At this point we apply the transpose of the connection matrix M^i and continue to propagate to the bottom half space with the compound matrices again.
The explicit expressions for the propagator and compound propagator matrices can be found in (19).
The numerical stability of the modified method and its applicability to layered structures thick compared to the wavelength is due to the fact that the above determinants are computed directly from the analytical expressions for the elements of the compound matrices.
The problems in the earlier Thomson-Haskell formulation are caused by the direct numerical evaluation of these determinants in terms

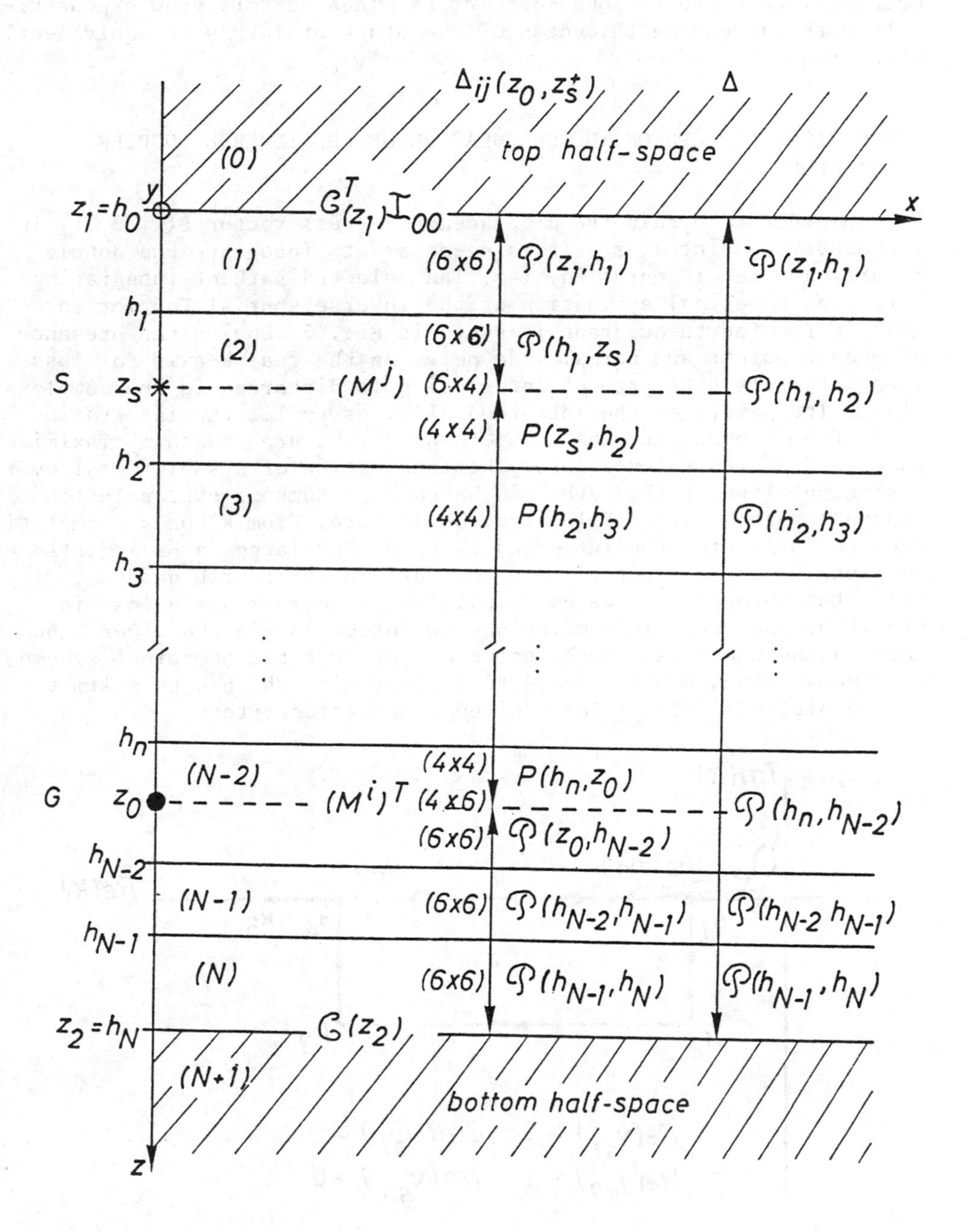

Fig. 4 Illustration of the assembly of the expressions for the determinants Δ_{ij} and Δ.

of the propagator matrix elements. It was shown in (19) that for P-SV waves terms which are theoretical identical zero do not cancel out completely due to round-off errors. These errors grow exponentially with increasing thickness of the stack of layers or equivalently with frequency.

3 NUMERICAL ASPECTS OF THE CALCULATION OF THE INVERSE FOURIER TRANSFORM

In order to obtain the displacement-stress vector $B(x_o,z_o,t)$ in the observer point x_o,z_o, it is necessary to integrate the double inverse Fourier integral Eq. (4). The selected path of integration Γ for the numerical evaluation of the inverse spatial Fourier integral in the fourth quadrant is shown in Fig. 5. Due to the presence of branch points and normal mode poles on the real k-axis for lossless materials, the path of integration is distorced in the complex plane. The parity of the integrand allow us to let the integration start from the origin. The parameters $k_1 \ldots k_5$ are chosen to maximize the speed of the calculations. The integration of a subintegral over a straight line of the path Γ is based on a Romberg extrapolation scheme together with a Filon type quadrature. From k_5 on a asymptotic representation for the integrand is used. For large $|x_o-x_s|$ distances one must be careful not to move too far in the fourth quadrant since the integrand grows exponentially for $k_i \to -\infty$ and a loss in precision can occur in summing the subintegrals. On the other hand integrating too close to the poles can prevent the numerical scheme from converging. One is forced to a compromise which sets a limit on the distance $|x_o-x_s|$ for a given computation effort.

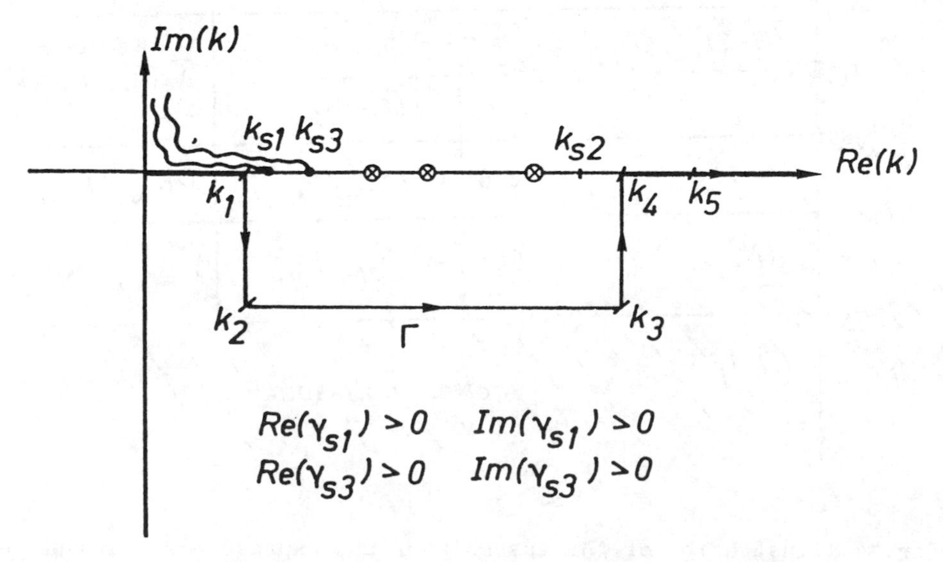

Fig. 5 The path of integration Γ in the complex wavenumber plane.

Other authors make the materials weakly attenuative so as to shift the singularities off the real k-axis in the first quadrant and the integration then takes place entirely on this axis. To assure the convergence of the integration in the latter case one is obliged to introduce sufficient loss which also sets a limit on the lateral distance for which the effect of the attenuation can be neglected. At the point of failure of these algorithms normal mode programs can take over since the response then consists only of guided mode contributions.
While performing the k-integrations for each frequency the amplitude spectra of the source components are kept constant. The spectrum so obtained is the frequency response of the medium for the given observer and source positions. This response is then multiplicated with the spectrum of the source. The remaining frequency to time inversion can be done with a Fast Fourier Transform routine since the spectrum to be inverted is a bandlimited spectrum without singularities.

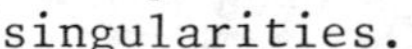

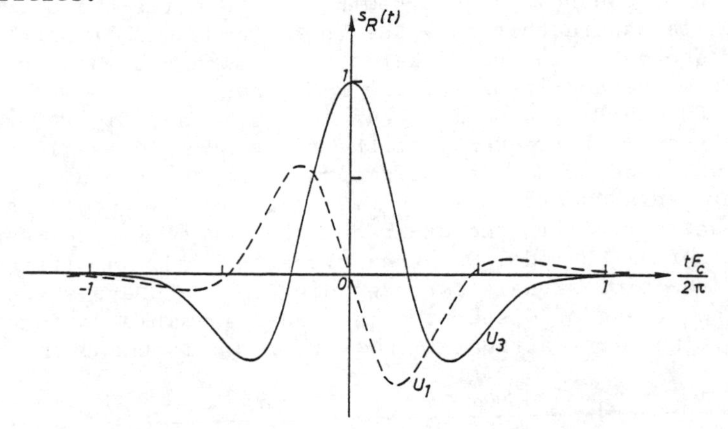

Fig. 6 The Ricker wavelet used as the source time function.

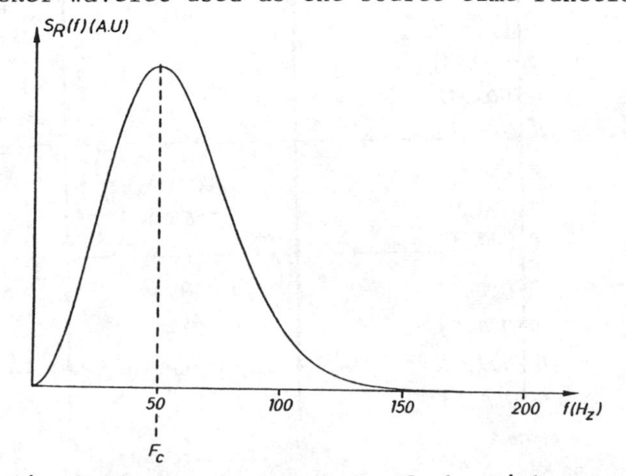

Fig. 7 The amplitude frequency spectrum of the Ricker wavelet.

4 SYNTHETIC SEISMOGRAMS

Synthetic seismograms have been calculated for a number of geometries. In each of the following examples a Ricker wavelet has been used as the source time function. This time signal and its frequency amplitude spectrum is shown in Fig. 6 and Fig. 7 respectively.

4.1 Seismograms in a Monomode One-layer Waveguide

The configuration is shown in Fig. 8. The waveguide consists of a coal layer sandwiched between two elastic half spaces. The elastic constants are given in the figure. The source is placed in the middle of the coal layer and 12 equidistant geophones are arranged in three vertical planes A, B and C respectively 100, 200 and 300 m away from the source plane. Seismograms have been calculated in the frequency range 0-400 Hz for SH waves and 0-200 Hz for P-SV waves to assure that only the lowest mode was propagative. The centre frequency F_c in the Ricker spectrum was chosen to be 100 Hz for the SH waves and 50 Hz for the P-SV waves.
The first SH and P-SV mode is shown in Fig. 9 and Fig. 10 respectively. The cut-off frequency of the first SH mode is 110 Hz and for the P-SV mode 70 Hz. The seismograms for the three displacements in the three planes are shown in Fig. 11. The time scale is 0.32 s. The dominant arrival in the u_2-traces (SH waves) is the response of the first normal SH mode. One recognizes the modal amplitude distribution of the time signals corresponding to the shape shown in Fig. 9. The effect of dispersion is clearly visible as function of the range. The first arrival in the u_2-traces is the transverse

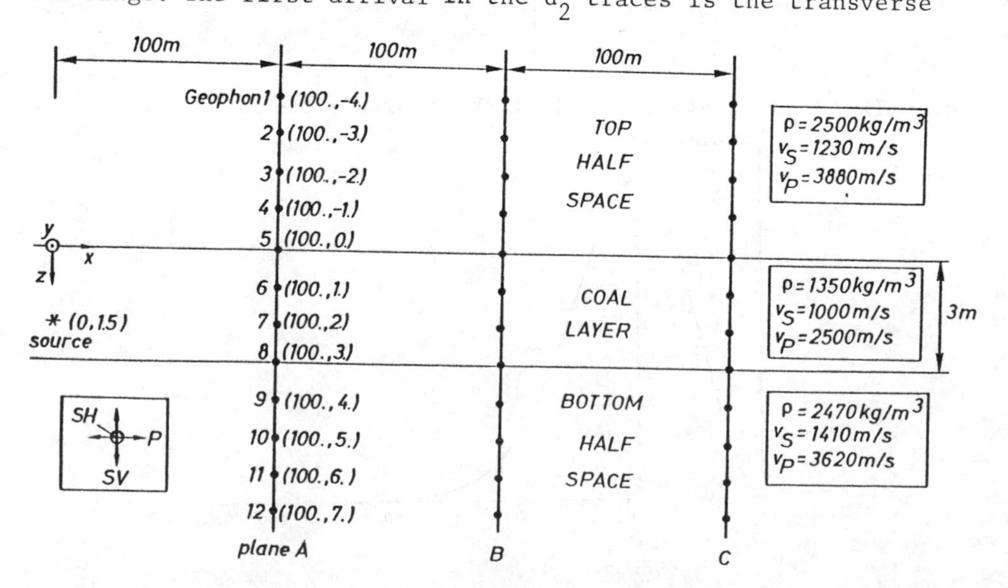

Fig. 8 The source and geophone positions in a monomode coal guide.

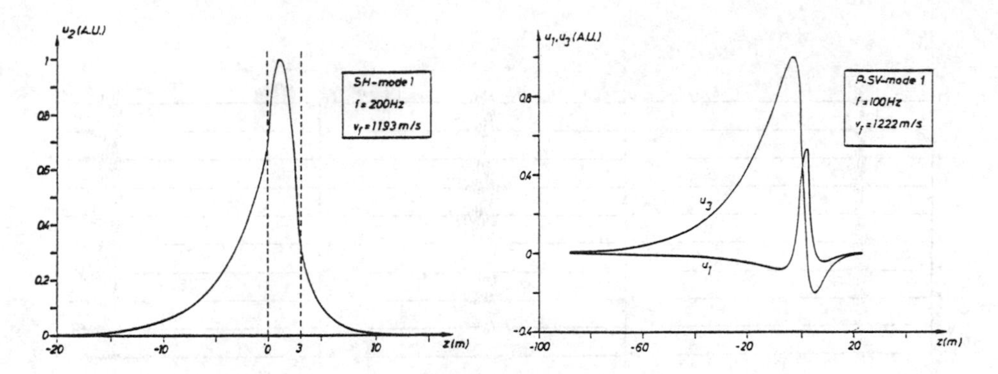

Fig. 9 The first normal SH mode. Fig. 10 The first normal P-SV mode.

bulk wave propagating with the local shear velocity of the half spaces. The first arrival in the u_1- and u_3-traces (P-SV waves) is caused by the longitudinal wave propagating with the local v_P of the half spaces. The strong motions in the u_1-traces is the projection of the displacement vector on the x-axis and the weak pulse in the u_3-traces is the projection of this polarization on the z-axis. As expected, the latter arrival vanishes in the middle of the layer. The last arrival is the response of the normal P-SV mode. The absence of any noticeable dispersion is due to the limited frequency range. One sees again the corresponding amplitude behavior over the z-axis as expressed by the modal shape in Fig. 10. A final pulse can be resolved immediately before the P-SV mode arrival. The propagation velocity suggests that this motion is again the response of the transverse volume wave.

4.2 Seismograms in a Multimode One-layer Waveguide

The coal layer shown in Fig. 12 is a multimode guide for the frequency range 0-400 Hz considered here. In this range 15 SH modes and 16 P-SV modes are propagative. The elastic constants can be found in the figure. The source S_1 is placed in the layer, the source S_2 is located in the top half space. A number of 12 equidistant geophones are placed in a borehole 70 m apart from the source plane. The centre Ricker frequency is 120 Hz. The time signals for the three displacement components and for the two source positions are shown in Fig. 13. The time scale is 0.16 s. The traces generated by the source S_1 show clearly the diffracting longitudinal and transversal wavefronts, marked with the capitals P and S respectively. These fronts propagate with the local v_S and v_P velocities. Later on, the arrivals of the normal modes can be seen in the traces belonging to the geophones in the layer. The signals produced by the source S_2 have a different nature. Due to the position of the source no guided modes can efficiently be excitated. One sees the reflected and refracted waves which come from the interface. This picture is

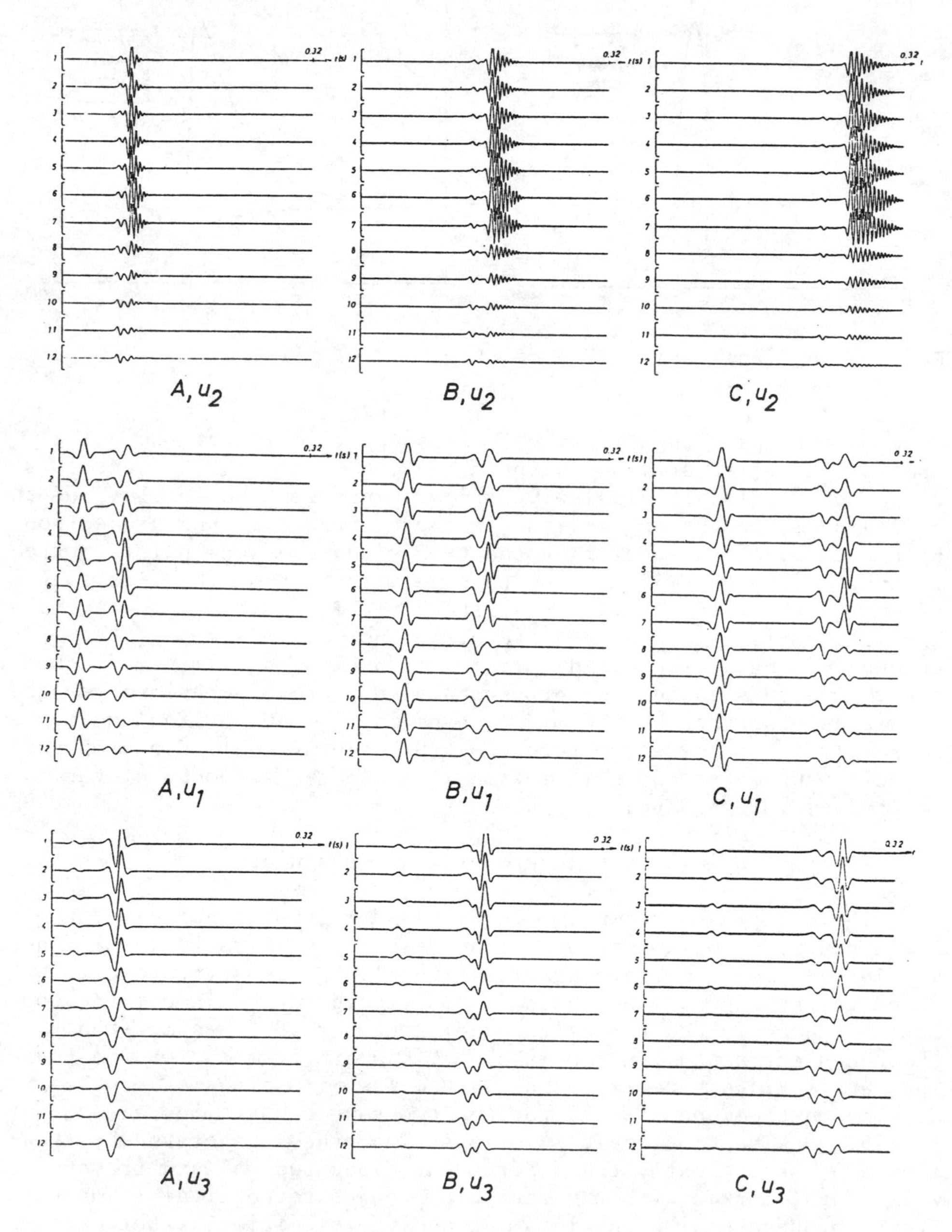

Fig. 11 The synthetic seismograms for the configuration of Fig. 8.

less clear for the P-SV waves because of the coupling of the P and S waves at each interface.

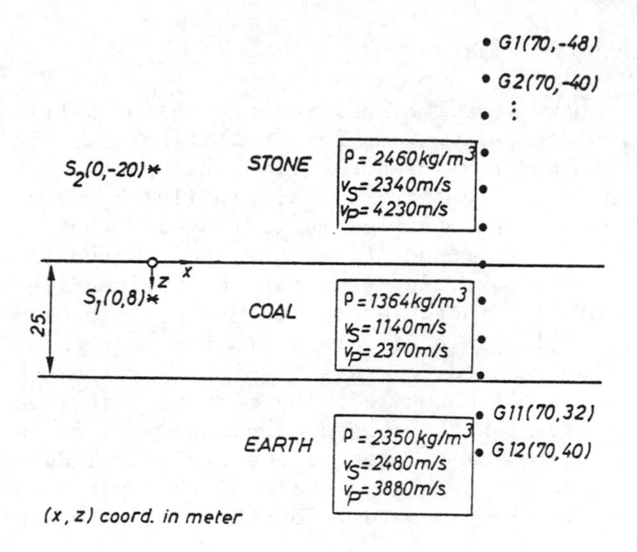

Fig. 12 Source and geophone positions in a multimode coal waveguide.

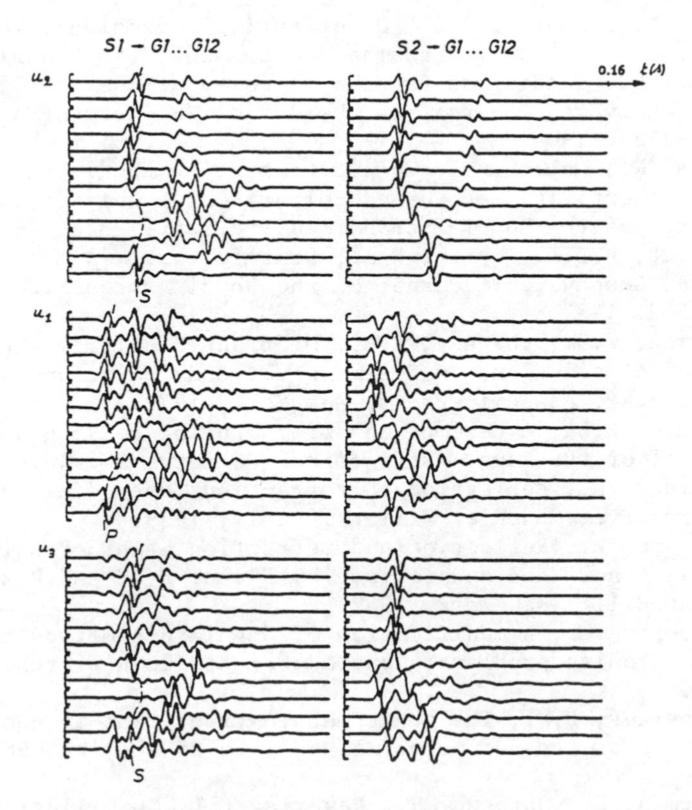

Fig. 13 The seismograms obtained from the configuration of Fig. 12.

5 CONCLUSIONS

It was shown that the modified propagator matrix method can be used to calculate seismograms for stratifications thick compared to the wavelength and for which the classical propagator matrix method could not be used. A comparison made in (19) between the propagator matrix method and the modified method reveals that the latter method is less sensitive to round-off errors and can run in single precision (8 decimals). The matricial structure of the modified method makes this method highly tractable even when a great number of layers are present and results in a efficient programmable code. The present method can also be used for anisotropic layers for which the SH and P-SV waves are still uncoupled. The main task is then to find the analytical expressions for the propagator matrices and the compound matrices. This problem reduces to the analytical determination of the eigenvalues and the eigenvectors of the system matrix A (with dimensions 4x4 for P-SV waves) for the considered anisotropy.

1. Aki, K., Richards, P.G., Quantitative Seismology, Theory and Methods, vol I & II, Freeman and Company, San Francisco, 1980.
2. Lalor, E., Conditions for the validity of the angular spectrum of plane waves, Journal of the Optical Society of America, 58, 1235-1237, 1968.
3. Fuchs, K., Muller, G., Computation of synthetic seismograms with the reflectivity method and comparison of observations, Geophys. Journal of the Royal Astronomical Society, 23, 417-433, 1971.
4. Kennett, B.L.N., Kerry, N.J., Seismic waves in a stratified half space, Geophysical Journal of the Royal Astronomical Society, 57, 557-583, 1979.
5. Chapman, C.H., Phinney, R.A., Diffracted seismic signals and their numerical solution, in Seismology : Body Waves and Sources, Ed. Bolt, B.A., Academic Press, New York, 1972.
6. Chapman, C.H., A new method for computing seismograms, Geophys. Journal of the Royal Astronomical Society, 54, 481-518, 1978.
7. Cagniard, L., Reflextion et refraction des ondes seismiques progressive, Gauthier-Villars, Paris, 1939.
8. Cagniard, L., Reflection and refraction of progressive seismic waves, translated and revised by Flinn, E.A. and Dix, C.H., McGraw-Hill , New York, 1962.
9. De Hoop, A.T., A modification of Cagniard's method for solving seismic pulse problems, Applied Scientific Research, B8, 349-356, 1960.
10. Helmberger, D.V., The crust-mantle transition in the Bering Sea, Bulletin of the Seismological Society of America, 58, 179-214, 1968.
11. Alterman, Z., Jarosch, H., Pekeris, C.L., Oscillations of the Earth, Proc. Royal Society London, A252, 80-95, 1959.

12. Takeuchi, H., Saito, M., Seismic surface waves, in Seismology : Surface waves and Earth Oscillations, Ed. Bolt, B.A., Academic Press, New York, 1972.
13. Gilbert, F., Backus, G., Propagator matrices in elastic waves and vibration problems, Geophysics, 31, 326-332, 1966.
14. Thomson, W.T., Transmission of elastic waves through a stratified solid, Journal of Applied Physics, 21, 89-93, 1950.
15. Haskell, N.A., The dispersion of surface waves in multilayered media, Bulletin of the Seismological Society of America, 43, 17-34, 1953.
16. Knopoff, L., A matrix method for elastic wave problems, Bulletin of the Seismological Society of America, 54, 431-438, 1964.
17. Gantmacher, F.R., The theory of matrices, vol I & II, Chelsea Publishing Company, New York, 1959.
18. Woodhouse, J.H., Efficient and stable methods for performing seismic calculations in stratified media, in Physics of the Earth's Interior, Eds. Dzeiwonski and Boschi, Elsevier, New York, 1981.
19. Franssens, G.R., Calculation of the elasto-dynamic Green's function in layered media by means of a modified propagator matrix method, Accepted for publication in the Geophysical Journal of the Royal Astronomical Society, 1983.

THE BEAM PROPAGATING METHOD IN INTEGRATED OPTICS

P.E. Lagasse, R. Baets

Lab. for Electromagnetics and Acoustics
Sint-Pietersnieuwstraat 41 B 9000 Gent, Belgium

ABSTRACT

The BPM algorithm is described with the emphasis on the various explicit and implicit assumptions made in this method. Starting from practical experience with the BPM, its limitations and possible extensions are discussed, as related to integrated optics.

1. INTRODUCTION

The Beam Propagation Method (BPM) is basically a numerical modeling method for the scalar wave equation in the paraxial and low contrast approximation. It has been introduced in the field of underwater acoustics since 1973 as the Parabolic Equation (PE) method (1,2) and has been reviewed recently for the same area of application (3). The above-mentioned conditions are usually fulfilled in the case of fiber optic or integrated optic structures. For this reason the BPM has been applied to a number of problems in those areas over the past few years (4-14). The BPM has been analysed and described in several ways. In this paper we would like to derive the BPM in a manner being simple and straightforward, but clearly showing the basic assumptions and limitations of the method.

Since the basic algorithm of the BPM is very easy to implement on a computer, one might be tempted to use the method without any further thought. Due to the very good numerical stability of the BPM, one always obtains a more or less plausible result. In order to obtain correct results, especially in the case of new applications, considerable care needs to be taken. The limitations of the BPM and possible extensions of the method will therefore be discussed in some detail. Methods of testing the obtained results will also briefly be described. Since numerous examples of the application of the BPM have already been published, the results presented here will be limited to a few illustrative and didactic examples.

For the sake of clarity, all derivations will be done in two dimensions. As will be explained, the extension to the three-dimensional case is trivial.

2. DERIVATION OF THE BPM

The problem we want to solve is the calculation of the propagation of a given input beam $E_o(x,z)$ through a medium with a refractive index n,x,z). From the start, the following assumptions are made :

1) the wave propagation is described by the scalar wave equation :

$$\nabla^2 E + k^2 n^2(x,z)E = 0 \tag{1}$$

   This assumption is restrictive since Maxwell's equations lead only to a scalar wave equation for one of the field components in a limited number of geometries. However, in a considerable number of cases the field consists of a summation of quasi-TE and quasi-TM modes. The propagation of each of those groups can be analysed in a scalar way but the coupling between them is neglected.

2) the refractive index variation can be written as :

$$n(x,z) = n_o(x) + \Delta n(x,z) \tag{2}$$

   where $\Delta n << n_o$ and $n_o(x)$ has been chosen thus that the solutions of :

$$\nabla^2 \phi + k^2 n_o(x)\phi = 0 \tag{3}$$

   are known eigenfunctions :

$$\phi_n(x)e^{-jk_n z} \tag{4}$$

3) the variation of n along the z-direction is such that, for an input beam containing only components propagating in the positive z-direction, one can neglect the influence of the reflected fields on the forward propagating beam. This means that there can be no large abrupt changes of n(x,z) as a function of z, or no periodic reflections that add up coherently as in a grating.
   The mathematical consequence of this assumption is that the solution on a line z = constant can be written as an eigen-

function expansion containing only components propagating in the positive z-direction :

$$E = \sum_{n=1}^{\infty} B_n \phi_n e^{-jk_n z} \tag{5}$$

One should realize that by excluding half of the eigenmode components, only one boundary condition is required instead of two. Basically this assumption transforms a boundary value problem into an initial value problem, so that a stepwise solution becomes feasible. This last assumption is very important because it makes a numerical solution much easier. It is not completely equivalent to the parabolic approximation where one assumes :

$$E(x,z) = G(x,z)e^{-jk_o z}$$

with G(x,z) a slowly varying function. Neglecting $\frac{\partial^2 G}{\partial z^2}$, one obtains then the classic parabolic wave equation for G. This is a first order differential equation in z and therefore also an initial value problem, but the difference with the BPM lies in the inherent paraxial approximation of this equation. This can be seen when drawing the wave vector diagram (fig. 1). The wave

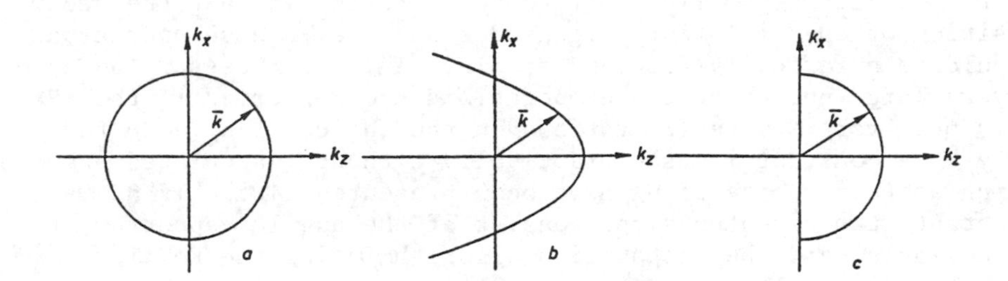

Fig. 1. Wave vector diagram : a. circular, b. parabolic approximation, c. half circular approximation

equation has a circular (fig. 1a) and the parabolic equation a parabolic diagram, whereas the approximation of the BPM yields a half circle. In the BPM paraxiality is not introduced at this stage. It is not used for the propagation in the unperturbed $n_o(x)$ space, but only in the calculation of the correction term.

It should be noted that other numerical methods than the BPM can be developed starting from this last assumption, namely that reflections can be neglected. In the field of seismology, a finite difference scheme based on this assumption is widely used (15).

For this finite difference scheme, it is necesarry to derive a first order wave equation in z having a half circle as wave vector diagram. In the case of the BPM, it is also possible to derive a first order differential equation in z by means of a Green's function approach. This method of analysis, which was presented in a previous paper (10) gives an interesting insight in the BPM, but is more complicated than the present derivation. Since in the case of the BPM the propagation in the unperturbed n_o space is calculated by means of the eigenfunction approach, the selection between the forward and backward propagating waves is done in the summation of equation (5) and no first differential equation needs to be derived.

Having looked carefully at the initial assumptions of the BPM, the general outlines of the method become apparent. The solution proceeds in a stepwise fashion from the given input beam towards increasing z-values. The arbitrary refractive index profile is split up into an unperturbed part n_o and a small perturbation Δn. Propagation in the unperturbed medium n_o is done by means of an eigenfunction expansion (5) and for each step, a correction due to Δn is calculated.

The choice of n_o is very important. The first criterion is that Δn should be small. The second criterion is that the decomposition of the field into eigenmodes and the inverse operation should be numerically fast and stable. This is necessary because a very large number of those operations are performed in the BPM. This puts very severe limitations on the choice of n_o. So far, only n_o = constant is used practically although calculated examples using another choice of n_o have been presented (10). If n_o is constant, the eigenfunctions consist of the angular spectrum of plane waves, and the propagation algorithm using the Fourier Transform is well known (16).

The problem we wish to solve can now be stated as follows : given the incident field $E(x,z_o)$, we want to calculate the field $E(x,z_o+\Delta z)$ where Δz remains to be chosen. In the BPM, $E(x,z_o+\Delta z)$ is written as the product of the field $\varepsilon(x,z_o+z)$ propagated in the unperturbed medium n_o and a correction factor e^{Γ} :

$$E(x,z_o+\Delta z) = \varepsilon(x,z_o+\Delta z).e^{\Gamma} \tag{6}$$

where $\varepsilon(x,z)$ satisfies

$$\nabla^2 \varepsilon + k^2 n_o^2 \varepsilon = 0 \tag{7}$$

with

$$\varepsilon(x,z_o) = E(x,z_o) \tag{8}$$

and

$$\varepsilon(x,z) = \sum_{n=1}^{\infty} B_n^o \, \phi_n(x).e^{-jk_n z} \tag{9}$$

$$B_n^o = \int_{-\infty}^{+\infty} E_o(x,z_o) \, \phi_n^*(x)dx \tag{10}$$

In the case n_o = constant, equations (9) and (10) represent Fourier Transforms. In the BPM, a multiplicative correction is chosen although an additive correction is also possible as will be discussed later. The exponential form of the correction factor allows one to easily distinguish between phase and amplitude corrections.

Substitution of (6) into (1), and taking (7) into account, yields :

$$\varepsilon \, \nabla^2\Gamma + \varepsilon|\text{grad}\Gamma|^2 + 2 \, \text{grad}\varepsilon \, \text{grad}\Gamma + k^2(n^2-n_o^2)\varepsilon = 0 \tag{11}$$

Since Δz can be chosen small, following series expansion for Γ is possible

$$\Gamma(x,z) = \sum_{n=1}^{\infty} A_n(x)z^n \tag{12}$$

Substitution of (12) into (11) and equating the coefficients of z^n to zero yields a set of equations for the A_n. The first equation is :

$$A_1^2 + \frac{2}{\varepsilon}\frac{\partial\varepsilon}{\partial z} \, . \, A_1 + k^2\Delta n^2 + 2A_2 = 0 \tag{13}$$

where $\quad \Delta n^2 = n^2 - n_o^2$

In order to obtain a simple form for A_1, a number of approximations are now required.
First we assume that :

$$A_2 << \frac{1}{2} k^2 \, \Delta n^2$$

This condition can be verified by calculating A_2 from the second equation resulting from the series expansion. (13) is then reduced to a quadratic equation which can easily be solved :

$$A_1 = \frac{1}{\varepsilon}\frac{\partial\varepsilon}{\partial z}\left[-1+\left[1-\left(\frac{1}{\varepsilon}\frac{\partial\varepsilon}{\partial z}\right)^{-2}.k^2\Delta n^2\right]^{1/2}\right] \tag{14}$$

In principle the calculation of (14) presents no special difficulty. Since ε is obtained by propagating the angular spectrum of plane waves, it is easy to also calculate $\frac{\partial\varepsilon}{\partial z}$ in the spectral domain.

Expression (14) can be further simplified if we assume

$$k^2 \Delta n^2 \left(\frac{1}{\varepsilon} \frac{\partial \varepsilon}{\partial z}\right)^{-2} << 1$$

so that

$$A_1 = -\frac{k^2}{2} (n^2 - n_o^2) \frac{\varepsilon}{\frac{\partial \varepsilon}{\partial z}} \tag{15}$$

Introducing paraxiality in the calculation of A_1, one can assume that ε consists of a plane wave travelling along the z-axis :

$$\varepsilon = e^{-jkn_o z}$$

With this last approximation, one obtains the very simple phase correction term of the BPM :

$$A_1 = -jk \, \Delta n \tag{16}$$

The complete algorithm for the case n_o = constant can then be summarized as follows : each step consists of the propagation in homogeneous space by means of the angular spectrum of plane waves, followed by a phase correction term $e^{-jk \, \Delta n(x) \, \Delta z}$ such as used for thin lenses.

3. LIMITS OF APPLICABILITY OF THE BPM

Starting from the various inequalities assumed in the previous paragraph, one can derive a set of conditions for the applicability of the BPM. One obtains then the set of conditions published in reference (10). Basically they amount to a limitation of Δn and grad n. Some of the conditions can be met by decreasing the step size Δz, although other conditions are independent of Δz.

When practically using the BPM, however, many other considerations should be taken into account. Based on our experience with the BPM, we will discuss some of the most relevant problems encountered.

As a first simple example we would like to point out that, although the expressions (14) and (15) for A_1 contain less approximations and require only 50 % more computer time than (16), their application is of little practical interest. The problem of numerical stability by far outweighs the wider angular spectrum that becomes available.

The Fourier Transforms required by the BPM are computed by means of the FFT algorithm. This algorithm satisfies the requirements of numerical efficiency and stability perfectly. It also means that the field and the refractive index profile need to be discretised in a finite number of points. The fields are only calculated over a limited extent of the x-axis. The rest consists of a periodic extension of the considered region. In order to obtain meaningful results, one should make sure that the complete region of interest lies within the chosen section of the x-axis, and that aliasing effects do not occur. This really amounts to the classic problem of analysing wave propagation in an infinite space by means of a finite computational domain. In the BPM, the radiation condition is simulated by introducing absorbing boundary conditions. The imaginary part of the refractive index profile, plotted in fig. 2, forms an elegant solution for this problem. The simplicity of implementation of this boundary con-

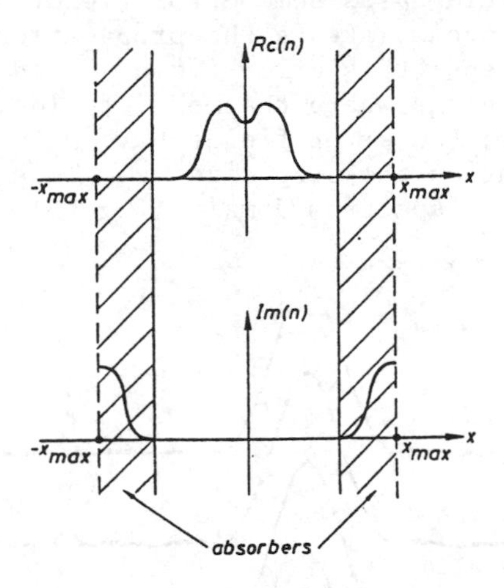

Fig. 2. Real and imaginary part of refractive index variation

dition is an advantage for the BPM compared to finite difference formulations, where it is notably difficult to obtain absorbing boundary conditions for a wide range of the angular spectrum.

When using the BPM to calculate the eigenmodes of a waveguide, an absorbing region too close to the waveguide can obviously cause errors. Another cause of errors is related to the discretisation of the refractive index profile. Consider for example a waveguide making a small angle with the z-axis. A simple rounding off of the index profile to the nearest discretisation point results in a staircase shape of the waveguide. Obtaining the same, correct, result for the eigenmode of a waveguide parallel to the z-axis and for the same waveguide making an angle with respect to the z-axis, requires very careful discretisation of the refractive index profile (14). This can for example be done by Fourier

Transforming this index profile with respect to the x-coordinate and operating the shift in this Fourier domain.

Since one of the main conditions for the application of the BPM is a limitation of grad n, step index guides can in principle not be modeled. It is, however, very instructive to attempt to use the BPM to determine the eigenmode of a simple step index waveguide. One learns immediately that the conditions mentioned in this paragraph fail to take into account the discretisation and numerical aspects of the BPM. Normally, the discretisation in the x-direction is sufficiently fine to model a grad n that is too large for the BPM, so that each propagation step generates the correct modal field plus some error field. One would expect this error field to accumulate as the propagation progresses. After a number of steps, this error field should become apparent, especially in the regions where the modal field is practically zero. In practice, however, the error fields have a special shape that is strongly dependent on the step size z. This can most clearly be seen in the angular spectrum domain (fig. 3). The central part

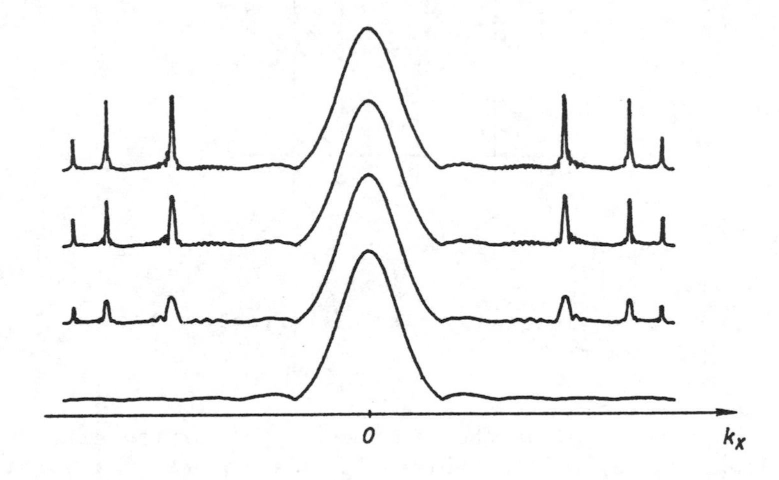

Fig. 3. Angular spectrum domain showing selective accumulation of errors

is the angular spectrum of the mode. The error field consists clearly of a set of plane waves propagating at an angle with respect to the z-axis. A qualitative explanation for this can easily be found. Since the modal field remains constant, the error field can only accumulate constructively, if its phase change is equal to $2n\pi$. In the spectral domain, one can immediately see (fig. 4) that those parts of the angular spectrum of the error field for which

$$(k_o - k_z) n_o \Delta z = 2n\pi \qquad (17)$$

add constructively. If one takes $\Delta z < \frac{2\pi}{kn_o}$ or $\Delta z < \lambda_o$, condition (17) can no longer be met. Error fields generated at each step add destructively so that good results can be obtained for step index guides using the BPM (fig. 5). Even the numerical attenuation of the eigenmode, introduced by the error fields generated in the absorbing region, is negligible for most practical purposes.

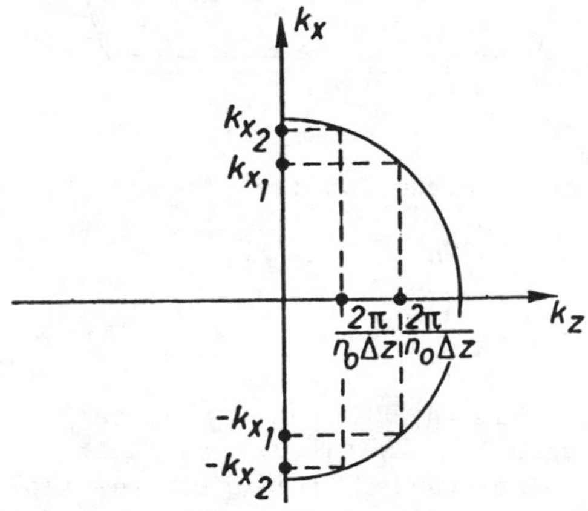

Fig. 4. Wave vector explanation of constructive error accumulation

One should be careful when using the calculation of eigenmodes of a waveguide as a test for the BPM. The fact that a stable mode pattern is obtained, is not sufficient. A comparison with the analytically calculated eigenmode is necessary. The numerical operations performed in a single step of the BPM can be written as the multiplication of the vector of the discretised field values with a matrix [A] containing the discrete Fourier Transform, the propagator, the inverse transform and the "lens law". For a z-independent waveguide, this matrix remains the same for each step, so that a stable modal field simply represents an eigenvector of this matrix [A] :

$$[A] \; [\phi] = e^{-j\beta\Delta z} \; [\phi]$$

Only if the BPM is applied in the correct conditions, this numerical eigenmode will be close to the analytical solution. A complete analysis of the BPM should therefore study the relationship between the original problem and the repeated application of the matrix multiplication used in the BPM.

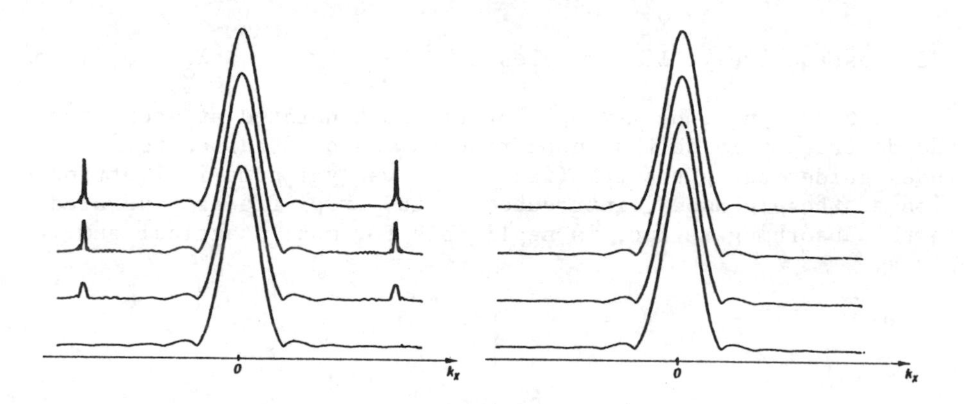

Fig. 5. Angular spectrum for : a. $\Delta z > 2\pi/kn_o$ and b. $\Delta z < 2\pi/kn_o$

4. EXTENSION OF THE BPM

It is obvious that the BPM can immediately be extended to three-dimensional problems by calculating a three-dimensional angular spectrum of plane waves. This requires replacing the one-dimensional FFT by a two-dimensional FFT. One drawback is that the computation time increases by an order of magnitude. Other small problems are that a three-dimensional discretisation of the refractive index profile is needed as input and that one has to interpret the three-dimensional output data. The coupling of a fibre to a planar waveguide could be modeled in this way. In many cases though, we have found it very useful to combine the two-dimensional BPM with the effective refractive index method to deal with three-dimensional problems [13] , [14] . Especially for cases in which a field is guided strongly in a single-mode fashion along one dimension, as is often encountered at an air-waveguide interface in integrated optics, this technique can offer satisfactorily accurate results at drastically reduced computation times. In this method the field is written as

$$E_x(x,y,z) = F(x,y,z).G(x,z)$$

in which F(x,y,z) represents the single mode profile in the y-direction, slowly varying in the other two directions due to slight waveguide alterations. G(x,z) is found by applying a two-dimensional BPM-calculation in a medium which takes into account the effective refractive index of the mode structure F(x,y,z). In this way, coupling of power to radiation modes in the y-direction is entirely neglected. Fig. 6 illustrates schematicly the method

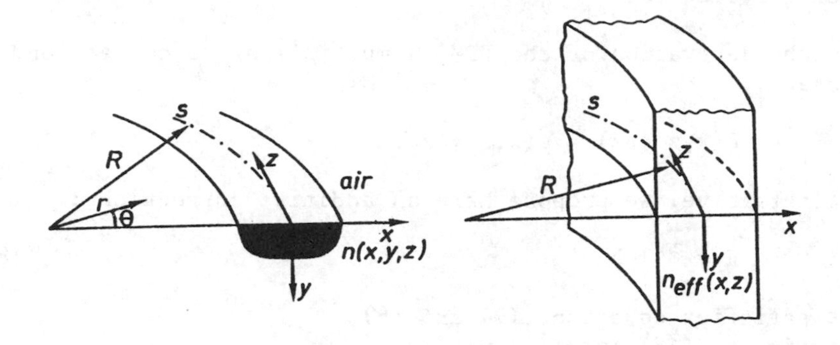

ig. 6. Illustration of the effective index method for a curved waveguide

for a curved planar waveguide.

As was explained in paragraph 2, the BPM fundamentally does not require n_o to be constant. Examples have been published [19], [17], where guides containing a large step in refractive index were modelled by including the step variation in $n_o(x)$. Since one can no longer use the FFT, the required computer time increases considerably and rounding off errors become a problem. This makes this extension of the BPM somewhat less attractive to use. An alternative and approximate way of dealing with large refractive index variations at one plane interface is to assume that the field becomes zero at the index step and stays zero at the lower side. In this case an antisymmetric field distribution together with an antisymmetric Fourier Transform (sine transform) can be used to simulate the zero boundary condition at the interface.

From the modeling of the wave equation by the BPM to that of coupled wave equations, there is only one step. Examples of this extension of the BPM have already been published [11], [12]. The main advantage lies in the possibility to use the BPM for the calculation of reflected waves, provided those can be described by coupled wave equations. This allows one for example to analyse gratings including the diffraction effects. The difficulties do not come from the modification of the BPM, but from the boundary conditions. Depending on the direction of the reflected waves, one can obtain a situation in which the input beams of incident and reflected wave are not known on the same line. This is a boundary value problem, which can only be solved by an iterative application of the BPM through the reflecting structure.

In the derivation of the BPM, a multiplicative correction was chosen :

$$E(x,z_o+\Delta z) = \varepsilon(x,z_o+\Delta z).e^{\Gamma}$$

As an alternative, we propose here an additive correction :

$$E = \varepsilon + E' \qquad (18)$$

where ε satisfies equations (7) and (8).
Substitution of (18) into (1) yields :

$$\nabla^2 E' + k^2 n^2(x,z)E' = -k^2 \Delta n^2(x,z)E_o \qquad (19)$$

where $\qquad \Delta n^2(x,z) = n^2(x,z) - n_o^2$

For the additive correction field E', we obtain a wave equation (19) with source term. For the solution of this equation, the BPM with a multiplicative correction is now used.

$$E' = \eta\, e^{\Gamma} \qquad (20)$$

where η satisfies

$$\nabla^2 \eta + k^2 n_o^2 \eta = -k^2 \Delta n^2(x,z)E_o \qquad (21)$$

Using again the series expansion (12) for Γ, one obtains the same result :

$$\Gamma = -jk\ \Delta n.\Delta z$$

If n_o was chosen properly, η can easily be computed from equation (21). Normally, n_o is taken constant, so that the solution of (21) can immediately be written in terms of a Green's function. The most interesting aspect of this additive correction is that so far, no assumption was made about the absence of reflections. The source term in equation (21) generates waves both in the positive and in the negative direction. This means that by combining a BPM for the positive and one for the negative z-direction, one can obtain a system that can handle reflections within the Born approximation. Various algorithms based on this idea are currently under investigation for application in the area of elastic wave propagation [18].

5. CALCULATION EXAMPLES

As has been mentioned in the introduction, the BPM has already been applied to a variety of problems in integrated optics and fibre optics. Each of these problems demands a particular approach. The choice of the discretisation step and the propagation step together with the choice of an appropriate absorber depend on the nature of the specific problem involved.
The examples described hereafter all make use of the effective refractive index method to reduce the problem to two dimensions. This means that the index profiles used for the calculation are in fact effective refractive index profiles.

As a first example, planar waveguide tapers are considered (fig. 7).

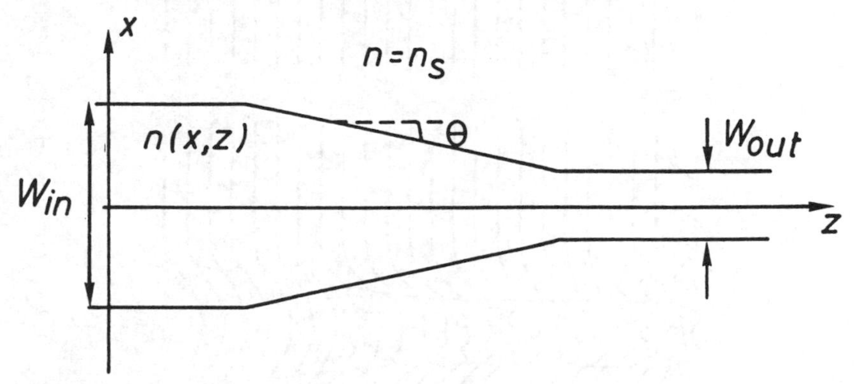

Fig. 7. Geometry of a planar waveguide taper

The refractive index profile is a step-index profile. The transmission losses and mode conversion properties can be analysed by introducing the modes of the input waveguide into the taper and calculating the expansion of the output field into the modes of the output waveguide. Fig. 8 shows a case in which the output guide has half the width of the input guide and is able to guide 3 modes. The fundamental normalised mode of the input waveguide is introduced and propagated through the taper both for a taper angle of 0.5° and 4°. The power in the zeroth and second order mode of the output waveguide is indicated in both cases. One can see that for $\Theta = 0.5°$ the taper transfers the fundamental mode adiabatically whereas the case with $\Theta = 4°$ is hardly better than an abrupt waveguide transition as shown in fig. 8c. More complete data on tapers both for single mode and multimode applications can be found in ref. 11. It appears that the transmission loss for the lowest-order mode in step-index tapers only depends on the normalized width of input- and output waveguides and on $\sin \Theta_{eff}$, defined as

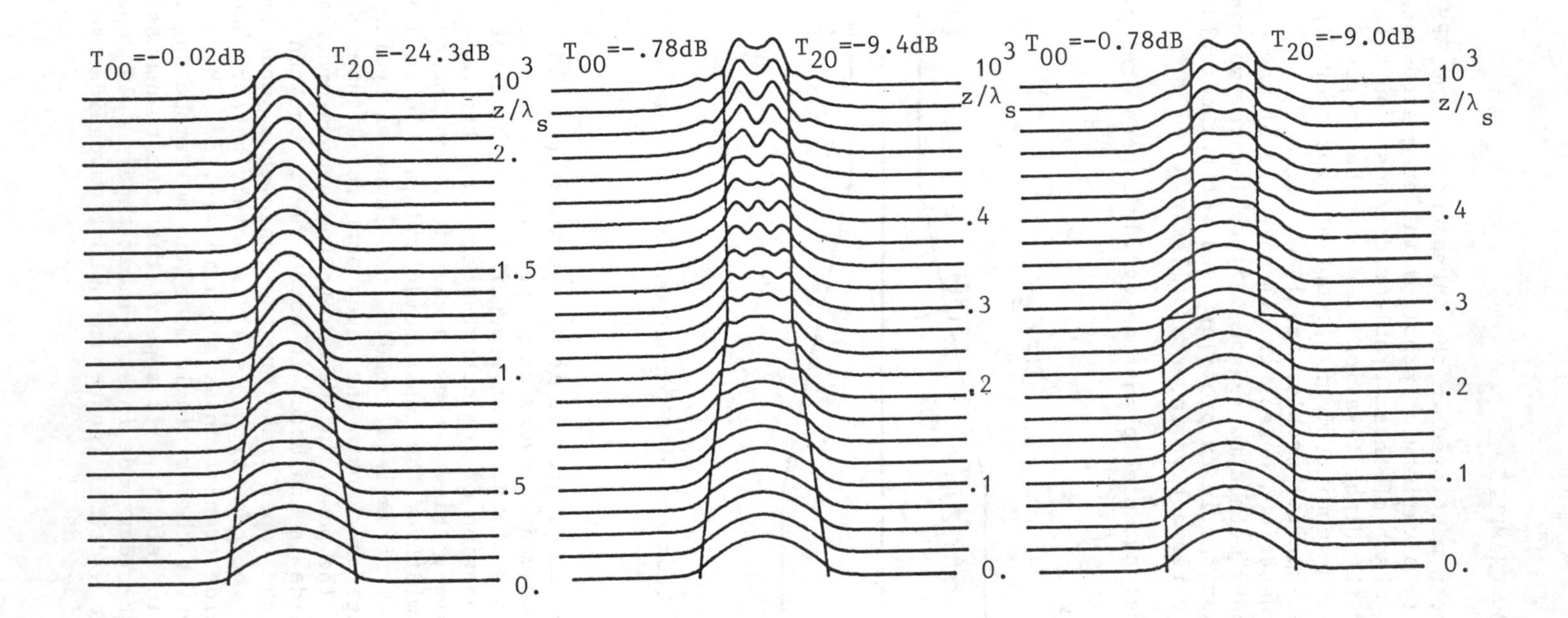

Fig. 8. Example of taper with input width of $40.\lambda_s$, output width of $20.\lambda_s$ and $\Delta n/n_s$ = 0.1 %
a. Θ = 0.5°, b. Θ = 4°, c. Θ = 90°

$$\sin \Theta_{eff} = \frac{\sin\Theta}{N.A.}$$

where N.A. is the numerical aperture of the waveguides in the medium.

The second example shows the propagation through a Y-junction (fig. 9), consisting of two single-mode input ports and one single-mode output port. In this case the refractive index was chosen to be a step profile filtered with a gaussian in the x-direction. In fig. 10 the propagation from each input arm is

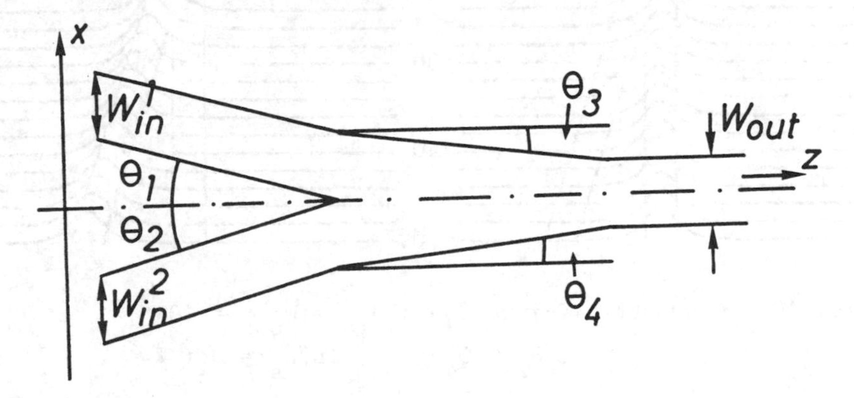

Fig. 9. Geometry of a planar Y-junction

shown for $\Theta_1 = 0$, $\Theta_2 = 2°$, $\Theta_3 = 0$, $\Theta_4 = 2°$. The attenuation for the mode is indicated for each input. It can be shown that the average of these attenuation values (in dB) always exceeds 3dB loss (ref. 11).
The power transfer behaviour of waveguide branches has been analysed in ref. 19. Depending on one waveguide parameter, a linear branch from one bimodal guide to two single mode guides can act as a power divider (e.g. a symmetrical branch) or as a mode splitter. When it is used reverse as a junction, the power divider will always excite both output modes, whereas the mode splitter will excite either the lowest order or the first-order mode, depending on the excited input guide. This phenomenon can easily be studied by means of the BPM. Some results are shown in fig. 11.

Finally, an example of curved waveguides is presented. Bends normally pose a problem to the Beam Propagation Method, because the limitation of paraxiality is exceeded in many geometrics. It would of course be possible to subdivide the structure into smal-

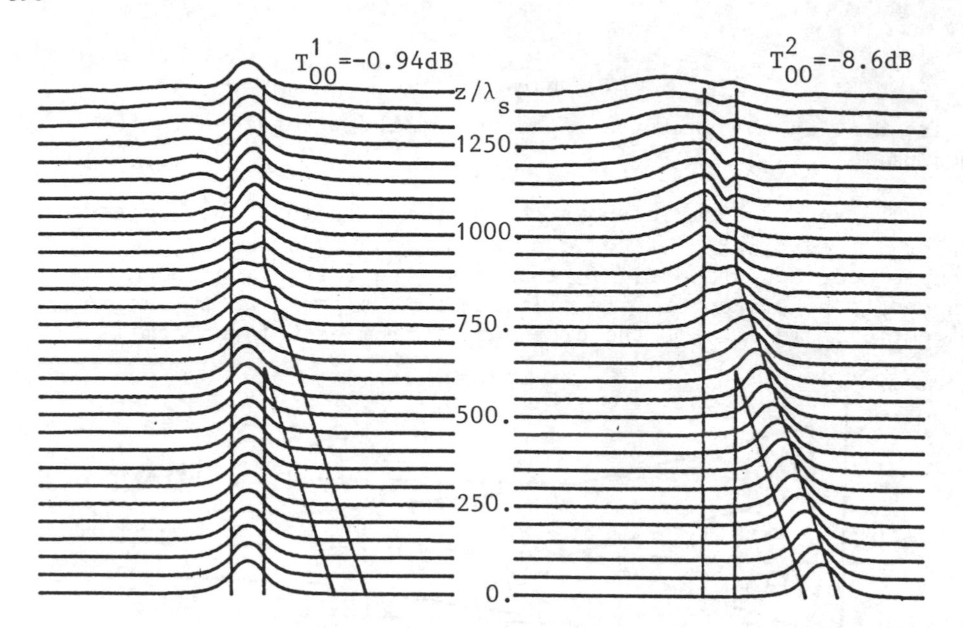

Fig. 10. Asymmetric Y-junction : $W_{in} = W_{out} = 10\lambda_s$, $\Theta_1 = \Theta_3 = 0$, $\Theta_2 = \Theta_4 = 2°$, $\Delta n/n_s = 0.1$ %

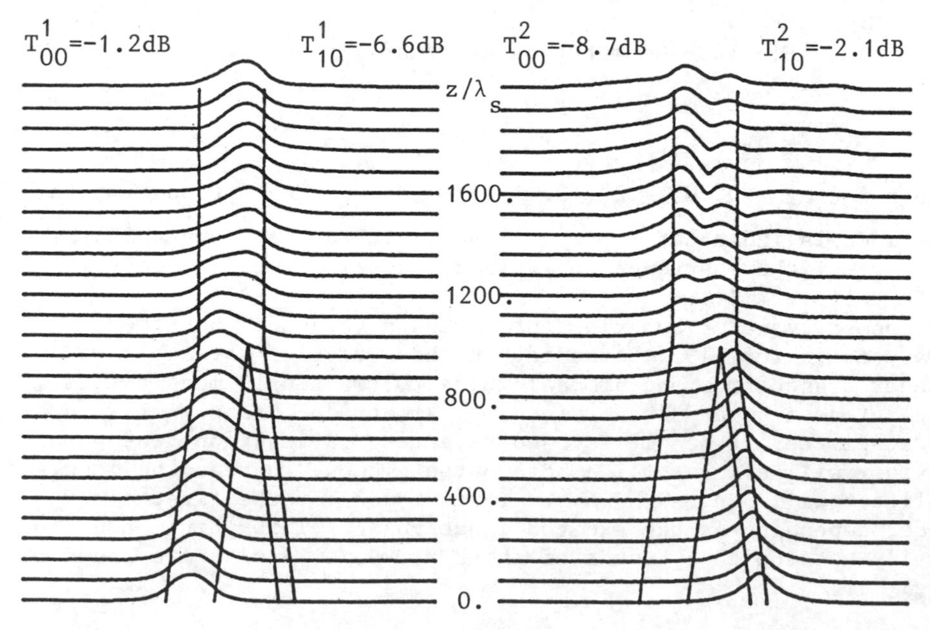

Fig. 11. Mode splitter : $W^1_{in} = 15\lambda_s$, $W^2_{in} = 5\lambda_s$, $W_{out} = 20\lambda_s$, $\Theta_1 = \Theta_s = .58°$, $\Delta n/n_s = 0.1$ %

ler parts, each having a different propagation direction so as to make the power distribution in the spatial frequency domain centralised about the propagation axis. This however leads to problems at the interfaces between the subdomains. For moderate radii of curvature, it is possible though to make use of a conformal mapping technique which transforms the original curved waveguide into a straight waveguide with a modified refractive index profile (ref. 20). In this way the transition losses at interfaces between straight and curved waveguides have been analysed as well as the losses in S-curved waveguides (ref. 13). The field pattern of uniform bend modes can also easily be calculated. Since these modes constantly radiate out of the waveguide bend, special care has to be taken for the choice of the absorber region, especially for sharp bends. Fig. 12 shows both the amplitude and the phase

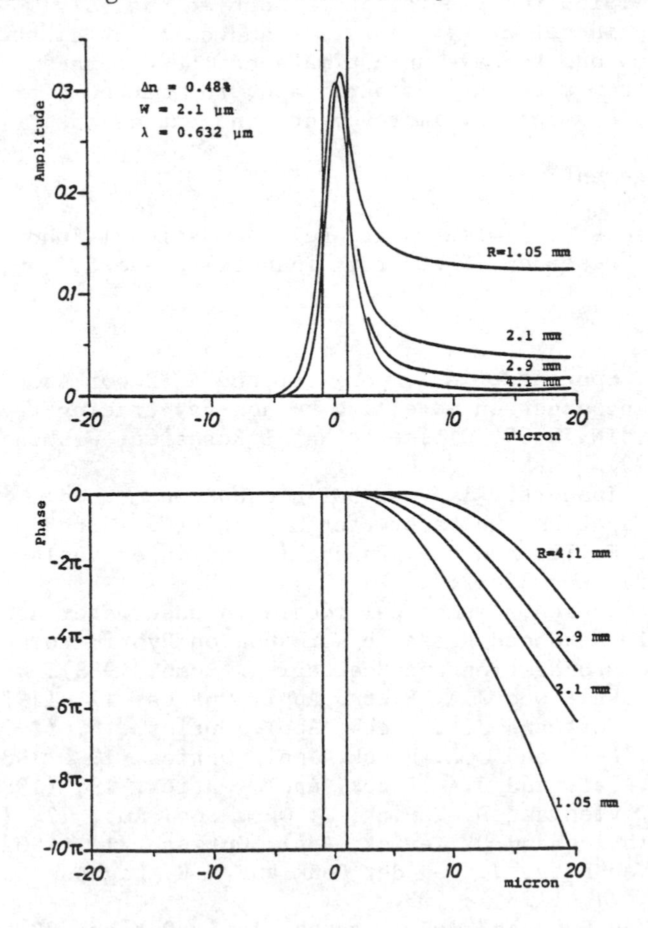

Fig. 12. Amplitude and phase of curved waveguide mode for several values of the radius of curvature

of uniformly curved waveguide modes for step index profiles. It is clear that the mode shifts away from the bend center as the radius of curvature decreases.

6. CONCLUSION

The BPM is a powerful tool for the numerical modeling of scalar wave propagation through media with arbitrary but slow and small variations in the refractive index. The basic algorithm is simple to implement, but care is required in order to obtain correct results. The BPM is relatively new and a lot of work remains to be done to fully understand the possibilities and limitations of the method. An error analysis that takes into account the discretisation and numerical aspects of the BPM remains to be developed. Such an analysis would be a great help in selecting step sizes and sampling distances, and would allow one to make an estimate of the accuracy. As is apparent from the previous paragraph, the method seems to offer considerable scope for improvements and extensions.

Acknowledgement

R. Baets acknowledges the Belgian National Fund for Scientific Research (NFWO) for financial support.

References

1. C.W. Spofford, 'A symposis of the AESD workshop on acoustic propagation modelling by non-ray-tracing techniques', Rpt. TN-73-05, Office of Naval Research, Washington D.C. (1973).
2. F.D. Tappert, 'The parabolic equation method', in 'Wave propagation and Underwater Acoustics', edited by J.B. Keller and J.S. Papadakis (Springer-Verlag, Berlin, 1977), pp. 224-287.
3. F.B. Jensen, 'Numerical models in underwater acoustics', (Nato Advanced Research Workshop on Hybrid Formulation of Wave Propagation and Scattering, Rome, 1983).
4. M.D. Feit and J.A. Fleck, Appl. Optics, 17, (1978), 3990.
5. M.D. Feit and J.A. Fleck, Appl. Optics, 18, (1979), 2843.
6. M.D. Feit and J.A. Fleck, Appl. Optics, 19, (1980), 1154.
7. M.D. Feit and J.A. Fleck, Appl. Optics, 19, (1980), 2240.
8. L. Thylen and D. Yevick, J. Opt. Soc. Am., 72, (1982), 1084.
9. L. Thylen and D. Yevick, Appl. Optics, 21, (1982), 2751.
10. J. Van Roey, J. van der Donk and P.E. Lagasse, J. Opt. Soc. Am., 71, (1981), 803.
11. J. Van Roey and P.E. Lagasse, Appl. Optics, 20, (1981), 423.
12. J. Van Roey and P.E. Lagasse, J. Opt. Soc. Am., 72, (1982), 337.

13. R. Baets and P.E. Lagasse, J. Opt. Soc. Am., 73, (1983), 177.
14. R. Baets and P.E. Lagasse, Appl. Optics, 21, (1982), 1972.
15. J.F. Claerbout, 'Fundamentals of Geophysical Data Processing", (McGraw-Hill, 1976).
16. J.W. Goodman, 'Introduction to Fourier Optics', (McGraw-Hill, 1968).
17. J. van der Donk, 'Toepassing van de bundelpropagatiemethode op problemen van de geïntegreerde optica', Ph.D. thesis, University of Gent, Belgium, 1982.
18. P. Kaczmarski, Internal Report, Lab. Electromagnetism and Acoustics, University of Gent, Belgium.
19. W.K. Burns, A.F. Milton, 'An analytic solution for mode coupling in optical waveguides branches', IEEE J. of Quantum Electronics, QE16, (1980), 446-454.
20. M. Heiblum, J. Harris, 'Analysis of curved optical waveguides by conformal transformation', IEEE J. Quantum Electronics, QE11, (1975), 75-83.

TIME DOMAIN SOLUTION OF NONLINEAR PULSE PROPAGATION

B. E. McDonald and W. A. Kuperman

Numerical Modeling Division
Naval Ocean Research and Development Activity
NSTL, Mississippi 39529 USA

ABSTRACT

This paper develops a two-dimensional (range and depth) theoretical-numerical model for the simulation of nonlinear acoustic pulses and weak shocks in a refracting-diffracting medium which can be range dependent. Nonlinearity is most efficiently treated in the time domain; i.e., the signal is not Fourier analyzed into discrete frequencies. Beginning with a progressive wave equation retaining linear and lowest order nonlinear terms, a nonlinear time domain parabolic equation (PE) is derived in a pulse-following frame of reference. The nonlinear PE gives a natural separation of terms governing the processes of refraction, diffraction, and nonlinear steepening. When the nonlinearity is absent and the signal is taken to be monochromatic, the usual linear frequency domain PE emerges. Thus the model is capable of producing wideband solutions without Fourier synthesis. Results from the model are in excellent agreement with one-dimensional analytic solutions for nonlinear pulse development. Preliminary numerical experiments with the two-dimensional model for an "N" wave incident on a caustic indicate that nonlinearity can weaken the reflected wave and alter its shape.

1. INTRODUCTION

This work is motivated by the desire to model the effects of weak nonlinearity upon ocean acoustic wave propagation in two and three dimensions. We hope to provide some answers to the following question: What are the theoretical and computational ingredients necessary to follow the propagation of a nonlinear pulse in the ocean acoustic waveguide (sound speed profile)? In the past decade the propagation of linear time harmonic signals has been treated successfully using the parabolic approximation to the wave equation [1]. The parabolic equation (PE) assumes small angle scattering of a single frequency wave and is basically a high frequency limit. Finite bandwidth linear pulses may be propagated by separate treatment of all component frequencies. The introduction of nonlinearity into the problem results in interaction among all frequencies. In the time domain, however, the interaction is confined to spatially neighboring points at a single time level. Thus the time domain is highly preferred for calculation of nonlinear effects in broadband pulses.

We seek to construct a model suitable for future investigations of the following problems: (1) Nonlinear pulse propagation in the ocean acoustic waveguide. It is well known that acoustic signals originating in the waveguide undergo an approximate focusing at "convergence zones" separated in range by 30-40 km [2]. The effect of nonlinearity on this focusing is not known; (2) Focusing of a finite amplitude pulse by a lens. This problem has been addressed when the input signal is sinusoidal, using successive corrections to the Lighthill equation [3]. It remains to be seen, however, what happens if a shock discontinuity is present in the incoming wave or focused field; (3) The reflection of a finite amplitude pulse from a caustic surface (see Figure 2). This problem has been addressed theoretically for sonic booms in the atmosphere [4], resulting in some general knowledge of the signal away from the caustic. At the caustic, however, the properties of the signal are not well understood. The model to be described in the next section has been applied to the case of a discontinuous plane wave pulse incident on a caustic. Results will be discussed in Section 3.

2. MODEL FORMULATION

2.1 Governing Equations

We begin with a progressive wave equation [5] retaining lowest order nonlinearity and neglecting attenuation:

$$\frac{\partial^2 \rho'}{\partial t^2} = \nabla^2 C^2 \left(\rho' + \beta \rho'^2 / \rho_o\right) \tag{1}$$

Here ρ' is the acoustic perturbation to the local density ρ_o, c is the local sound speed and β (≈ 3.5 for ocean application) is the dimensionless coefficient of nonlinearity. For a medium with an adiabatic equation of state $p = p(\rho)$ for the pressure, one has

$$C^2 = \frac{\partial p}{\partial \rho} \tag{2}$$

and

$$\beta = 1 + \frac{1}{2} \rho_o \frac{\partial^2 P}{\partial \rho^2} / \frac{\partial P}{\partial \rho} . \tag{3}$$

Implicit in (1) are assumptions of weak nonlinearity and propagation within a small angle cone about a preferred direction. Equation (1) may be tidied up slightly by the introduction of a dimensionaless variable

$$R \equiv \rho' / \rho_o . \tag{4}$$

Taking ρ_o constant, (1) becomes

$$\frac{\partial^2 R}{\partial t^2} = \nabla^2 C^2 (R + \beta R^2) . \tag{5}$$

For the purpose of following the evolution of a pulse propagating through the medium, it is computationally inefficient to use (5) in the rest frame of the medium. Time resolution requirements would be determined by the timescale of the pulse's passage over a fixed point, whereas the timescale of interest is that of the gradual development of the pulse itself as it traverses large distances. Thus (5) should be recast in a frame moving with some average speed C_o descriptive of the pulse's propagation. This is achieved by the substitution

$$D_t \equiv \frac{\partial}{\partial t} + C_o \frac{\partial}{\partial r} , \tag{6}$$

where r is the range variable in the primary direction of propagation. We assume azimuthal symmetry and take z to be the variable transverse to r. With the substitution

$$C(r, Z) \equiv C_o + C_1(r, Z), \tag{7}$$

where C_1 is presumably small, (5) becomes

$$\left(D_t - C_o \frac{\partial}{\partial r}\right)^2 R = \left(\frac{\partial^2}{\partial r^2} + \frac{1}{r} \frac{\partial}{\partial r} + \frac{\partial^2}{\partial Z^2}\right) (C_o + C_1)^2 (R + \beta R^2). \tag{8}$$

$$\uparrow \qquad\qquad\qquad\qquad \uparrow \qquad \uparrow \qquad\qquad \uparrow \qquad \uparrow$$

The arrows under (8) indicate terms within each bracket which are assumed to be small. Proceeding from left to right in (8), the smallness assumptions have the following physical interpretation: (a) the evolution of the pulse is gradual compared to its transit time over a given point; (b) the gradient scale size in range is much smaller than the total range; (c) the variation in the transverse direction is more gradual than that in range; (d) spatial fluctuations in the sound speed are small; and (e) the nonlinearity is weak. Expanding (8) and keeping only the lowest order in "small" terms yields an equation which may then be integrated with respect to r. The result is

$$D_t R = -\frac{\partial}{\partial r}\left(C_1 R + \frac{C_o}{2}\beta R^2\right) - \frac{C_o}{2r}R - \frac{C_o}{2}\int^r \frac{\partial^2 R}{\partial Z^2}\,dr. \tag{9}$$

We refer to (9) as the time domain parabolic equation. Connection with the more familiar frequency domain PE will be demonstrated below. Each term on the right hand side of (9) describes a distinct physical process. Proceeding left to right these are refraction, nonlinear steepening, radial spreading and diffraction. The extraction of equation (9) from (5) has not only separated out distinct physical processes, but has also reduced the problem from second order in time to first order. This simplifies calculations, and corresponds physically to the selection of the relevant propagation characteristic (forward going as opposed to backward going).

A few points concerning the nonlinear time domain PE should be stressed. Equation (9) illustrates how the presence of a finite amplitude disturbance affects the medium: the nonlinearity can be thought of as effective perturbation in the local sound speed. That is, the substitution $C_1 \rightarrow C_1 + 1/2\,\beta C_o R$ achieves the same result. Where the disturbance is positive (negative) the local propagation speed is increased (decreased). As illustrated below this can lead to the well known steepening of initially smooth pulses into shock fronts.

For linear propagation, an obvious feature of the time domain PE is that it permits calculation of wideband pulses without Fourier analysis. This may or may not be an advantage, depending upon the problem under consideration and the level of access one has to existing propagation codes.

2.2 Relation to the Frequency Domain PE

The properties of equation (9) for a linear monochromatic wave are illustrated by making the following far-field substitution:

$$R(r,Z) = r^{-1/2}\, e^{i(kr-\omega t)}\, f(r,Z), \tag{10}$$

where

$$\omega = k\, C_o \tag{11}$$

Taking the limit of large k after substituting (10) into (9) with $\beta = 0$ gives

$$\frac{\partial f}{\partial r} = -ik\frac{C_1}{C_o}\, f + \frac{i}{2k}\,\frac{\partial^2 f}{\partial Z^2}\,. \tag{12}$$

This is the usual frequency domain PE. Thus we are assured that (9) is a proper generalization of the single frequency PE since it contains no reference to a frequency or wavenumber.

2.3 Numerical Methods

The individual terms in (9) are treated by timestep splitting; i.e., each term is integrated separately by a scheme with appropriate stability and accuracy. The total timestep is accurate as long as the changes within each sub-step are "small".

The radial spreading term in (9) is simplest to account for. Denoting radial and transverse grid points by i and k respectively, we have

$$R_{i,k} \longrightarrow \left(\frac{r_i}{r_i + C_o\,\delta t}\right)^{1/2} R_{i,k} \tag{13a}$$

and

$$r_i \longrightarrow r_i + C_o\,\delta t \tag{13b}$$

The arrows represent replacement of an old value by the indicated new value.

The diffraction term in (9) is integrated by the following finite difference techniques: trapezoidal rule for the r integral, second order centered differences for the second z derivative, and a Crank-Nicholson representation for the time derivative. The resulting process for the diffraction step is implicit, second order in space and time, requiring a tridiagonal matrix inversion at each successive r_i. The combination is stable for all space and time steps. The use of an explicit time differencing scheme would appear an attractive alternative to Crank-Nicholson for integration on a vector computer, but a severe stability limit on the timestep offsets most of the gain in efficiency.

The terms in (9) requiring the most careful numerical treatment are refraction and steepening. Both are integrated in a single advective step using a new scheme [6] of the flux correction type [7,8]. During the past decade such methods have been developed for dealing with Gibbs' oscillations which form when a highly structured profile is advected in a fixed Eulerian grid. These methods [7,8] effectively use a nonlinear filter to switch locally and conservatively from a high order finite difference scheme to a low order one in the vicinity of artificial oscillations that may try to form.

Our scheme [6] defines the direction of information flow to be that of the relevant propagation characteristic. This direction is used in two upwind schemes (one first order and the other second order) which are combined in a monotonicity-preserving hybrid scheme by the method of flux correction.

3. RESULTS

3.1 Comparison with 1D Analytic Solution

In one dimension the time domain PE reduces to

$$D_t R = -\frac{\partial}{\partial X}\left(C_1 R + \beta \frac{C_o}{2} R^2\right), \tag{14}$$

where the cartesian coordinate x has replaced r. When C_1, β, and C_o are constant, (14) may admit closed form analytic solution, depending on initial conditions. When the initial condition has R quadratic in x, the profile at later times is given by solution of a quadratic equation with time dependent coefficients. Details are given in [6]. Such a case is shown in Figure 1 where two parabolic lobes are used to approximate one cycle of a sinusoid. Normalizing R_{max} to unity in the initial condition parameters are chosen so that $\beta C_o \delta t / \delta x = 0.25$ and $C_1 \delta t / \delta x = -.0875$. The characteristic speed changes sign where R = .35, providing a test of stability against formation of expansion shocks at that location. The integration proceeds using the scheme [6] resulting in excellent agreement between numeric and analytic results. The numerical grid consists of 202 equally spaced points. Periodic boundaries are used in this test, resulting in a reentry of the left shock front as shown at step 601. Shock development and propagation are predicted accurately without the use of Lagrangian markers to define the shock location. These properties follow from an accurate treatment of fluxes [6].

3.2 Reflection of a Finite Amplitude Pulse From a Caustic

The reflection of a plane wave pulse from a caustic is inherently a two dimensional problem. Thus we retain all terms in (9) except radial spreading and replace r by x. The geometry of the problem is shown in Figure 2. The sound speed in the medium is taken to be constant below some depth Z_o and to increase linearly above Z_o. Linear ray theory has the rays being straight below Z_o and circular arcs above Z_o. The radius of the arcs is $C_o/(dc/dz)$, where C_o is the sound speed at the apex. In this geometry two adjacent parallel rays with positive vertical

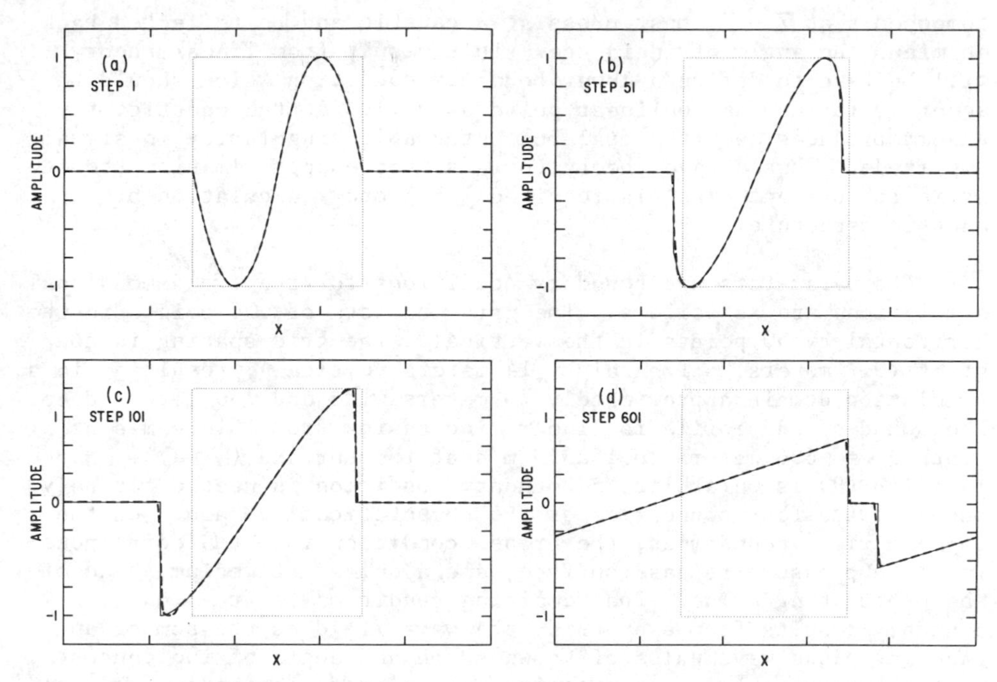

Figure 1. Development of a one dimensional pulse from a smooth initial condition into an aging "N" wave shock. Solid line: analytic solution; dashed line: numerical results from our algorithm [6]. A leftward drift and periodic boundaries are added to demonstrate stability against formation of expansion shocks.

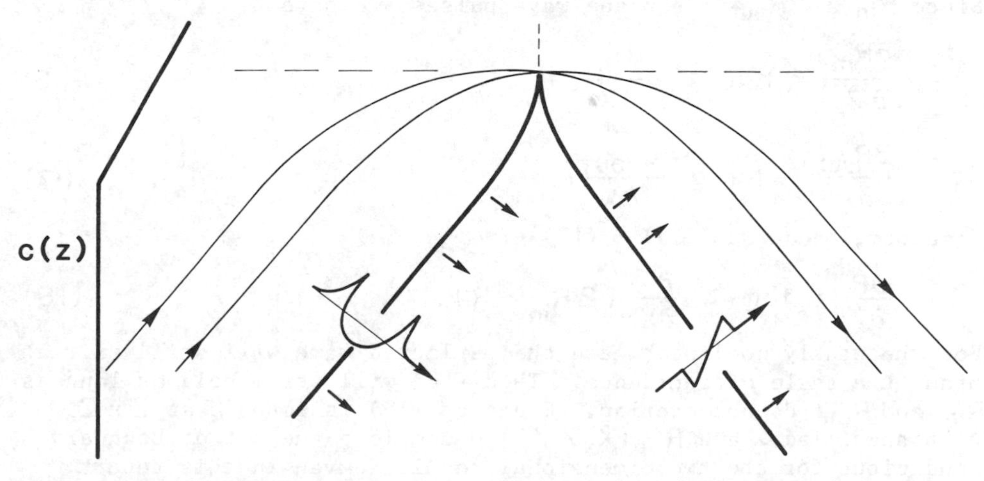

Figure 2. Wavefronts (heavy lines) and rays (thin lines) incident on a caustic indicated by horizontal dashed line. Arrows indicate direction of propagation. The calculation zone for the present model ranges upward from the breakpoint in C(z).

components at $Z = Z_o$ must cross at a caustic and be reflected back at minus the angle of incidence. This result from linear theory will be used to define a lower boundary condition which should be accurate for a weak nonlinear pulse as well. At the caustic, ray theory produces an artificial but integrable singularity in signal amplitude. Thus a wave theory (or a suitable approximation thereto as in the present case) is required for proper calculation of caustic structure.

The parameters and boundary conditions for the two dimensional simulations are as follows. The grid consists of 100 points in the horizontal by 50 points in the vertical. The grid spacing is constant (0.7 meters horizontal by 14 meters vertical), resulting in a simulation domain approximately 70 meters wide and 700 meters deep. The sound speed profile is linear, increasing from 1516.4 m/s at a depth Z_o = -686 meters to 1523.5 m/s at the surface ($Z = 0$). Since equation (9) is parabolic, a boundary condition is needed for only three of the four boundaries of the computational domain. On the top and right boundaries, the proper condition is $R = 0$ corresponding to a pressure-release surface, and a quiescent medium ahead of the propagating pulse. The remaining condition for the lower boundary results from expressing the wave field as the sum of an incoming plane wave pulse of known shape and angle of incidence α to the horizontal plus an outgoing wave propagating at an angle $-\alpha$ to the horizontal. Let the known incident wave be $R_{in}(X,Z,t)$ and the unknown outgoing wave be $R_{out}(X,Z,t)$ for $Z \leq Z_o$. Then for linear pulses

$$R(X, Z_o, t) = R_{in}(X, Z_o, t) + R_{out}(X, Z_o, t). \tag{15}$$

Since R_{in} and R_{out} are plane wave pulses, we have

$$\frac{\partial R_{in}}{\partial Z} = \tan\alpha \frac{\partial R_{in}}{\partial X} \tag{16}$$

$$\frac{\partial R_{out}}{\partial Z} = -\tan\alpha \frac{\partial R_{out}}{\partial X}. \tag{17}$$

Together, equations (15) - (17) give

$$\frac{\partial R}{\partial Z} = \tan\alpha \frac{\partial}{\partial X}(2R_{in} - R). \tag{18}$$

For the weakly nonlinear case the reflected wave will still exit at minus the angle of incidence. Then (18) will still hold as long as R_{in} and R_{out} do not overlap. Equation (18) is imposed at $Z = Z_o$ with specified α and $R_{in}(X, Z, t)$ to complete the set of boundary conditions for the two dimensional results given in this report. For a range independent sound speed profile, we take

$$\sec\alpha = C_o / C(Z_o) \tag{19}$$

in accordance with Snell's law. Taking C_o = 1520 m/s and $C(Z_o)$ = 1516.4 m/s gives α = 3.94° and results in a caustic at a depth of 336 meters, where $C(Z) = C_o$. With $X_i = i\,\delta X$ and δX = 0.7 m, the incoming wave is taken to be

$$R_{in}(X_i, Z_o, t) = 3.7 \times 10^{-5} \times \begin{cases} i-82.5, & 72.5 \le i \le 92.5 \\ 0 & \text{otherwise} \end{cases} \qquad (20)$$

From (2), this results in a peak pressure of approximately 8.5 atmospheres. This large value was chosen somewhat arbitrarily to illustrate nonlinear pulse development. The form of the incoming wave (20) is a so-called "N wave", which could result from an explosive disturbance a large distance away. Figure 3 shows the transient behavior as the wave enters the calculation zone, turns over at the caustic, and is reflected. Shown together are results from a linear case ($\beta = 0$ in (9)) and a nonlinear case (β = 3.5). By step 401 the wave at the caustic in both cases has reached a steady state. This is confirmed by calculations (not shown) in which the solutions were carried out an extra 400 timesteps. A steady state is possible only because the form of the incident wave is held fixed in time and the radial spreading term is omitted from (9). The results are therefore intended to be illustrative, and do not claim to represent an actual oceanic event. In the atmosphere, however, a steady state shock incident on a caustic could be generated by the flight of a supersonic aircraft [4].

The results given in Figure 3 contain several points of qualitative agreement with expectations to be discussed briefly. Statements concerning the quantitative accuracy of the results will await validation studies involving the sensitivity of the results to temporal and spatial resolution.

The linear and nonlinear calculations in Figure 3 are not significantly different for the first 100 steps. The disturbance imposed on the lower boundary must enter the region via the diffraction term in (9). Only after a substantial amount of disturbance has entered can nonlinear steepening take place. The difference becomes apparent at step 201 where positive (negative) contours in the nonlinear case have shifted forward (backward) relative to their position in the linear case. By step 401 both linear and nonlinear cases have reached a steady state. The approximate symmetry of contours just below the caustic (depth 400 m) in the linear case has been replaced in the nonlinear case by a pair of shock waves: a Mach stem on the left at a depth of approximately 350 m, and a weaker shock on the right at a depth of approximately 550 m. The points of qualitative agreement with expectations are: (a) The outgoing waves exit at minus the angle of incidence. This is imposed on the lower boundary, but in fact holds true in the region of integration as well. (b) A Mach stem exists at the juncture of shocks in the nonlinear case. (c) The

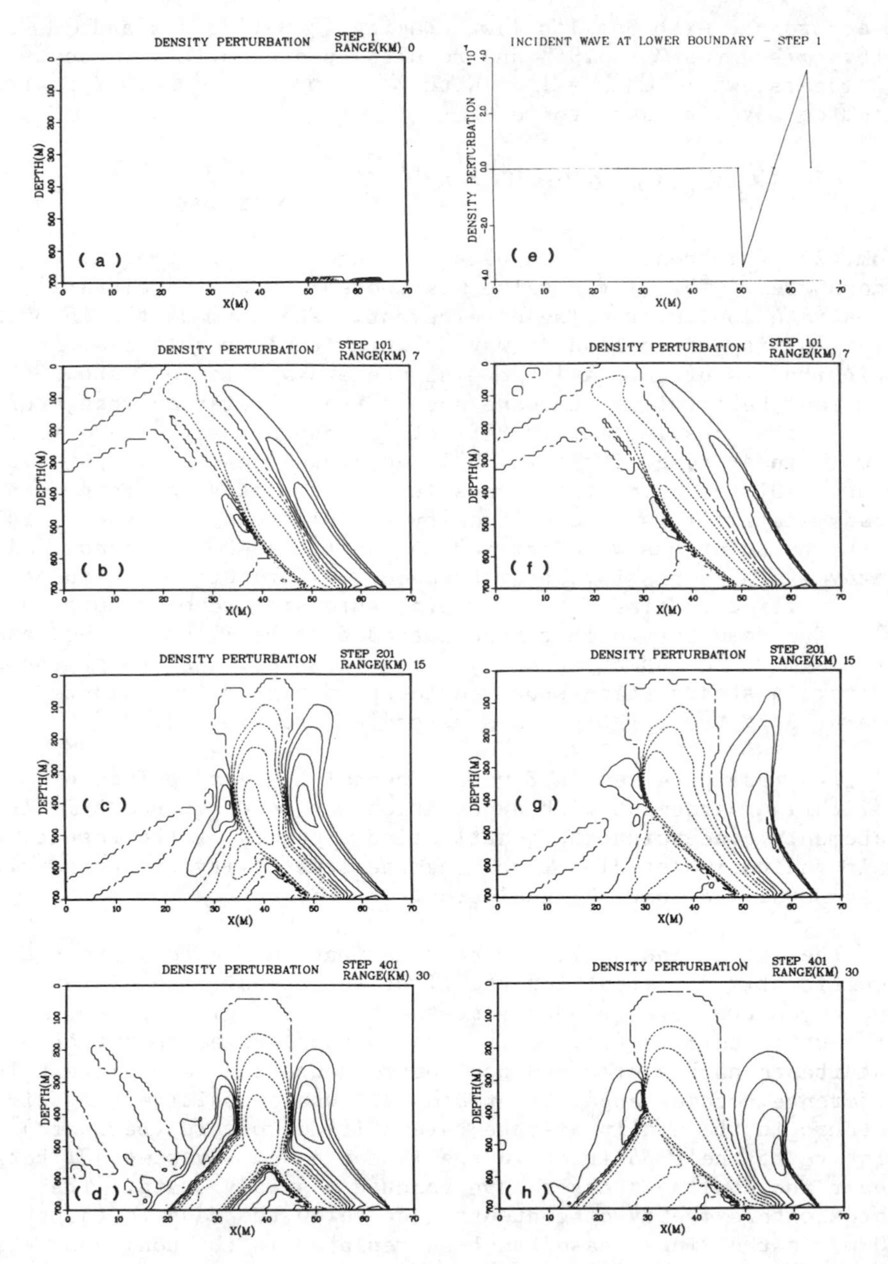

Figure 3. R(x,z) at successive times for a plane "N" wave incident on a caustic. (a)-(d): linear case, $\beta = 0$. (f)-(h): nonlinear case, $\beta = 3.5$. Contour values ($\times 10^{-4}$): -4, -2, -1, -.5, 0, .5, 1, 2, 4. Dashed contours: negative; dot-dash: zero; solid: positive. Time has been converted to range by $C_o = 1520$ m/s.

sonic points (location of vertical contours) are displaced properly in the nonlinear case: downward for the positive portion of the incoming wave ($\sim$550 m) and upward for the negative portion ($\sim$350 m) relative to the linear case (both $\sim$420 m). This reflects the amplitude dependence of propagation speed in the nonlinear case. (d) In the linear case, an incoming "N" wave is reflected as a "U" wave [4] (two positive maxima surrounding a negative minimum). This final point is illustrated in Figure 4a.

Qualitative features of the results which may require further investigation are as follows: (a) The injection of the incoming wave is accompanied by a transient evident in the contours of Figure 3 near the injection region. This is probably a result of differences between the first order numerics used to represent the boundary condition (18) and the second order numerics used in the interior. (b) The reflected nonlinear signal is considerably weaker than the reflected linear signal (Figure 4). This is evident in the amplitude of the shock outbound from the Mach stem (Figures 3h and 4b). (The expected outbound wave from the sonic point at 550 m depth in Figure 3h falls outside the calculation region.) A physically plausable explanation is as follows. The linear case in our model is free of dissipation, so the energy in the reflected wave should equal that of the incident wave. The shocks formed in the nonlinear case, however, transform wave energy into entropy and thus represent a loss mechanism. (c) The "U" wave reflection in Figure 4a is free from the logarithmic spikes predicted from linear theory for an N wave incident on the caustic at a small angle α to the horizontal. This probably results from a combination of finite α and numerical resolution effects. In the present case $\alpha \approx .07$ radians, whereas the smallness criterion used in deriving the logarithmic spike result is $\alpha \ll (\lambda dc/dz)/C_0 \approx 10^{-4}$. Here λ = 20 m is the largest wavelength descriptive of the pulse, and the length by which it is normalized is the ray radius of curvature.

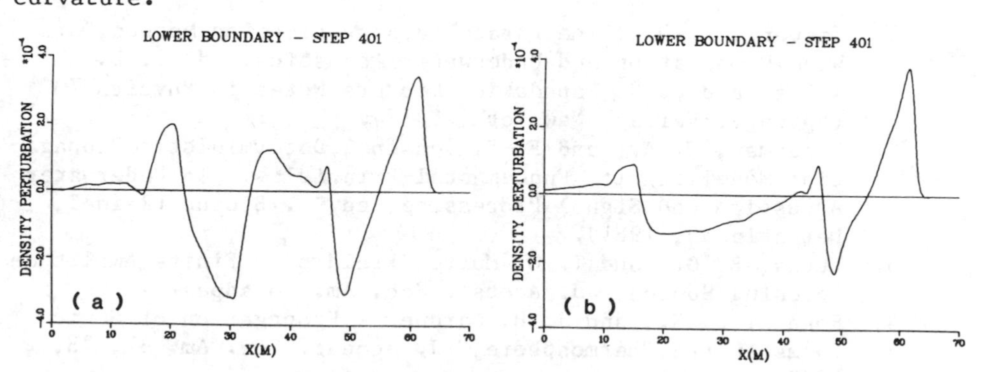

Figure 4. R vs. x near the lower boundary (specifically, line 3, where the incoming transient has died away) at step 401. (a): linear case, β = 0. (b): nonlinear case, β = 3.5.

4. SUMMARY

The time domain parabolic equation (9) provides a mechanism for (a) incorporating nonlinearity into a multidimensional propagation model and (b) calculating broadband pulse development without Fourier analysis. The time domain PE (9) contains separate terms for physical processes of refraction, nonlinear steepening, spreading and diffraction. A numerical model has been constructed for the time integration of (9) and a new algorithm proposed for the refractive and nonlinear terms [6].

Results in one dimension are in excellent agreement with analytic solutions describing formation of shock discontinuities from smooth initial conditions. Results in two dimensions for an "N" wave incident on a caustic are in qualitative agreement with expectations (analytic solutions are not available) on a number of points. Further investigation is needed to determine the quantitative accuracy of the two dimensional results.

The most important single implication of the numerical results is that weak nonlinearity at a caustic can have a significant effect on the form and amplitude of a reflected pulse, with noticeable energy loss due to shock formation. This means, for example, that a nonlinear calculation which may be valid within ray tubes [4] up to a caustic may not be able to rely on a linear connection procedure between incoming and outgoing waves.

5. ACKNOWLEDGMENT

This work was supported by the Office of Naval Research.

6. REFERENCES

1. Tappert, F. D., "The Parabolic Approximation Method," in Wave Propagation and Underwater Acoustics, ed. J. B. Keller and J. S. Papadakis, Lecture Notes in Physics 70 (Springer-Verlag, New York, 1977).
2. Kuperman, W. A., and F. B. Jensen, "Deterministic Propagation Modelling I: Fundamental Principles," in Underwater Acoustics and Signal Processing, ed. L. Bjorno (Reidel, Netherlands, 1981).
3. Lucas, B. G., and T. G. Muir, "Field of a Finite-Amplitude Focusing Source," J. Acoust. Soc. Am. to appear.
4. Rogers, P. H., and J. H. Gardner, "Propagation of Sonic Booms in the Thermosphere," J. Acoust. Soc. Am. 67, 78, 1980.
5. Hamilton, M. F., "Diffraction in Beams," in lecture notes on nonlinear acoustics, D. Blackstock, Univ. of Texas (unpublished).

6. McDonald, B. E., and J. Ambrosiano, "High-Order Upwind Flux Methods for Hyperbolic Conservation Laws," J. Comp. Phys. to appear.
7. Boris, J. P., and D. L. Book, "Flux-Corrected Transport I: SHASTA-A Fluid Transport Algorithm that Works," J. Comp. Phys. 11, 38, 1973.
8. Zalesak, S. T., "Fully Multidimensional Flux-Corrected Algorithms for Fluids," J. Comp. Phys. 31, 335, 1979.

STRATIFICATION METHODS IN WAVEGUIDING STRUCTURES

G. COPPA, P. DI VITA, M. POTENZA, U. ROSSI

CSELT, Centro Studi e Laboratori Telecomunicazioni S.p.A.
Via G. Reiss Romoli, 274 - 10148 TORINO (Italy)

ABSTRACT

Stratification methods to solve waveguiding problems in planar and circular structures are presented. They are based on the approximation of the actual index profile of the guide by means of a wide class of analytically solvable curves (quartic profiles for circular fibres; quartic, Epstein and linear profiles for planar guides). The method allows to deduce accurately the propagation characteristics in the least computing time, particularly for guides with non-monotonic index profiles.

1. INTRODUCTION

Waveguiding phenomena take place in several branches of physical sciences and especially in acoustics and electromagnetics [1]. A lot of these problems can be analyzed in terms of a scalar wave (Helmholtz) equation. The aim of this work is to discuss the various problems to which this kind of equation can be applied, with boundary conditions suitable to describe confined propagation, and to propose an improved stratification method for arbitrary profiles (see below), both for planar and circular structures. The stress will be put on optical applications of guided propagation in dielectric structures, but it is evident that the present method of solution can prove useful to analyze similar waveguiding problems in other physical situations. Stratification methods were originally proposed in [2] for circular guides, while applications to planar structures had already been discussed, for example in [3]. These early methods were based on a "staircase" approximation to the real refractive-index profile of the guide: the interval of variability of the profile is divided into q subintervals (of different lengths) where the approximating function is assumed piecewise constant and where the waveguiding problem is analytically solvable. It is clear that in the limit as $q \to \infty$, as the thickness of each subinterval goes to zero, the exact profile is reproduced. Such a method allows to calculate dispersion curves, group velocities and so on. The authors claimed that a few layers were quite enough to calculate all relevant quantities with sufficient accuracy. In [4] however, the method was criticized as far as its computational stability and computing time were concerned, showing that also for q as high as forty, there were anomalies in the calculation of group-delay values. Similar considerations were done in [5]. Starting from these observations, comparisons among the staircase stratification method and the direct Runge-Kutta numerical solution of the scalar wave equation for the given profile were made [6], indicating a significant superiority of the direct method for comparable orders of magnitude of the error in the propagation characteristics. This was due to the fact that the stratification method requires generation of redundant information with respect to more direct methods of solution for the scalar wave equation [7]. So at that stage staircase stratification methods remained useful only for profiles which were actually multistep [8, 9] or to obtain some analytical results [10, 11]. However, the capabilities of this approach were notably improved by the use of a more suitable set of approximating functions [12].

In fact a generic refractive-index profile n(r) is better approximated by a sequence of parabolas than by a staircase; it is even more so for optimised profiles, which are nearly parabolic. Of course, also in this case the solution is known analytically over each subinterval into which the multi-parabolic (MP) approximation is split [12, 13]. Comparisons of accuracy and computing time to evaluate propagation constants and group delays between the MP-method and the Runge-Kutta (RK) one have been performed for various index profiles and modes [12]: in each case a significant superiority of the MP-stratification method over RK-one has been obtained. Finally a more refined version of this stratification method has been recently proposed [14], which enlarges the attractive capabilities of the MP-technique to more complex, non-monotonic refractive-index profiles, often encountered in particular lightguides as monomode fibres. This is the multi-quartic (MQ) method in which the actual profile is piecewise approximated by suitable quartic functions which allow a good reproduction of particulars in the index profile such as dips, bumps and ripples. In all these cases a proper choice for the approximating functions is very effective in reducing the number q of stratification sublayers and then the computing time required by the method. A comparison with direct methods proves the superiority of the MQ-approach.

As already pointed out, staircase stratification methods have been applied to planar structures as well [3, 15], also in the case of anisotropic media [16, 17]. The same argument which has led from the original staircase approximation to the MQ-one for optical fibres can be applied to planar guides too. In this latter case at least three families of analytical solutions which can be employed for piecewise approximation of an arbitrary profile are available: the quartic one (as in the previous case), the linear one and the Epstein [18] one. The first case obviously includes the parabolic variation, while the Epstein layer covers, as particular cases, the exponential variation, the Morse-type profile and the inverse squared hyperbolic cosine function (see [19] and references quoted therein). All these analytical solutions can be utilized as "bricks" to approximate in the best way (the meaning of this statement will be made clear below) a given refractive-index profile.

2. THE HELMHOLTZ SCALAR EQUATION AND THE STRATIFICATION METHODS

It is well known that in dielectric structures the application of Maxwell equations leads to a vector wave equation for the guided field components. If the guide index profile varies slowly over the distance of the wavelength, then the vector wave equation for time-harmonic fields reduces to a scalar Helmholtz equation for the transverse components, both for planar and cylindrical structures [20, 21 and references therein]. Then in both cases we can write:

$$\Delta \quad \phi(\mathbf{R}) + k^2(\mathbf{R})\,\phi(\mathbf{R}) = 0 \tag{1}$$

R is the position vector inside the guide whose longitudinal axis is the z-axis; in the transverse plane a pair of cartesian coordinates (x, y) for planar structures, or a pair of polar coordinates (r,

ψ) for circular guides are used. The function $\phi(\mathbf{R})$ is a transverse field component (longitudinal components can be obtained, if needed, from these through Maxwell's equations), while $k(\mathbf{R})$ is the space dependent wavenumber given by

$$k^2(\mathbf{R}) = k_0^2 \, n^2(\mathbf{R}) = \left(\frac{\omega}{c}\right)^2 n^2(\mathbf{R}) \tag{2}$$

in terms of the refractive-index profile $n(\mathbf{R})$ (which may be a complex-valued function if gain/loss phenomena are present in the dielectric medium) and of the vacuum wavenumber $k_0 = 2\pi/\lambda_0$. The vacuum speed of light is indicated by c while ω is the angular frequency of the guided optical field whose vacuum wavelength is λ_0. Having in mind guided wave solutions to Eq. (1), the transverse optical field is usually looked for in the form

$$\phi(\mathbf{R}) = \begin{cases} \varphi(x, y)\, e^{-i\beta z} & \text{for planar structures} \quad (3a) \\ \varphi(r)\, e^{-i(\beta z + \nu\psi)} & \text{for circularly symmetric guides} \quad (3b) \end{cases}$$

β is the (complex) axial propagation constant while ν is the azimuthal mode number for circular guides. In this last case Eq. (3b) reduces Eq. (1) directly to a one-dimensional radial waveguiding problem. The case of planar structures can be similarly reduced to one-dimensional problems when the refractive-index distribution $n^2(x, y)$ (we suppose the guide translationally invariant along z) is separable in the transverse coordinates. However, even when the separation does not occur exactly, one can use approximate methods such as the effective dielectric constant procedure [19, 22 and references therein quoted] or similar ones [23] to obtain a "pseudo-separation" of the transverse effects: both the rectangular waveguides and the two-dimensional active guides present in semiconductor laser devices can be treated in this way. In the former case a dielectric material of rectangular cross-section is surrounded on its four sides by other materials of lower refractive-indices. In the second example the waveguiding is obtained by means of a sharp index variation along a transverse (say y) direction and a weaker, smooth variation in the orthogonal, lateral (say x) direction of refractive index or optical gain. In both cases the field amplitude $\varphi(x, y)$ of Eq. (3a) can be reasonably approximated [19, 24] by a product of the form $X(x)\,Y(y)$ (or more generally by a superposition of such terms). The waveguiding mechanism in the lateral direction can then be expressed through the so called effective-index profile $n_{eff}(x)$ where both x-and y-confining actions fall in suitably [19, 22, 25, 26].

Summarizing, the analysis of a wide class of waveguiding phenomena in planar structures leads to solving the following equation:

$$\varphi''(x) + [k^2(x) - \beta^2]\,\varphi(x) = 0 \tag{4}$$

while in circular structures we deal with

$$\varphi''(r) + \frac{1}{r}\varphi'(r) + [k^2(r) - \beta^2 - \frac{\nu^2}{r^2}]\,\varphi(r) = 0 \tag{5}$$

obtained by substituting (3) in (1).

3. ANALYTICAL SOLUTIONS FOR THE ONE-DIMENSIONAL FIELD EQUATIONS

We first consider the quartic profile in circular fibres and then in planar guides together with the linear and the Epstein profiles.

3.1 Circular fibres

The most general refractive-index profile that can be solved in the case of circular fibres is the quartic one [14], see Fig. 1a:

$$k^2(r) = A - B\,r^2 - \frac{C}{r^2}\ . \tag{6}$$

Eq. (15) with this profile becomes:

$$\varphi''(r) + \frac{1}{r}\varphi'(r) + [A - \beta^2 - B\,r^2 - \frac{C+\nu^2}{r^2}]\,\varphi(r) = 0 \tag{7}$$

whose general solution is [14]:

$$\varphi(r) = r^{\sqrt{C+\nu^2}}\ e^{-\frac{1}{2}\sqrt{B}\,r^2} \cdot \Big\{ D \cdot M\left(\frac{1+\sqrt{C+\nu^2}}{2} + \frac{\beta^2 - A}{4\sqrt{B}},\ 1+\sqrt{C+\nu^2},\ \sqrt{B}\,r^2\right)$$

$$+ E \cdot U\left(\frac{1+\sqrt{C+\nu^2}}{2} + \frac{\beta^2 - A}{4\sqrt{B}},\ 1+\sqrt{C+\nu^2},\ \sqrt{B}\,r^2\right)\Big\} \tag{8}$$

M and U are the confluent hypergeometric functions [27]. The integration constants D, E are determined by imposing continuity of the logarithmic derivative of the field at the interface of each layer and the convergence of the field for $r \to 0$ and for $r \to \infty$. The quartic profile (6) includes as particular cases the parabolic (C = 0) and constant profiles (B = C = 0).

3.2 Planar guides

The following three families of profiles can be analytically solved in the case of planar structures: the quartic, the Epstein and the linear profiles. All of these can be given a further parameter after the translation $x \to x - x_0$ [cf. Eqs. (11), (19)], which will not be considered explicitly for sake of brevity.

The *quartic profile*

$$k^2(x) = A - B\,x^2 - \frac{C}{x^2} \tag{9}$$

has a solution directly related to (8):

$$\varphi(x) = x^{\sqrt{C+1/4}}\ e^{-\frac{1}{2}\sqrt{B}\,x^2} \cdot \Big\{ D \cdot M\left(\frac{1+\sqrt{C+1/4}}{2} + \frac{\beta^2 - A}{4\sqrt{B}},\ 1+\sqrt{C+1/4},\ \sqrt{B}\,x^2\right)$$

$$+ E \cdot U\left(\frac{1+\sqrt{C+1/4}}{2} + \frac{\beta^2 - A}{4\sqrt{B}},\ 1+\sqrt{C+1/4},\ \sqrt{B}\,x^2\right)\Big\}; \tag{10}$$

as in the case of Eq.(8) similar boundary conditions specify the integration constants D, E. Also in this case the profile (9) can be applied, with $C \neq 0$, only on intervals not containing the origin x = 0. The case C = 0 gives the parabolic profile and for B = C = 0 one has a step profile.

The general form of the ***Epstein profile*** [18], is (see Fig. 1b):

$$k^2(x) = A + B\,\frac{e^{x/a}}{1+e^{x/a}} + C\,\frac{e^{x/a}}{(1+e^{x/a})^2}\ ; \tag{11}$$

its asymptotic behaviour is asymmetrical ($a > 0$):

$$\lim_{x \to -\infty} k^2(x) = A\ ,\quad \lim_{x \to +\infty} k^2(x) = A + B\ . \tag{12}$$

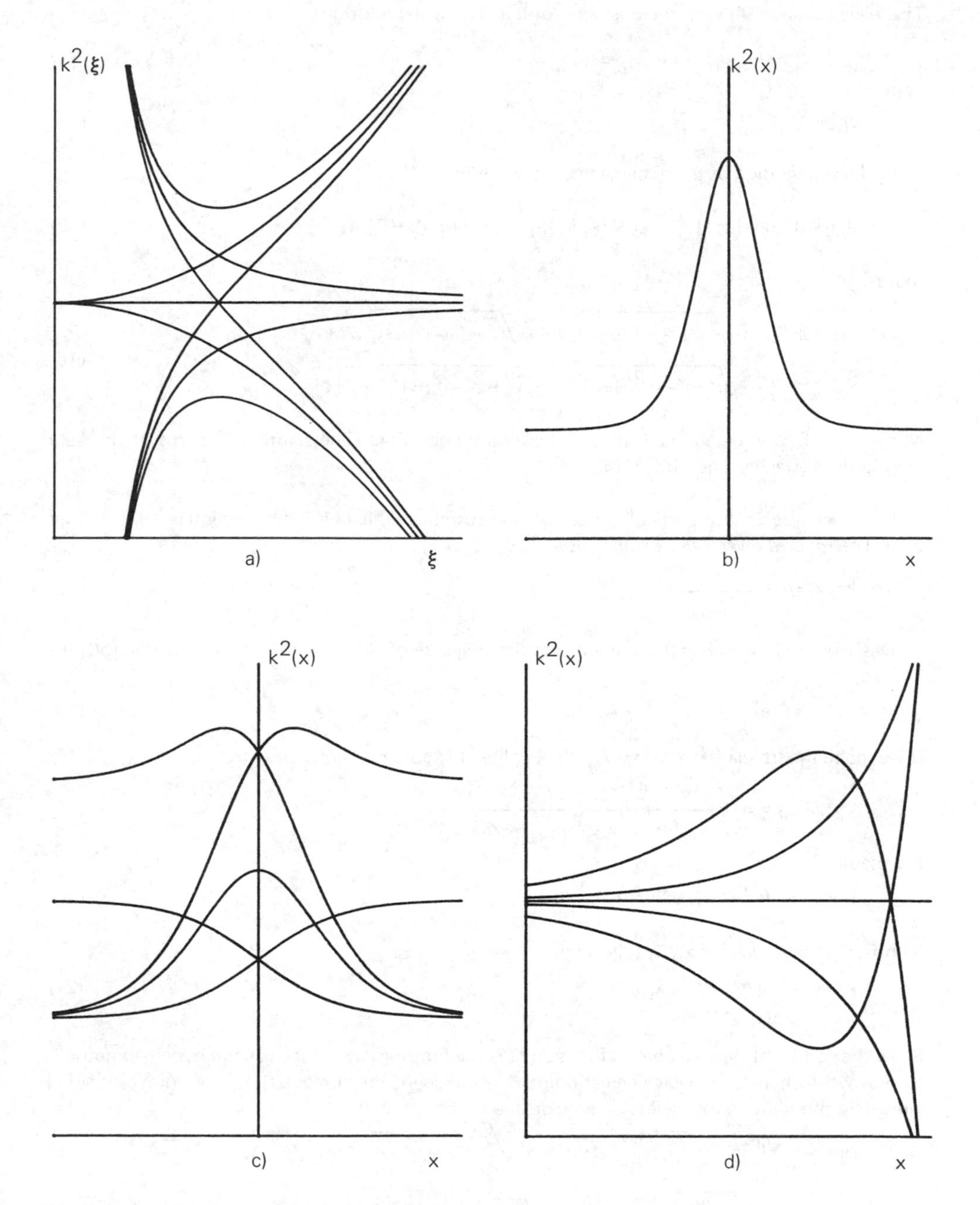

Fig. 1 *Sketch of the analytically solvable profiles: a) sample of quartic curves that can be obtained from Eq. (6) for some values of the parameters A, B, C; b) the Epstein profile of Eq. (19); c) the squared inverse hyperbolic cosine (symmetric Epstein) profile of Eq. (17); d) the Morse-type profile of Eq. (18) (upon translation $x \rightarrow x - x_0$)*

The wave equation (4) with the Epstein profile (11) has the solution:

$$\varphi(x) = e^{sx/a}(1+e^{x/a})^r f(-e^{x/a}) \tag{13}$$

with

$$r(r-1) = a^2 C, \; s^2 = a^2 \cdot (\beta^2 - A). \tag{14}$$

and f (ξ), is a solution of the hypergeometric equation [27]

$$\xi(1-\xi) f''(\xi)+[2s+1-(2s+2r+1)\xi] f'(\xi)-[r(2s+1)+a^2(B+C)] f(\xi)=0 \tag{15}$$

that is:

$$\begin{aligned} f(\xi)=&D\cdot F(r+s-\sqrt{r^2+s^2-r-a^2(B+C)}, r+s+\sqrt{r^2+s^2-r-a^2(B+C)}; 2s+1; \xi) \\ &+E\cdot \xi^{2s} F(r-s-\sqrt{r^2+s^2-r-a^2(B+C)}, r-s-\sqrt{r^2+s^2-r-a^2(B+C)}; 1-2s; \xi) \end{aligned} \tag{16}$$

where D, E are determined by the boundary conditions. The solution in terms of the field amplitude is given by Eqs. (13), (16).

There are various particular cases of the Epstein profile (11) which are interesting on their own. The *squared inverse hyperbolic cosine* [28, 29], Fig. 1c:

$$k^2(x)=A + \frac{C}{4\,ch^2\left(\frac{x}{2a}\right)} \tag{17}$$

derives from (11) with B = 0. An important limiting case of (11) is the *Morse-type profile* [30] see Fig. 1d:

$$k^2(x)= G\, e^{x/a} + H\, e^{2x/a} + A. \tag{18}$$

If we make the translation $x \to (x - x_0)$ the profile (11) goes over into

$$k_E^2(x;x_0) = A+B\frac{e^{(x-x_0)/a}}{1+e^{(x-x_0)/a}}+C\frac{e^{(x-x_0)/a}}{[1+e^{(x-x_0)/a}]^2}. \tag{19}$$

If we put

$$B = F\, e^{2x_0/a}, \; C=G\, e^{x_0/a}-H\, e^{2x_0/a} \tag{20}$$

with F, G independent of x_0, it follows that

$$k^2(x) = \lim_{x_0 \to +\infty} k_E^2(x;x_0). \tag{21}$$

Since the confluent hypergeometric function M is the limiting case of the general hypergeometric F whose two (canonical) regular singular points 1, ∞ merge into the irregular singular point ∞ of M [1] we obtain the solution for the Morse-type profile in the form:

$$\begin{aligned} \varphi(x)=&e^{x\sqrt{\beta^2-A}\, a\sqrt{-H}\, e^{x/a}}[D\cdot M(\tfrac{1}{2}-a\sqrt{\tfrac{-G^2}{4H}}+a\sqrt{\beta^2-A}; 1+2a\sqrt{\beta^2-A}; 2a\sqrt{-H}\, e^{x/a}) \\ &+E\cdot U\left(\tfrac{1}{2}-a\sqrt{\tfrac{-G^2}{4H}}+a\sqrt{\beta^2-A}; 1+2a\sqrt{\beta^2-A}; 2a\sqrt{-H}\, e^{x/a}\right)] \end{aligned} \tag{22}$$

The *exponential profile*

$$k^2(x) = A+G\, e^{x/a} \tag{23}$$

is in turn a particular case of the profile (18) with H = 0. The limiting relations existing among confluent hypergeometric and Bessel functions [27] allow to obtain the pertaining solution as

$$\varphi(x)=D\cdot J_{2a\sqrt{\beta^2-A}}\left(2a\sqrt{G}\,e^{\frac{x}{2a}}\right)+E\cdot Y_{2a\sqrt{\beta^2-A}}\left(2a\sqrt{G}\,e^{\frac{x}{2a}}\right). \tag{24}$$

Finally, the *linear profile*

$$k^2(x) = P-Q\,x \tag{25}$$

particularizes Eq. (4) in the form:

$$\varphi''(x)+[P-\beta^2-Qx]\varphi(x)=0 \tag{26}$$

whose solution is

$$\varphi(x)=D\cdot Ai\left(\frac{\beta^2-P+Qx}{Q^{2/3}}\right)+E\cdot Bi\left(\frac{\beta^2-P+Qx}{Q^{2/3}}\right) \tag{27}$$

where Ai, Bi are the Airy functions [27].

Moreover, it can be straightforwardly shown that the asymmetrical hyperbolic tangent profile discussed by various authors in different situations [1, 19, 28] is a particular case of Eq. (19).

4. CHOICE OF THE APPROXIMATING FUNCTIONS

Let us consider a circular or planar guide with profile $k^2(\xi)$ (ξ = r or x, respectively). The ξ - domain will be divided into n intervals where the actual profile is approximated by means of the functions considered in Sect. 3. These last are characterized by a set $\{A_\sigma\}$of parameters (e.g. {A, B, C}for the quartic profile and $\{A, B, C, x_0, a\}$for the Epstein model) which allow to obtain the best fit to the actual $k^2(\xi)$. This best fit on the i-th interval $[\xi_{i-1}, \xi_i]$ (with $\xi_0 < \xi_1 < ... < \xi_{n-1} < \xi_n$) is found by minimizing the weighted square deviations of the real profile $k^2(\xi)$ from the approximating one $k^2(\xi;\{A_{\sigma i}\})$ [14]

$$F_i(\{A_{\sigma i}\})=\int_{\xi_{i-1}}^{\xi_i}[k^2(\xi;\{A_{\sigma i}\})-k^2(\xi)]^2\,\varphi^2(\xi)\mu(\xi)d\xi \tag{28}$$

as functions of the parameters $\{A_{\sigma i}\}$. The quantity $\mu(\xi)$ is a metric factor

$$\mu(\xi)=\begin{cases}\xi\equiv r & \text{for circular fibres}\\ 1 & \text{for planar guides}\end{cases} \tag{29}$$

while $\varphi(\xi)$, the field amplitude, is not known. It can be replaced by a suitable trial function or a field function obtained from the previous iteration. Since both the Epstein and the quartic profile have an additive parameter (which we generally indicate by $A_{\sigma i}$ on the i-th subinterval) a necessary condition to minimize $F_i(\{A_{\mu i}\})$ is

$$\frac{\partial F_i(\{A_{\sigma i}\})}{\partial A_{\sigma i}}=\int_{\xi_{i-1}}^{\xi_i}[k^2(\xi;(A_{\sigma i}\})-k^2(\xi)]\varphi^2(\xi)\,\mu(\xi)d\xi=0 \tag{30}$$

which ensures that the actual profile and the approximating one have the same propagation constant to the first order in the "perturbation"

$$\delta k^2(\xi; \{A_{\sigma i}\}) = k^2(\xi; \{A_{\sigma i}\}) - k^2(\xi), \ \xi_{i-1} \leqslant \xi \leqslant \xi_i \tag{31}$$

At the second order the difference in the propagation constants is less than a quantity proportional to $\sum_{i=1}^{n} F_i(\{A_{\sigma i}\})$ [31]. Since the various F_i of Eqs. (29) must be minimized, the second order difference in the β's is minimized too and this is the reason why the present method is very accurate.

5. RESULTS

We have applied the discussed method to the calculation of the propagation constant for the fundamental HE_{11} mode of circular optical fibres with three values of the normalized frequency V (V = 1, 2, 3)

$$V = a\,k_0\sqrt{n_0^2 - n_1^2} \tag{32}$$

where a is the fibre core radius, k_0 the vacuum wavenumber and n_0, n_1 the maximum and the cladding refractive-indices respectively. Three refractive-index profiles were chosen in the form (r is normalized to the core radius a) [14]:

$$\begin{aligned} &a)\ k^2(r) = k_0^2 n_1^2 + k_0^2(n_0^2 - n_1^2)\cdot \sin^2(\pi r) \\ &b)\ k^2(r) = k_0^2 n_1^2 + k_0^2(n_0^2 - n_1^2)\cdot \cos^2\left(\frac{3\pi}{2} r\right) \\ &c)\ k^2(r) = k_0^2 n_1^2 + k_0^2(n_0^2 - n_1^2)\cdot 0.74\,(1-r^2)[1-0.4|\sin(5\pi r)|]-0.5\,e^{-30r} \end{aligned} \tag{33}$$

The profile a) has a noticeable central dip, while the b)-one is of the W-type and the last one is a parabolic profile with central dip and various ripples. The fundamental eigenvalue β has been calculated both by the present improved stratification method (IS) with multiquartic approximations (Sect. 3.1) and by the usual Runge-Kutta (RK) procedure, requiring in each case a precision of 10^{-7} on the normalized transverse propagation constant in the cladding

$$w = a\sqrt{\beta^2 - k_0^2 n_1^2} \tag{34}$$

The IS is faster than the RK one for all the considered V-values and for the whole set of profiles, as is shown in Table 1, in which a comparison among the computing times (on a desk-top computer) required for the various applications by both methods is shown. The IS is three times faster on the average, up to a maximum of five times and more. Similar advantages of the IS can be verified also for the calculation of group velocities and dispersion. Moreover, further tests confirm the superiority of the present approach for higher-order modes, particularly for complicated index profiles and at low V-values.

V	$s_a(r)$		$s_b(r)$		$s_c(r)$	
	(IS)	(RK)	(IS)	(RK)	(IS)	(RK)
1	2.2	8.0	3.4	9.8	4.4	24.7
2	2.4	8.1	3.4	8.0	5.7	18.4
3	2.4	8.4	3.8	9.6	8.0	17.8

Tab. 1 - Computing time (s) to obtain a precision of 10^{-7} on w with improved stratification (IS) and Runge-Kutta (RK) methods

6. CONCLUSIONS

We have presented an improved stratification method for the solution of the one-dimensional scalar wave equation both in circular and planar waveguides. This approach is particularly effective with respect to its former versions, since the actual refractive index profile is no longer approximated by means of a staircase function, but by a set of analytically solvable profiles which make the stratification method much more powerful and flexible, even with respect to direct solution methods such as the Runge-Kutta procedure. The set of approximating functions is made up by the quartic profile (which generalizes the step and parabolic ones) for circularly symmetric guides, and by the quartic, the Epstein and the linear profiles for planar structures. Particular attention has been devoted to the analysis of the Epstein profile, showing that the squared inverse hyperbolic cosine profile, the Morse profile and the exponential profile are particular cases of the Epstein profile.

Then the actual refractive-index distribution is approximated, according to its shape, by a suitable sequence of the above mentioned "basis" functions, whose parameters are chosen in such a way as to minimize the weighted square deviations among the approximating functions and the given profile. This choice proves to be very effective in the applications shown: in fact the improved stratification method is always much faster than the Runge-Kutta one.

The procedures here described have been applied to practical problems encountered in the optical communication field. It is worth noticing that a similar formalism can be used in different physical environments, such as electromagnetics, acoustics, elasticity, seismology, since the basis equations which rule the phenomena can be treated with formally analogous procedures.

REFERENCES

1. Morse, P.M. and H. Feshbach. Methods of theoretical physics (New York, Mc Graw-Hill, 1953).
2. Clarricoats, P.J.B. and K.B. Chan. Electron. Lett. 6, 22 (1970) 694-695.
3. Brekhoskikh, L.M. Waves in layered media (New York, Academic Press, 1960).
4. Arnaud, J.A. and W. Mammel. Electron. Lett. 12, 1 (1976) 6-8.
5. Bianciardi, E. and V. Rizzoli. Opt. Quantum Electron. 9, 2 (1977) 121-133.
6. Arnold, J.M., G.A.E. Crone and P.J.B. Clarricoats. Electron. Lett. 13, 10 (1977) 273-274, and Errata on Electron. Lett. 13, 13 (1977) 390.
7. Arnold, J.M. Electron. Lett. 13, 22 (1977) 660-661.
8. Kawakami, S. and S. Nishida. IEEE J. Quantum Electron. QE-10, 12 (1974) 879-887.
9. Yeh, P., A. Yariv and E. Marom. J. Opt. Soc. Am. 68, 9 (1978) 1196-1201.
10. Di Vita, P. and U. Rossi. CSELT Rapp. Tecn. 6, 3 (1978) 213-215.
11. Di Vita, P. and U. Rossi. Opt. Acta 27, 8 (1980) 1117-1125.
12. Coppa, G. and P. Di Vita. Electron. Lett. 17, 23 (1981) 896-897.
13. Yamada, R., T. Meiri and K. Okamoto. J. Opt. Soc. Am. 67, 1 (1977) 96-103.
14. Coppa, G. and P. Di Vita. Electron. Lett. 19, 11 (1983) 430-432.
15. Suematsu, Y. and K. Furuya. IEEE Trans. Micr. Th. and Tech. MTT-20, 8 (1972) 524-531.
16. Vassel, M.O. Journ. Opt. Soc. Am. 64, 2 (1974) 166-173.
17. Yamanouchi, K., T. Kamiya and K. Shibayama. IEEE Trans. Micr. Th. and Tech. MTT-26, 4 (1978) 298-305.

18. Epstein, P.S. Proc. Nat. Acad. Sci. USA 16 (1930) 627-637.
19. Adams, M.J. An Introduction to Optical Waveguides (New York, Wiley, 1981).
20. Marcuse, D. Light Transmission Optics (New York, Van Nostrand Reinhold, 1972).
21. CSELT Tech. Staff. Optical Fibre Communication (New York, Mc Graw-Hill, 1981).
22. Buus, J. IEEE J. Quantum Electron. QE-18, 7 (1982) 1083-1089.
23. Payne, F.P. Opt. Quantum Electron. 14, 6 (1982) 525-537.
24. Paoli, T.L. IEEE J. Quantum Electron. QE-13, 8 (1977)662-668.
25. Streifer, W. and E. Kapon. Appl. Opt. 18, 22 (1979) 3724-3725.
26. Thompson, G.H.B. Physics of Semiconductor Laser Devices (Bath, Wiley, 1980).
27. Abramowitz, M. and I.A. Stegan, eds. Handbook of Mathematical Functions (New York, Dover, 1970).
28. Nelson, D.F. and J. Mc Kenna. J. Appl. Phys. 38, 10 (1967) 4057-4074.
29. Kogelnik, H. Theory of Dielectric Waveguides, in T. Tamir, ed., Integrated Optics (Berlin, Springer-Verlag, 1975) pp. 31-81.
30. Landau, L.D. and E.M. Lifchitz. Mécanique quantique (Moscou, Mir, 1974).
31. Cohen-Tannoudji, C., B. Diu and F. Laloë. Mécanique quantique (Paris, Hermann, 1973).

CONCLUSIONS

L.B. Felsen

Assessments of the Workshop emerged from the individual working groups which dealt with what transpired in each of the four principal areas. Their reports, presented at the final panel discussion, form the basis for the observations to follow. In this description, reference to the relevant workshop papers is made by citing the first-named author in parentheses. As noted in the Preface, to avoid delay, no attempt was made to render verbatim transcripts. However, each of the group leaders had an opportunity to read the final manuscript and include modifications deemed appropriate. I would like to express my appreciation to the working groups and to the group leaders for their efforts.

Concerning Rays and Beams, the Geometrical Theory of Diffraction (GTD) remains a most useful tool for attacking a broad class of high frequency propagation and diffraction problems. Continuing modifications via uniform asymptotic theory make accessible an ever larger category of structures and configurations that give rise to previously unexplored transition phenomena where simple GTD fails. Spectral techniques play an important role in the construction of uniform transition functions, both in the time-harmonic and transient domains. Thus, our understanding of scattering by edge discontinuities on plane and curved impedance surfaces and (or) in the presence of penetrable media has been improved substantially (Tiberio, Idemen), but more work remains to be done. The goal is to assemble more and more entries in the catalog of GTD launching and diffraction coefficients, uniformized for transition regions, in order to steadily increase the utility of this basic technique. Here, emphasis should be not only on analytical methods but also on numerical tabulations, if an analytical approach proves to be intractable.

A relatively recent and highly promising development is the use of complex rays for the description of propagation and scattering involving beam type fields. Results for beam type excitation can be generated from those for line or point source excitation by analytic continuation of the initial data (isolated source locations or aperture distributions) into a complex coordinate space. Considerable progress has been made in structuring complex ray tracing programs, thereby generalizing the real ray tracing of GTD. In this endeavor, the selection of physically relevant complex rays and the proper complex analytic extension of surfaces (especially of edges) is of paramount importance (Ghione). A number of practical problems in beam fed large reflector antennas and integrated optical devices (Jacob) have already been analyzed in this manner. An important and quite recent application concerns the modeling of non-tapered initial amplitude distributions as a superposition of Gaussian beams. Use of dynamic (paraxial) beam tracing furnishes the field at any exterior point in a complicated medium. The advantage of using Gaussian beams instead of real rays is the avoidance of singularities at ray caustics. This procedure has been introduced and applied for seismic propagation but much work remains to be done to remove some ambiguities which arise from the free choice of the beam parameters (Hanyga).

To sum up, ray methods remain a most versatile tool for dealing with high frequency propagation and diffraction, and also with ray guiding (Scheggi), in complicated environments. Real ray theory is well developed but requires continuing extension to new problems. Complex ray theory and its relation to beam-type fields is a newer discipline that promises to provide analytical and numerical versatility for propagation, guiding and diffraction problems involving not only highly collimated incident fields but also incident fields with wide angle coverage. Here, some basic work, through comparisons with exact solutions of canonical problems, is still required to remove certain ambiguities in the theory at present. Also, comparisons should be made between the complex ray approach and the more conventional parabolic equation treatment, as exemplified here for guiding applications (Kuester).

The papers in the Rays and Modes group provide an up-to-date view of the interplay between (local) ray fields and (global) mode fields in waveguide configurations with rather general characteristics. Here, ray descriptions as pursued under Rays and Beams, which provide essential physical insight, become complicated by the possibility of many multiple reflections due to the guiding environment, while mode descriptions may be deficient when an excessively large number of modes is required. For this reason, hybrid methods combining rays and modes, wherein groups of the one are expressed collectively as groups of the other, have been applied most extensively here (Felsen, Weston). While the relation between rays and modes has a rigorous spectral formulation in waveguides with coordinate separability along the guiding direction (Felsen, Shang), the spectra can be manipulated to effect conversion from ray fields to mode fields also for configurations that depart weakly from separability (Arnold, Harrison, Weston). In this extension, geometrical or adiabatic invariants and the concept of flux (Weston) play a central role. The invariants ensure that the essential features of an individual modal field remain locally intact as it propagates through its range-dependent three-dimensional environment. They also furnish understanding of the behavior of rays and modes that can be exploited in attempts at further generalizations of guided ray and local mode concepts. As in other formulations of approximate theories, reliable solutions of canonical problems for special configurations serve as benchmarks with which approximate results can be compared. Being one of the simplest but non-trivial canonical structures, the wedge-shaped waveguide with penetrable boundaries has been explored extensively in this context (Buckingham), with recent progress made to chart the progress of rays and modes uniformly through critical regions related to transitions from the trapped to the leaky regime (Arnold).

As analytical models and numerical algorithms are extended to successively more complicated environments, it may be anticipated that the physically motivated ray-mode equivalence and flux concepts will continue to guide the formulations so as to render numerical evaluations more efficient. The flexibility afforded by choosing different models poses a challenge to construct numerical algorithms that are suited best for the study of particular types of wave phenomena.

While time-dependent wave processes entered into discussions throughout the workshop, special emphasis on problems arising in this context is focused in the Transient Propagation and Scattering section. When the transient signal has a broad frequency spectrum and the response is desired over a long time interval, the problem becomes complicated by the need to incorporate both weakly dispersive (high frequency) and strongly dispersive (low frequency) effects. By the traditional route of passing via Fourier inversion from the time-harmonic to the time-dependent regime, numerical evaluation of the transform integral is inefficient at the high frequency end due to rapid oscillation of the integrand. Therefore, attention has been given recently to inversion schemes that overcome these difficulties by separating the weakly dispersive portions from the strongly dispersive ones and inverting the former explicitly by temporal spectral integration _before_ spatial integration in a plane wave spectral representation of the transient field. Moreover, for layered media and finite scatterers, alternative formulations in terms of the complex resonances (complex frequency poles) of the structure have been exploited for construction of the transient response. This procedure, known as the Singularity Expansion Method (SEM), has been reviewed in detail (Pearson). The complex resonance expansion is deficient at early observation times due to difficulties in synthesizing the abrupt first arrival fields by resonance superposition. This feature can be clearly understood by characterizing the early time response in terms of wavefronts that describe the phenomenology there in its most fundamental form. There can then emerge a hybrid combination of wavefronts and resonances for a physically incisive and numerically stable description (Heyman[1]). While SEM has been principally applied to impenetrable scatterers, some recent efforts have focused on layered media with and without dispersion. The previously mentioned reversed spectral inversion provides here an effective means for extracting the weakly dispersive response more efficiently

(Heyman[2], Kuester). The dynamic ray tracing based on Maslov theory (Hanyga) should also be mentioned in this connection.

The spectral options implied by wavefronts, resonances and weakly vs. strongly dispersive regimes have only begun to be exploited systematically. Much activity may be anticipated here for constructing efficient numerical algorithms that take advantage of these models.

None of the discussions throughout the workshop was divorced from the numerical implementation of various analytical models but such observations were formalized in the session on Numerical Modeling. Comparisons are made here between various algorithms (Jensen), especially those that are "robust" in the sense that they can handle propagation environments of complexity and broad generality. Algorithms may be based on rays, modes, parabolic equation approximations, beam tracing (Lagasse), or spectral inversion (Franssens, Schmidt). Much discussion ensued on the advantage or not of employing algorithms that are matched efficiently to a particular type of problem vs. robust algorithms that accommodate many diverse problems. The robustness is achieved by access to advanced computational facilities and ample computer time. The decision here must clearly be influenced by the facilities available to the user, by cost and by other constraints that may make a tradeoff between specificity and robustness a serious option. Even the presently available robust algorithms are constrained to two-dimensional environments. Approaches to their extension to three dimenions may be aided through incorporation of previously discarded options. Inclusion of nonlinear effects in the propagation model, when required (McDonald), poses challenges that are beginning to receive attention. Proper discretization of continuum models continues to be an area of future endeavor for all of these applications (Coppa).

It is not inappropriate to end this discussion on a philosophical note. While everyone agreed that physical insight to interpret observed phenomena is desirable and even essential, there was much debate over what constitutes such insight. Numerical modeling enthusiasts tended toward the opinion that the results themselves furnish the insight while those with more analytical inclinations felt that identification of relevant wave phenomena through study of canonical problems is essential for basic understanding of previously unexplained features in actual or synthetic data. One thing is certain: because of the rapid development of computer technology, numerical considerations will play an ever more important role in the study of propagation and diffraction in complicated environments. In this pursuit, the combination of analysis and numerical modeling will provide options to challenge the talents of all who work in this fascinating field. As this workshop has demonstrated, similar concerns beset various disciplines, and the options in one discipline can be enlarged by considering those employed in another. It is hoped that this interdisciplinary exposure will be maintained among the workshop participants.

LIST OF PARTICIPANTS

Arnold, J.M. — Department of Electrical Engineering
University of Nottingham
University Park
Nottingham NG7 2RD
England

Bardati, F. — Universita di Roma
via Eudossiana 18, 00184
Roma
Italy

Bassi, P. — Universita di Bologna
Dep. Elettronica, Informatica Sistemistica,
viale Risorgimento 2,
40100 Bologna
Italy

Buckingham, M.J. — Naval Research Laboratory
Acoustic Divsion (Code 5170)
Washington, D.C. 40375, U.S.A.

Coppa, G. — CSELT
via G. Reiss Romoli 274
10148 Torino
Italy

Crone, G. — European Space Agency
Keplerlaan
2200 AG Noorwijk
Netherlands

DeVincenti, P. — Selenia S.p.A,
via Tiburtina Km 12,4,
00100 Roma
Italy

DiSalvo, E. — Istituto di Scienze Fisiche
viale Benedetto XV 5
16122 Genova
Italy

Falciai, R. — I.R.O.E. - C.N.R.
via Panciatichi 64
50127 Firenze
Italy

Felsen, L.B.	Polytechnic Institute of New York Route 110 Farmingdale, NY 11735 U.S.A.
Franssens, G.R.	University of Gent Lab. Electromagnetism and Acoustics St. Pietersnieuwstraat B-9000 Gent Belgium
Gerosa, G.	University of Rome via Eudossiana 18 00100 Roma Italy
Hanyga, A.	Institute of Geophysics Polish Academy of Sciences ul Pasteura 3 Warszawa 00973 Poland
Harrison, C.H.	CAP Scientific 233 High Holborn London England
Heyman, E.	Tel-Aviv University Department of Electrical Engineering Tel-Aviv Israel
Idemen, M.	Technical University of Istanbul Turkey
Jensen, F.	Saclant ASW Research Centre viale S. Bartolomeo 400 19026 La Spezia Italy
Krenk, S.	Risq National Laboratory 4000 Roskilde Denmark
Kuester, E.F.	Department of Electrical and Computer Engineering University of Colorado Campus Box 425 Boulder, Colorado 80309 U.S.A.

Kuperman, W.A.	NORDA (Code 320) NSTL STN MS 39529 U.S.A.
Lagasse, P.E.	University of Gent Lab. Elektromagnetism and Acoustics St. Pietersnieuwstraat B 9000 Gent Belgium
Mania, L.	Universita di Trieste Istituto di Elettrotecnica e di Elettronica via A. Valerio 10 34127 Trieste Italy
McDonald, B.E.	US Naval Ocean NORDA (code 320) NSTL STN MS 39529 U.S.A.
Montrosset, I.	Dipartimento di Elettronica Politecnico di Torino corso Duca Abruzzi 24 10129 Torino Italy
Nicolai, C.	Elettronica S.p.A. via Tiburtina Km 13,700 00131 Roma Italy
Pearson, L.W.	University of Mississippi Department of Electrical Engineering University, Mississippi 38677, U.S.A.
Pisani, C.	Istituto Universitario Navale via B. Bologna 20/4 Napoli Italy
Rossi, U.	CSELT via Reiss Romoli 274 10148 Torino Italy
Scheggi, A.M.	I.R.O.E. - C.N.R. via Panciatichi 64 50127 Firenze Italy

Schmidt, H.	Saclant ASW Research Centre viale S. Bartolomeo 19026 San Bartolomeo (La Spezia) Italy
Shang, E.C.	University of Wisconsin 1215 W. Dayton Street Madison, Wisconsin 53706 U.S.A.
Tiberio, R.	Universita di Firenze via S. Marta 3 50100 Firenze Italy
Unger, H.G.	Technical University of Braunschweig Postfach der TU 33 Braunschweig West Germany
Weston, D.E.	AUWE Portland Dosset DT 4 9QN England
Zich, R.	CESPA Politecnico di Torino corso Duca Abruzzi 24 10129 Torino Italy

SUBJECT INDEX